Introduction to Nucleic Acids

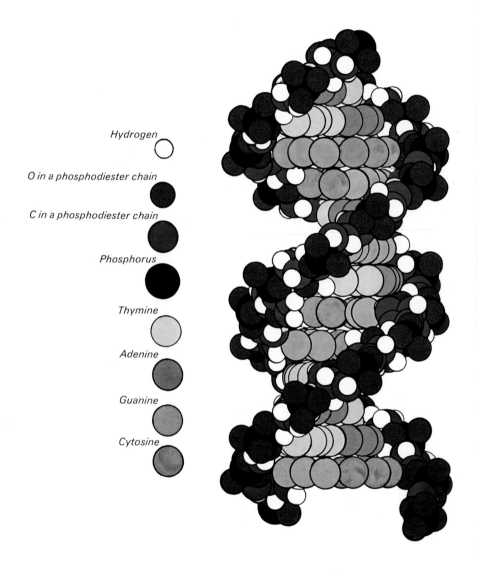

Hydrogen

O in a phosphodiester chain

C in a phosphodiester chain

Phosphorus

Thymine

Adenine

Guanine

Cytosine

Fig. 26. Molecular model of DNA-B. The base substituents have been omitted for the sake of clarity.

Introduction to Nucleic Acids
CHEMISTRY, BIOCHEMISTRY AND FUNCTIONS

EBERHARD HARBERS

Institute of Medical Physics and Biophysics
University of Göttingen
Göttingen, Germany

IN COLLABORATION WITH

GÖTZ F. DOMAGK

Institute of Physiological Chemistry
University of Göttingen
Göttingen, Germany

WERNER MÜLLER

Institute of Organic Chemistry
University of Göttingen
Göttingen, Germany

Translated by Scripta Technica, Inc.

REINHOLD BOOK CORPORATION
A subsidiary of Chapman-Reinhold, Inc.
New York   Amsterdam   London

Professor **KARL THOMAS**
Dedicated in honor of his 85th birthday

Preface
TO THE GERMAN EDITION

This book is based on a course of lectures that have been given regularly during the past years. Although still introductory, the content has been expanded considerably, not only to indicate the numerous biomedical problems in which nucleic acids have already been recognized as playing a major role, but also to help the student become familiar with recent developments. Therefore an extensive bibliography has been included.

The book also contains an appendix with some of the techniques used in this field. The terminology for enzymes is the traditional one, but footnotes have been included that refer to the newer terminology (including EC numbers) proposed by the "Report of the Commission on Enzymes for the International Union of Biochemistry 1961." The literature cited includes that published until the fall of 1963, but it does not pretend to be complete. Some newer references have also been included in the galley proofs.

We would like to thank Professors Rich, Sandritter and Wilkins for making original illustrations available to us. Special thanks are due our publisher, Dr. Günther Hauff, who met our wishes with great understanding and made it possible for the book to be printed rapidly and with care. Dr. Stamm helped in the production of the book and Mrs. Renate Kautz was responsible for preparing the many illustrations.

Göttingen
June 1964

E. Harbers

Preface
TO THE AMERICAN EDITION

Since the publication of the original German edition in 1964, progress in the study of nucleic acids has been very rapid and, consequently, the amount of literature has increased tremendously. In order to bring the book up to date for this edition, the contents had to be considerably revised. Some of the chapters on biochemical topics have been more or less completely rewritten and several new figures have been introduced, while some of the old ones have been omitted. Also the organization of the whole text is somewhat changed. Thus the appendix of the original edition containing various biochemical methods has been excluded.

As its title implies, this book is an attempt to summarize the essentials of an introductory course in nucleic acids covering most of the important facts of their chemistry and biochemistry. In view of the enormous scope of this field, a certain selectivity in choosing the material to be presented has been necessary, the choice being influenced to some degree by the authors' own viewpoints and interests. As far as possible, specific references to original publications have been provided so that the more interested readers may become familiar with the special types of research involved. However, it is neither possible nor desirable for such a book to include complete coverage of the literature. Numerous excellent papers have not been mentioned at all. The resulting selection of references does not always reflect their importance; rather, to a great extent it is the result of the availability of the publications to the authors.

We are indebted to Drs. R. W. Horne (Cambridge) and W. Zillig (Munich) for supplying original copies of electron micrographs, and to all the publishers who generously permitted the reproduction of several illustrations. Our special thanks are due to Dr. S. N. Alam for valuable comments when the revised manuscript was prepared, and to Mrs. U. Spaar for drawing all the new or altered figures.

Göttingen
May 1968

E. Harbers

Contents

CHAPTER I Introduction and Historical Review

"Two principles are necessary so that life shall succeed: One consists of proteins, the other of nucleic acids." These sentences, used by Professor Theorell in introducing the two Nobel prize winners in Physiology or Medicine in 1959, characterize very appropriately the key position of the two cell constituents which must be present for any living cell to function. While the importance of the proteins in cell metabolism has long been known, the role of nucleic acids has been appreciated only in recent times. Modern developments in the field of nucleic acid research, made possible by a variety of new experimental techniques, have attracted the attention of scientists from many previously disparate disciplines such as biochemistry, embryology, genetics, cancer research, microbiology, pharmacology, radiation biology, virology, and the like. The consequence of this community of interests has been the creation of a new branch of biological sciences termed *molecular biology* which, during the past few years, has grown with tremendous rapidity and has established a new era in scientific thought.

Nucleic acids were discovered by the Swiss Friedrich Miescher (1844–1895) who from 1868–1869 was an associate of Hoppe-Seyler's in Tübingen. Miescher isolated a material from pus-containing bandages which consisted largely of nuclei and which he designated as "nuclein." He found that it contained relatively large amounts of phosphates and was acid in character; he also noted that it was soluble in alkali but precipitated in acid (Miescher, 1871a and b). After returning to Switzerland, Miescher (1874) used Rhine salmon sperm as a starting material from which he isolated a basic protein called "protamine." Miescher's extensive studies were published posthumously in 1897 by his friends.

The term "nucleic acids" was introduced by Altmann in 1889. Subsequently some of the nucleic acid constituents were identified, particularly by Kossel and co-workers. Thus in 1879–1881, the purine bodies were isolated; later thymine (1884), cytosine (1894) and uracil (1909) were identified. A sugar was also found to be a nucleic acid constituent and, in yeast nucleic acid, was shown by Levene and Jacobs (1909) to be ribose. It was not until much later (1930) that Levene and collabora-

1

tors identified the sugar moiety of thymus nucleic acid as 2-deoxyribose. The *type of pentose* in nucleic acids was then used to *classify* these compounds into *ribonucleic acids* (commonly abbreviated as RNA) and *deoxyribonucleic acids* (DNA).   Because thymus was the most common starting material for the isolation of DNA, the latter was often designated as thymonucleic acid, while RNA was called yeast nucleic acid.

Pentose nucleic acid is another term that has been used for RNA, and deoxypentosenucleic acid, for DNA.   These terms, now outdated, were introduced when it was not known that the sugar moiety of nucleic acids was always either ribose or 2-deoxyribose.

At first it appeared that RNA was strictly of plant, and DNA of animal origin, but it was soon established that both types of nucleic acids occur in all cells.   This was demonstrated not only by biochemical methods but also by histological techniques which allowed intracellular localization of the nucleic acids.   The presence of DNA was established by means of a reaction first described by Feulgen and Rossenbeck in 1924.   Using a ribonuclease test, Brachet in 1940 was able to show that part of the basophilic material in the cell was RNA.   Caspersson developed a procedure whereby the presence of nucleic acids in histological preparations was established by measuring their absorption in the ultraviolet range.   Numerous studies of this type showed that DNA is an integral part of the chromosomes and that RNA plays a decisive role in protein biosynthesis (Caspersson, 1941; Brachet, 1942).   This led to some recognition of the role of nucleic acids, although the possibility that DNA might constitute the building block of the gene seemed remote to many investigators.   Nevertheless, the number of reports implicating DNA as the "information carrier" began to increase.   For example, the highest rate of mutations induced by ultraviolet light occurred at wavelengths which correspond to the absorption maxima of the nucleic acids and not those of the proteins (Knapp *et al.*, 1939; Hollaender and Emmons, 1941).   The first convincing evidence, however, for the biological role of DNA came from the work of Avery and co-workers in 1944; using pure DNA, they showed that it was possible to transfer the capacity to form capsules from one pneumococcal strain (S) to another (R) and that this property acquired by "transformation" was transmitted to the daughter cells.   Numerous virus studies further substantiated the biological role of DNA.   Thus, bacteriophages were shown to multiply in host cells only if the former contained DNA.   When DNA was removed from the phages, they were still capable of inducing an abortive infection, but no new phages arose (Herriot, 1951).   When *Escherichia coli* is infected with T phages, only the phage DNA enters the cell, while the protein portion remains outside (Hershey and Chase, 1952).   Concurrently with these biological studies, further chemical investigations were carried out, largely by Chargaff, Furberg, Gulland, Todd and Wilkins; these served as the basis for the structural model of DNA proposed by Watson and Crick in 1953. This model illustrates how, in the process of DNA replication (a necessary step in cell multiplication), exact copies of the DNA molecule can be formed.

Insight into the enzymatic steps in nucleic acid biosynthesis is of more recent origin. In 1955, Grunberg-Manago, Ortiz and Ochoa reported the synthesis of RNA-like polynucleotides from ribonucleoside diphosphates by means of an enzyme isolated from *Azotobacter vinelandii* and subsequently termed polynucleotide phosphorylase. Shortly thereafter, in 1956, Arthur Kornberg and co-workers succeeded in isolating from *E. coli* an enzyme, designated as DNA polymerase, which catalyzed the biosynthesis of DNA. Similarly RNA biosynthesis has been shown to proceed generally via ribonucleoside triphosphates. The histochemically observed relationships between RNA and protein synthesis have been confirmed in recent years by biochemical studies and have helped clarify the mechanism of protein biosynthesis (Hoagland, Lipmann, Watson, Zamecnik). They have also led to an understanding of the "code" that determines the synthesis of specific gene-dependent proteins (Nirenberg, Khorana).

With modern techniques, it is now possible to isolate undenatured nucleic acids which often retain their biological activity. Thus, pure viral RNA or DNA may induce an infection or the formation of neoplasms. Moreover, biologically active nucleic acids can be used to study the effects of interference with genetic information and, thus, to establish relationships between macromolecular structure and genetic function. This development marked the beginning of molecular genetics.

The problems which are discussed only briefly here and relevant experimental results will be treated in greater detail in the chapters to follow. The early work on nucleic acids has been discussed in German in a book by Feulgen (1923) and in English by Jones (1920) and by Levene and Bass (1933). A comprehensive treatise on this subject is the three-volume work "The Nucleic Acids," edited by Chargaff and Davidson. In addition, a number of introductory works (e.g., Potter, 1960; Steiner and Beers, 1961; Davidson, 1965) and more special accounts (e.g., Brachet, 1960; Jordan, 1960; Perutz, 1962; Chargaff, 1963) dealing with the biochemistry of the nucleic acids have been published. Current progress in this ever-expanding field is reported in the new series "Progress in Nucleic Acid Research and Molecular Biology" (edited by Davidson and Cohn).

Chemistry of Nucleic Acid Components and Precursors

CHAPTER II

A current method for the isolation of nucleic acids from biological material yields white amorphous products that decompose on heating without melting, contain about 9% phosphorus and are characterized by high molecular weights. These compounds are soluble in water, alkalies, formamide and dimethylsulfoxide; they may be reprecipitated from aqueous solution by the addition of acid or alcohol. If nucleic acids are treated vigorously with acid, certain pyrimidine and purine bases, along with D-ribose or 2-deoxy-D-ribose and phosphoric acid, occur as low molecular weight hydrolysis products. Nucleic acids which yield ribose on hydrolysis are termed ribonucleic acids (RNA); deoxyribose-containing nucleic acids are called deoxyribonucleic acids (DNA). On careful treatment with dilute alkali or suitable enzymes, both kinds of nucleic acids give rise to degradation products in which each pyrimidine and purine base is combined with a phosphorylated pentose residue. These so-called nucleotides or nucleoside phosphates constitute the monomeric units of the nucleic acids, which are, therefore, termed polynucleotides.

The following sections are concerned with the structure and synthesis of nucleotides, whose general composition and nomenclature can be expressed schematically as:

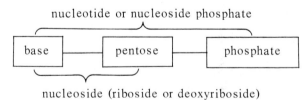

nucleotide or nucleoside phosphate

| base | pentose | phosphate |

nucleoside (riboside or deoxyriboside)

BASES

The base components of the nucleotides are derivatives of pyrimidine (I) or purine (II); formally speaking, they are derived from these two heterocyclic compounds by substitution in the 2-, 5-, and 6- or 2- and 6-positions, respectively. 2-Hydroxy-6-aminopyrimidine, cytosine, whose structure is shown in the keto form (IV), and 2,6-dihydroxypyrimidine, uracil (III, diketo form), constitute most of the pyrimidine bases of RNA. The analogous components of DNA are cytosine and thymine (V, 2,6-dihydroxy-5-methylpyrimidine); DNA generally contains no uracil.

		III	IV	V
I	II	Uracil	Cytosine	Thymine

In addition, small amounts of 5-methylcytosine (VI) occur in the DNA of most species. The DNA of T-even phages contains 5-hydroxymethylcytosine (VII; Wyatt and Cohen, 1952) in the place of cytosine (cf. p. 260). 1-Methyluracil and 1-methylcytidine have been found in soluble RNA (Hall, 1963c; cf. p. 101).

VI	VII	VIII	IX
5-Methylcytosine	5-Hydroxymethyl-	Adenine	Guanine
	cytosine		

Adenine (6-aminopurine, VIII) and guanine (IX) are the principal purine derivatives of both types of nucleic acids; in addition, small amounts of N- and C-methyl derivatives (Littlefield and Dunn, 1958) also occur (the so-called rare bases; see pp. 101 and 114.

Upon treatment with acid, the amino groups of guanine and cytosine are replaced by hydroxyl groups, yielding xanthine (XII in Fig. 2) and uracil, respectively. Until this reaction became known, it had been assumed that DNA contained some uracil. Since acid degradation was commonly used in the base analysis of the nucleic acids, it was not possible to exclude xanthine as a natural component of nucleic acids. However, with the availability of milder degradation methods, the base composition of the nucleic acids was established. Most of the bases referred to were isolated from nucleic acid hydrolysates by Kossel and co-

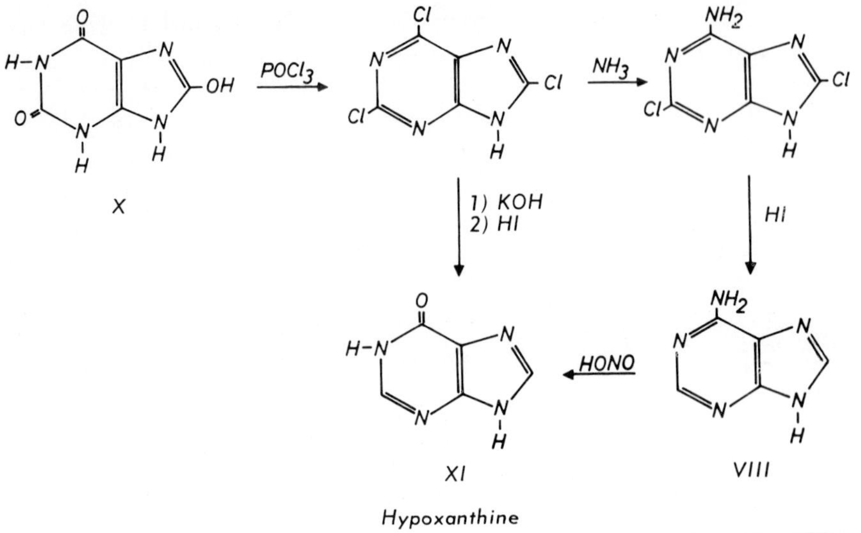

Fig. 1.   Conversion of uric acid to adenine and hypoxanthine.

workers (Kossel, 1889, 1899); uracil was found by Ascoli (1900–1901) in the degradation of yeast RNA. Clarification of the chemical structure of the purine bases became possible after Fischer's work on uric acid showed that adenine (VIII) and hypoxanthine (XI in Fig. 1) can be readily synthesized from uric acid. The synthetic pathway shown in Fig. 1 provided the evidence for the structure of the end products. The structure of guanine became known on the basis of its conversion to xanthine (XII in Fig. 2) and from the products of oxidative degradation, which Strecker identified as guanidine and parabanic acid (Fig. 2). Ultimately, E. Fischer (1897) partially synthesized this compound from uric acid via 2,6,8-trichloropurine.

Elucidation of the structure of the pyrimidine bases of the nucleic acids proved somewhat simpler, since similar pyrimidine derivatives were already known. Thus Steudel (1900) was able to assign the correct formula to thymine after oxidation to urea; this was later confirmed by synthesis (Wheeler and Merriam, 1903). Kossel and Steudel (1903) derived the structural formula of cytosine (IV) by isolating biuret, following the vigorous oxidation of cytosine and its conversion to uracil (III) by deamination with nitrous acid. The structure of uracil was already known, Fischer and Roeder having synthesized it a short time before. Final confirmation of the structure of cytosine came from chemical synthesis.

Of the various methods for preparing purine bases, Traube's method

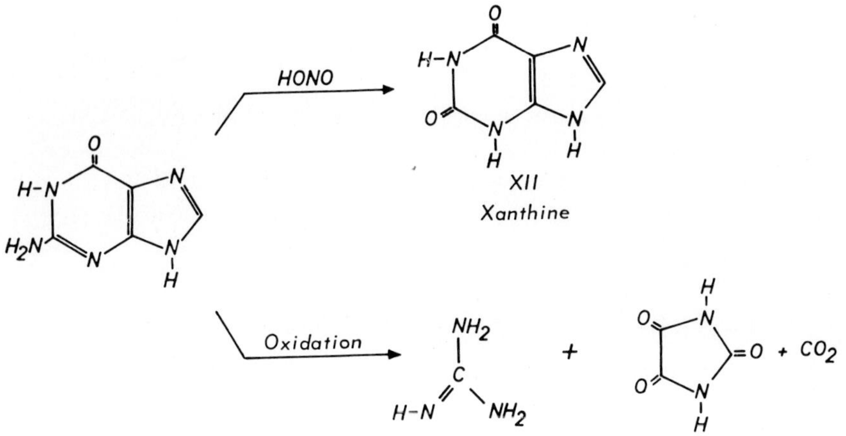

Fig. 2. Nitrosation and oxidative degradation of guanine (from Strecker, 1861).

for synthesizing adenine (Traube, 1904), hypoxanthine and guanine deserves mention, since it is a classic example of organic synthesis. More recently, preparative methods in this field have been greatly improved by the use of formamide (Brederick *et al.*, 1959). The procedure of Wheeler and Johnson (1903) for synthesizing cytosine is an example of the synthesis of pyrimidine bases (Fig. 3); the reaction is analogous to that

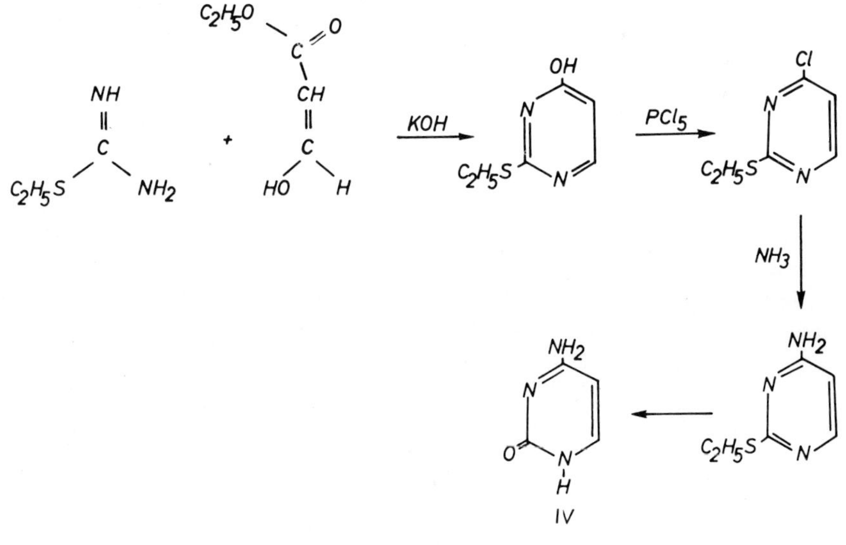

Fig. 3. Chemical synthesis of cytosine (from Wheeler and Johnson, 1903).

employed in the synthesis of thymine. The condensation of a urea derivative with a C_3 compound is also the basis for the uracil synthesis of
Fischer and Roeder (1901).

Most recently, it has been shown that purine and pyrimidine bases,
like amino acids, can be formed under the conditions of the so-called
primordial atmosphere (ammonia, methane, hydrogen and electric dis-
charges). These findings are of great interest in relation to the problem
of the origin of life.

THE SUGAR MOIETY OF THE NUCLEIC ACIDS

The sugars that occur in nucleotides are far more unstable chemically
than the bases. As early as 1891, Kossel recognized that the hydrolysis
products of yeast RNA contained a carbohydrate, and Hammarsten was
able to identify it as a pentose. However, not until 1908 were Levene
and Jacobs able to crystallize this sugar in pure form. Mild oxidation of
this substance yielded D-ribonic acid, an enantiomorph of a product obtained by Fischer and Piloty. Further oxidation produced the inactive
trihydroxyglutaric acid, so that structure XIII could be derived for the
sugar (Fig. 4). Van Eckenstein and Blanksma (1913) verified the struc-

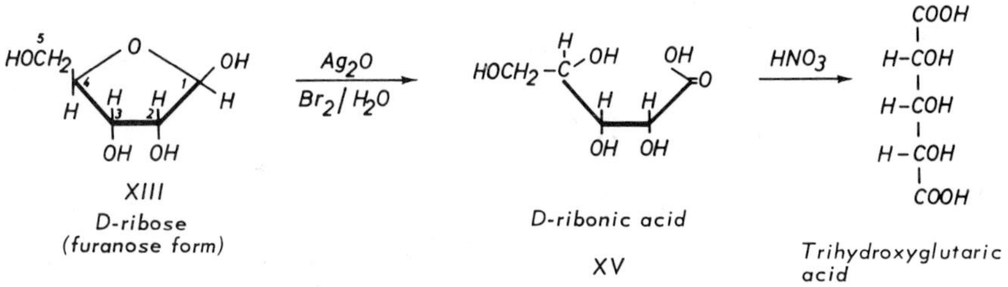

Fig. 4.   Oxidation of D-ribose (after Levene and Jacobs, 1909).

ture by synthesis. Until recently, D-ribose was the only reported sugar
in RNA nucleotides, but Hall (1963b) has detected traces of 2′-O-methyl-
ribonucleoside in various types of RNA; the methylated sugar was found
in pyrimidine as well as purine ribosides.

Initially, the isolation and study of the pentose portion of DNA or
deoxyribonucleotides met with many difficulties. While RNA can easily
be degraded with mild alkali to ribonucleotides, whose purine ribotides
yield D-ribose on mild acid hydrolysis, DNA resists the action of alkalies. Although Thannhauser and Ottenstein successfully employed picric

acid hydrolysis to liberate pyrimidine deoxyriboside diphosphate, only the enzymatic degradation to mononucleotides by Levene, Mikeska *et al.* (1930) produced deoxyribonucleoside phosphates in amounts sufficient to yield, on very mild acid hydrolysis, the already known 2-deoxy-D-ribose (XIV) in crystalline form.   Purine deoxyribosides yield the unchanged

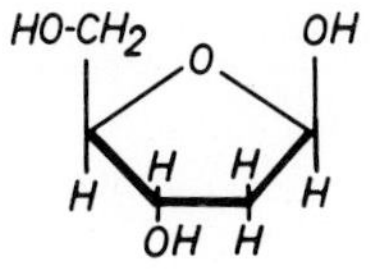

HO-CH$_2$ OH

H   H   H   H

OH H

XIV

2-Deoxy-D-ribose
(furanose form,
β-configuration)

sugar on acid treatment; acid hydrolysis of pyrimidine deoxyribosides caused deoxyribose to rearrange to levulinic acid.   Later it was shown that in order to remove the undamaged pentose portion from the pyrimidine nucleotides, it was necessary to hydrogenate the pyrimidine bases to the corresponding dihydropyrimidines.

In the preparation of 2-deoxy-D-ribose, the synthesis of Hardegger and collaborators (1957) is of greatest practical importance, since it has made this sugar more readily available.

STRUCTURE OF THE NUCLEOSIDES

The investigations outlined above provided reasonably accurate knowledge of the structure of the individual nucleotide components, but did not indicate the way in which these are joined together.   The solution of this problem was complicated by the fact that nucleotides prepared in various ways from different starting materials exhibited the same composition but different physical properties.   After the development of methods for the selective cleavage of the phosphate residue, it was shown that these nucleotides were isomers, differing in the position of the phosphate group.   Cleavage of the phosphate enzymatically or by chemical means (Levene and Jacobs, 1909; Carter and Cohn, 1949) converted the nucleotides into nucleosides which are composed of base and pentose; with one exception, no naturally occurring isomers of these products are known.

The names of the nucleosides are usually derived from the bases con-

TABLE 1. Major Nucleosides and Nucleotides and Common Abbreviations

Base	Sugar	Nucleoside[a]	Nucleotide	Abbreviation[b]
Adenine	Ribose	Adenosine (XVI)	Adenosine mono-, di- and triphosphate	AMP,ADP,ATP
	Deoxyribose	Deoxyadenosine	Deoxyadenosine mono-, di- and triphosphate	dAMP,dADP,dATP
Cytosine	Ribose	Cytidine (XIX)	Cytidine mono-, di- and triphosphate	CMP,CDP,CTP
	Deoxyribose	Deoxycytidine	Deoxycytidine mono-,di- and triphosphate	dCMP,dCDP,dCTP
Guanine	Ribose	Guanosine (XVII)	Guanosine mono-, di- and triphosphate	GMP,GDP,GTP
	Deoxyribose	Deoxyguanosine (XXI)	Deoxyguanosine mono-, di- and triphosphate	dGMP,dGDP,dGTP
Hypoxanthine	Ribose	Inosine	Inosine mono-, di- and triphosphate	IMP,IDP,ITP
	Deoxyribose	Deoxyinosine	Deoxyinosine mono-, di- and triphosphate	dIMP,dIDP,dITP
Thymine	Ribose	Thymine riboside	Thymine riboside mono-, di- and triphosphate	
	Deoxyribose	Thymidine (XX)	Thymidine mono-, di- and triphosphate	dTMP,dTDP,dTTP
Uracil	Ribose	Uridine (XVIII)	Uridine mono-, di- and triphosphate	UMP,UDP,UTP
	Deoxyribose	Deoxyuridine	Deoxyuridine mono-, di- and triphosphate	dUMP,dUDP,dUTP

TABLE 1. *(continued)*

Base	Sugar	Nucleoside[a]	Nucleotide	Abbreviation[b]
Xanthine	Ribose	Xanthosine	Xanthosine mono-, di- and triphosphate	XMP,XDP,XTP
	Deoxyribose	Deoxyanthosine	Deoxyxanthosine mono-, di and triphosphate	dXMP,dXDP,dXTP

[a]The Roman numerals in parentheses refer to structural formulas in the text.
[b]A number of authors prefer to have the abbreviations of thymidine phosphates preceded by a "d" so as to designate them clearly as deoxyribotides.

tained in them. Table 1 summarizes the usual designations for nucleosides and nucleotides and the internationally used abbreviations.

Acid hydrolysis of a nucleoside or nucleotide generally releases the base (Levene and Harris, 1933), while neutral or alkaline hydrolysis of the nucleotide liberates only the phosphate residues (Levene and Jacobs, 1909). This behavior in acids indicates that the two units of a nucleoside are not linked by a carbon-carbon bond, but by a glycosidic carbon-nitrogen bond. Proof that this kind of bond in fact occurs in the nucleoside and that pentose β-glycosidic bonds of the purine bases occur in the 9-position, and those of the pyrimidine bases in the 3-position, was obtained by a series of reactions, the most important of which will be summarized here.

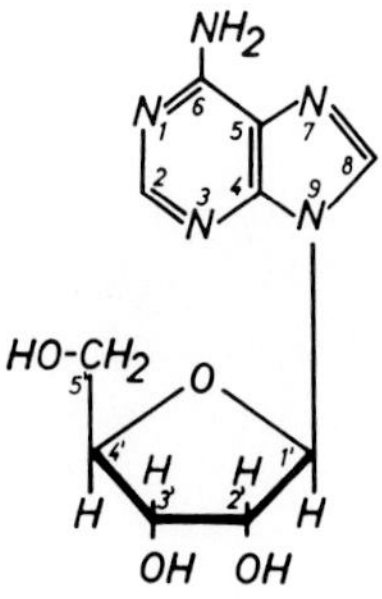

XVI

Adenosine

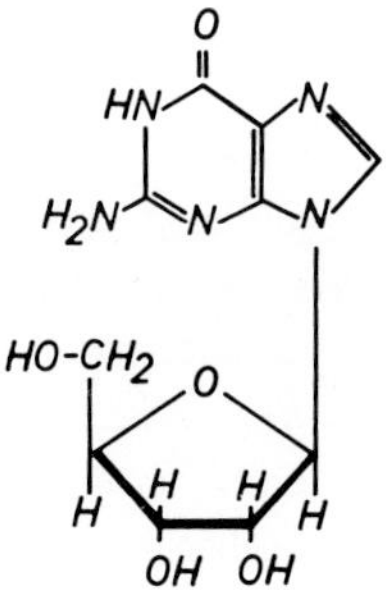

XVII

Guanosine

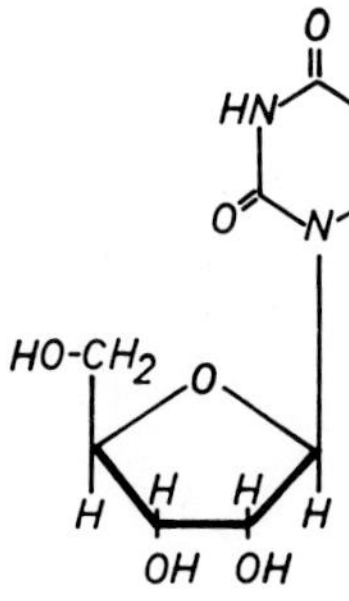

XVIII

Uridine

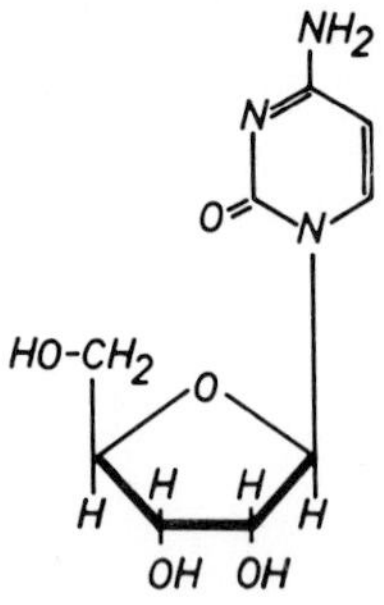

XIX
Cytidine

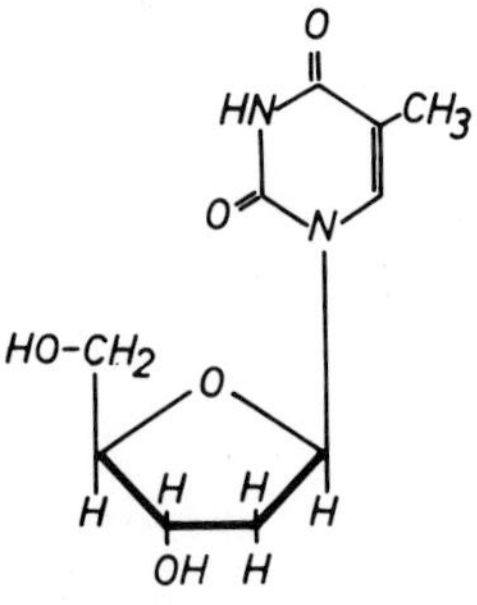

XX
Thymidine

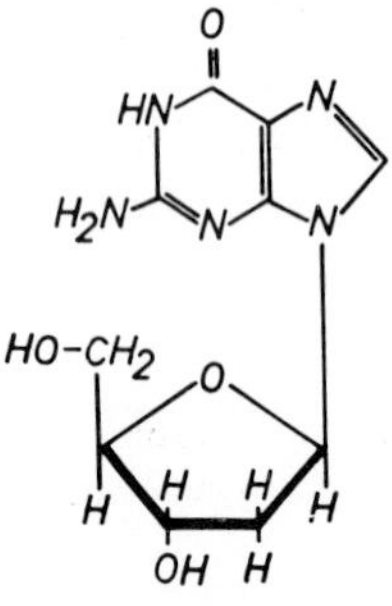

XXI
Deoxyguanosine

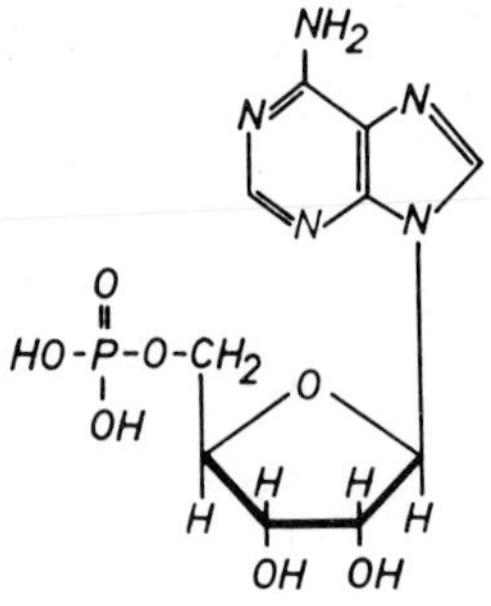

XXII
Adenosine-5'-phosphate
(Muscleadenylic acid)

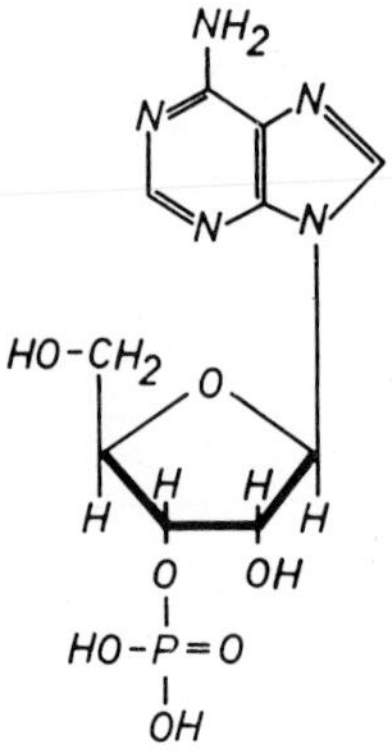

XXIII
Adenosine-3'-phosphate
(Adenylic acid b)

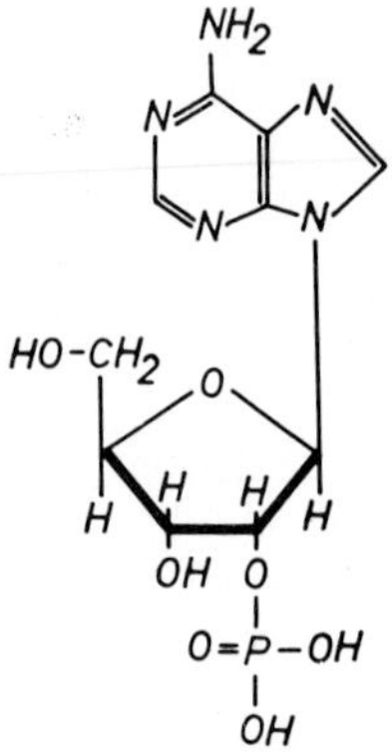

XXIV
Adenosine-2'-phosphate
(Adenylic acid a)

Levene and co-workers succeeded in converting adenosine into a permethyl ether, after preliminary N-acetylation. After hydrolytic cleavage of the N-acetyl residue and subsequent nucleotide hydrolysis with dilute mineral acid, 2,3,5-trimethylribofuranose was isolated. An analogous reaction sequence gave the same product from guanosine. According to these results, the bond between the actual purine base and the ribose residue could only originate from carbon 1 of this sugar, as is true for glycosides; moreover, the furanoid structure of the product indicated that ribose in the nucleosides exists in the furanose form. Since a pentose arranged in this cyclic form must carry a primary hydroxyl group, it was obvious that this might be detected in the nucleosides by suitably specific

etherification and esterification reactions. Indeed, with both ribonucleosides, and deoxyribonucleosides, such a reaction can be carried out by
converting nucleosides to trityl ethers with triphenylmethyl chloride
(trityl chloride) (Bredereck and Berger, 1940). This reagent preferentially
etherifies primary hydroxyl groups, but is not strongly specific. Therefore, the monotrityl ethers so obtained were esterified with p-toluenesulfonyl
chloride (tosyl chloride), and replacement of the tosyloxy groups by
iodine after reaction with sodium iodide in acetone was attempted (with
primary tosyloxy groups, this could be expected to be successful).
These reactions were never successful, however; therefore, one might
assume that the trityl residue had in fact previously etherified the primary hydroxyl groups. If, however, the nucleosides were first converted
with acetone and a dehydrating agent into acetonides or 2′,3′-isopropylidene
derivatives (Hampton, 1961), and the remaining free hydroxyl groups were
then esterified with tosyl chloride, the tosyloxy group so formed was
easily replaced by iodine. The only explanation of these results is the
existence of a primary hydroxyl group in the 5′-position.

Direct proof of the furanoid ring structure of the pentoses in the
ribonucleosides and deoxyribonucleosides came from their behavior on
periodate oxidation. In this reaction, 1,2-diols are broken down oxidatively with dialdehyde formation; 1,2,3-triols are similarly decomposed
to a dialdehyde, with the liberation of a molecule of formic acid. It was
found that all ribonucleosides are decomposed by periodate to dialdehydes without the liberation of formic acid; therefore, the ribonucleosides
must have a 1,2-diol grouping. The deoxyribonucleosides, on the other
hand, are not attacked by this reagent (see reaction scheme in Fig. 5).
Periodate oxidation of the sugars gave further information about the configuration of the glycosidic bond, as well as their position on the hetero-

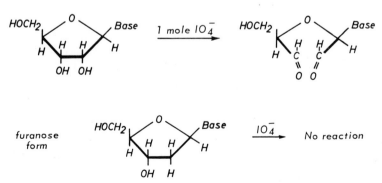

Fig. 5. The effect of periodate on ribonucleosides and 2′-deoxyribo-
nucleosides (from Lythgoe and Todd, 1944; Brown and Lythgoe, 1950).

cyclic bases. On periodate oxidation, 9-D-β-glucosyladenine, available
by an unambiguous synthetic route, yielded a dialdehyde which was
identical in all its properties to the corresponding oxidation product of
adenosine. Accordingly, the glycosidic linkage in adenosine must have
originated from the heterocyclic nitrogen atom in the 9-position and, with
reference to the ribofuranose ring, must be in the β-configuration. The
ultraviolet spectra of the purine nucleosides confirm this finding; for
they may be clearly associated with 9-methyladenine or 9-methylguanine,
respectively, and not with the corresponding 7-methyl derivatives (Gulland
and Story, 1938). With the 3-β-D-glucopyranosyl compounds of uracil
and cytosine, periodate treatment yielded dialdehydes which were identical with the corresponding degradation products of uridine or cytidine.
Therefore, the glycosidic bonds in the two pyrimidine nucleosides must
have been derived from the 3-position and likewise must have the
β-glycosidic configuration. Further confirmation of this result results
from the easy conversion of some ribonucleoside and deoxyribonucleoside
derivatives, tosylated in the 5-position, to cyclic nucleosides (see Fig.
6). For steric reasons, this reaction is only possible with a glycoside

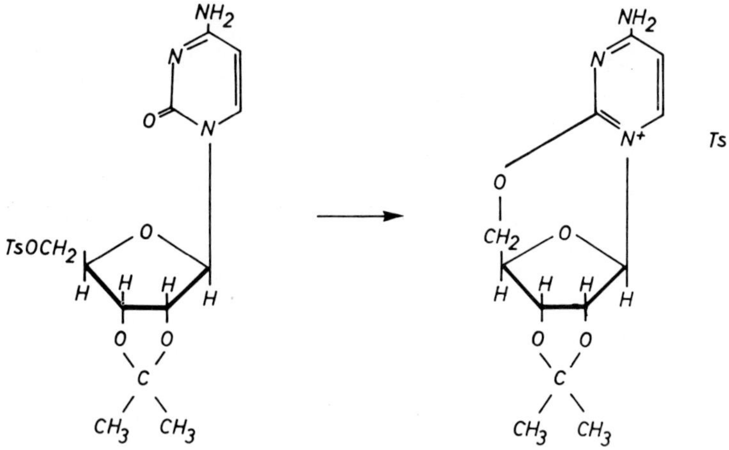

Fig. 6.   Formation of cyclic nucleosides (from Michelson and Todd, 1955).

bond in the β-configuration. An analogous structure may be demonstrated for thymidine by synthesis from the corresponding riboside
(Visser et al., 1948).

Pseudouridine (5-ribosyluracil), a nucleoside which occurs in transfer
RNA (see pp. 101 and 210), has a structure different from the usual nucleo-

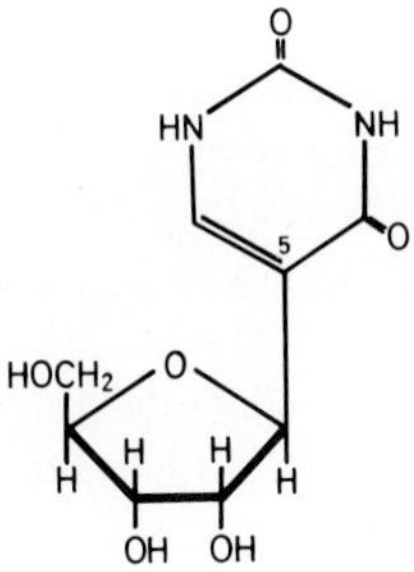

Fig. 7. Structure of pseudouridine (from W. E. Cohn, 1960). (Reproduced with permission of American Society of Biological Chemists).

sides. The structure given in Fig. 7 was derived from proton resonance spectra as well as from the products of periodate degradation (Cohn, 1960). The carbon-carbon linkage between the base and the sugar is consequently relatively insensitive to acids.

X-ray analysis of single crystals or of suitable halogen derivatives of the nucleotides has provided ultimate proof of the structure of nucleosides (Furberg, 1950; Furberg et al., 1956). These investigations have shown that the ribo- or deoxyribofuranose rings are roughly perpendicular to those of the heterocyclic bases. This made it easier to construct models of the secondary structure of polynucleotides. Moreover, data from X-ray analyses provided information on bond angles, bond lengths and the tautomeric forms of the nucleotide bases. Thus it became apparent that the oxygen atoms in the heterocyclic ring are in the keto rather than the enol form, since the small carbon-oxygen distances indicate a predominantly double-bond character. In contrast, it may be shown that the nitrogen atoms not in the ring belong to amino and not imino groups, because the carbon-nitrogen distance is practically normal. Detailed comparison of the infrared spectra of nucleotides, nucleosides and bases in heavy water showed that these substances exhibit the same tautomeric relations in solution as in crystalline form (Miles, 1958).

PREPARATION AND CHEMICAL SYNTHESIS OF NUCLEOSIDES

The completely synthetic production of nucleosides is usually intricate and tedious; moreover, the yields are often unsatisfactory. Therefore, it is generally preferable to degrade abundantly available natural polynucleotides and to separate the resulting nucleosides by ion exchange chromatography (see p. 192). The four ribonucleosides may be obtained from RNA in a one-step reaction after treatment for 3–5 hours with concentrated ammonia at 176–180 °C (Levene and Jacobs, 1910).

The preparation of deoxyribonucleosides is less simple, since DNA is not depolymerized by alkali treatment. Acid hydrolysis yields pyrimidine deoxyribotides (Levene, 1938; Dekker *et al.*, 1953), which may be converted enzymatically into the corresponding nucleosides (Heppel and Hilmoe, 1957); however, the purine deoxyribosides are lost and can be obtained only after enzymatic hydrolysis of DNA.

Although large amounts of nucleosides may be isolated as described, considerable effort has recently been focused on chemical methods to facilitate the production of isotopically labeled nucleosides and structural analogs suitable for chemotherapy (see p. 192). Attempts to synthesize these analogs have contributed substantially to perfecting nucleoside and nucleotide synthesis. Michelson (1961) has summarized the most significant synthetic approaches in his detailed review.

The following material will illustrate the principles involved in the syntheses of some analogs and suitable also for the production of deoxyribonucleosides, which can otherwise be obtained only by enzymatic methods.

Once a simple synthesis of 2-deoxy-D-ribose was achieved (Hardegger *et al.*, 1957), it seemed likely that deoxyribonucleosides could be made by direct linkage of deoxy sugars to heterocyclic bases. This can succeed only if a sufficiently reactive mercury or silver derivative of the intrinsically inert bases reacts with a diester of 2-deoxyribofuranosyl chloride. In this way, thymidine and deoxycytidine have been prepared in good yield from the corresponding monomercury compounds and 3,5-di-O-p-chloro-(or *p*-methyl)-benzoyl-2-deoxyribofuranosyl chloride (Hoffer *et al.*, 1959; see Fig. 8). Deoxyadenosine is obtained by the same principle via chloromercuri-6-benzamidopurine by reaction with 3,5-di-O-p-nitrobenzoyldeoxyribofuranosyl chloride, hydrolytic cleavage and separation from the α-anomer (Ness and Fletcher, 1960). If one starts with

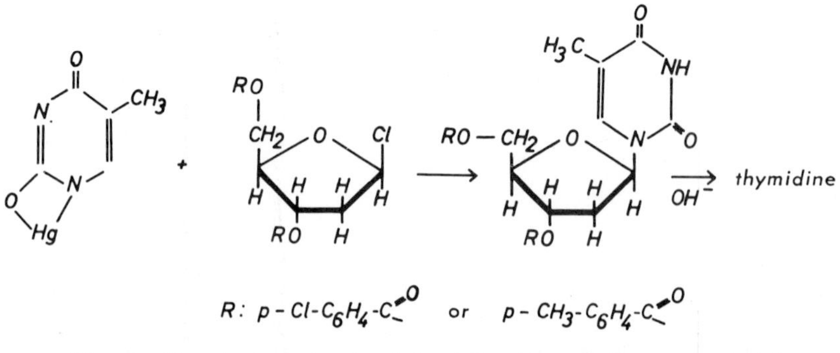

Fig. 8.   Chemical synthesis of thymidine (from Hoffer *et al.*, 1959).

the silver salt of 2,8-dichloroadenine, both deoxyguanosine and de-oxyadenosine can be obtained after combination with 3,5-di-O-acetyl-2-deoxyribofuranosyl chloride (Venner, 1960); the yield, however, is not satisfactory.

The properties of nucleosides and nucleotides will be discussed later (see p. 23).

STRUCTURE OF NUCLEOTIDES

The nucleoside monophosphates are the characteristic monomers of nucleic acids. As seen in structural diagrams, each ribonucleoside possesses three hydroxyl groups which may be esterified by phosphoric acid in the ribose residue (each deoxyribonucleoside has only two hydroxyls), so that according to the position of the phosphate residue, two or three isomeric nucleoside phosphates are possible. Identification of these isomers, which are denoted in the following as ribonucleoside-2'-, 3'- or 5'-phosphate and deoxyribonucleoside-3'- or 5'-phosphate, was achieved only after the development of suitable chromatographic separa-tion methods and after synthetically produced substances of defined structure were available for comparison.

Enzymatic degradation of DNA by pancreatic deoxyribonuclease (Volkin *et al.*, 1951) yields deoxyribonucleotides whose properties are identical with those of the readily synthesized deoxyribonucleoside-5'-phosphates (Hayes *et al.*, 1955). By the action of 5'-nucleotidase, an enzyme that selectively dephosphorylates 5'-nucleoside phosphates, deoxyribonucleo-sides are obtained (Carter, 1951). In contrast, calf spleen deoxyribo-nuclease depolymerizes DNA to deoxyribonucleotides which resist attack by 5'-nucleotidase (see p. 190), but have the same physical properties as synthetic 3'-deoxyribonucleotides. Depending on the kind of degradation used, one can therefore obtain 5'- or 3'-deoxyribonucleoside monophos-phates. When RNA is subjected to the action of snake venom diesterase (Cohn and Volkin, 1953), purine and pyrimidine nucleoside diphosphates are obtained, in addition to ribonucleoside-5'-phosphates. The 5'-position of the phosphate residues in these ribonucleotides was established by comparison with unambiguous synthetic compounds and, in some in-stances, by appropriate analytical reactions.

Adenosine monophosphate (AMP), obtained by the action of phos-phodiesterase on RNA, proved to be identical with adenylic acid from muscle (XXII), whose structure had already been elucidated by conver-sion to inosine-5'-phosphate. This nucleotide, also termed muscle inosinic acid, proved to be the 5'-phosphate; after acid treatment, both ribose-5-phosphate and hypoxanthine were liberated.

Uridine monophosphate (UMP), liberated from RNA after phosphodi-esterase treatment, reacted with a molecule of periodate with oxidative breakdown. Since this reaction does not take place if there is partial esterification of the 1,2-diol grouping, the phosphate residue must be in the 5′-position.

If RNA is degraded with pancreatic ribonuclease (Jones and Perkins, 1923) or with alkali (Levene, 1919), the nucleotides obtained show a free 5′-hydroxyl group and do not react with periodate. Accordingly, these nucleotides must carry the phosphate residue in the 2′- or 3′-position (Khym et al., 1953). Ion exchange chromatography of such hydrolysates suggested that both isomers (the so-called a- and b-forms, cf. XXIII and XXIV) are produced in these degradation proce-dures. This was confirmed by synthesis of a mixture of adenosine-2′- and 3′-phosphates (Brown et al., 1952) and comparison with the corre-sponding degradation products from RNA. However, the problem of which isomer was to be assigned the structure of the 2′-phosphate and which the 3′-phosphate remained unresolved. Because the two phosphates in-terconvert in dilute acid (Brown et al., 1952), comparison of the degrada-tion products with synthetic nucleoside-2′- or 3′-phosphates was not feasible, and their identification had to be attempted on the basis of physical data. The isomeric adenosine phosphates exist in solution as zwitterions, hence comparison of the pK values and densities allows one to estimate the charge separation and therefore the structure of the zwitterions. Studies of adenosine-2′- and 3′-phosphates and of the corresponding cytidine phosphates (Cavalieri, 1953) showed that solu-tions of the b-isomers had lower pK values and higher densities, imply-ing greater charge separation; consequently, these isomers must be re-garded as the 3′-phosphates. These conclusions have been confirmed by X-ray diffraction studies (Brown, Fasman et al., 1953; Alver and Furberg, 1959). Finally, Todd and his group achieved the unambiguous synthesis of adenosine-2′-phosphate and of uridine-2′-phosphate. The synthetic products coincided in all properties with the a-isomers from the alkaline RNA hydrolysates.

The existence in dilute acid of an equilibrium between nucleoside-2′- and 3′-phosphates also occurs in other cis-1,2-diol monophosphates; this rearrangement is clearly facilitated by a sterically favorable cyclic transition form (Fig. 9). The tendency to develop cyclic transition forms affects the speed of hydrolysis of phosphodiesters. While a diester of the form XXVI is sensitive to acids and alkali and is cleaved via XXVIII to the isomeric mixture XXIX (Bailly and Gaume, 1935; Baer and Kates, 1950), the diester XXVII is rather insensitive to hydrolysis. Accordingly, the hydroxyl group which is α to the phosphodiester group

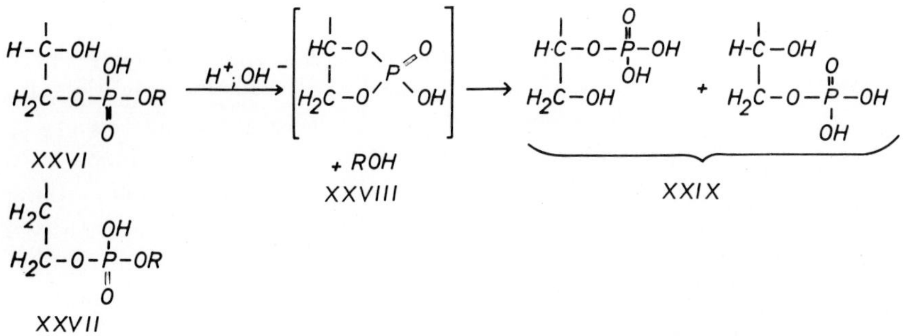

$$
\begin{array}{c}
\overset{|}{H-C-OH} \\
\overset{|}{H_2C-O}\underset{\overset{||}{O}}{-P}\overset{OH}{-OR}
\end{array}
\quad \xrightarrow{H^+,\,OH^-} \quad
\left[
\begin{array}{c}
\overset{|}{HC-O} \\
\overset{|}{H_2C-O}
\end{array}
\!\!\!\underset{O}{\overset{\nearrow O}{P}}\!\!\!{}^{O}_{OH}
\right]
\quad \longrightarrow \quad
\begin{array}{cc}
\overset{|}{H\cdot C-O-\overset{\overset{O}{||}}{P}-OH} & \overset{|}{H-C-OH} \\
\overset{|}{H_2C-OH}\;{}_{OH} & +\;\; \overset{|}{H_2C-O-\underset{OH}{\overset{\overset{O}{||}}{P}}-OH}
\end{array}
$$

XXVI   + ROH   XXIX
                              XXVIII

$$
\begin{array}{c}
\overset{|}{H_2C} \\
\overset{|}{H_2C-O}\underset{\overset{||}{O}}{-P}\overset{OH}{-OR}
\end{array}
$$

XXVII

Fig. 9. Hydrolysis of phosphodiester groups (from Bailly and Gaume, 1935).

can considerably increase the rate of hydrolysis. It was to this factor that Fono (1947) attributed the much higher alkali sensitivity of RNA as compared with DNA. Brown and Todd postulated that cyclic phosphates of type XXV must have occurred as intermediates in the alkali degrada-

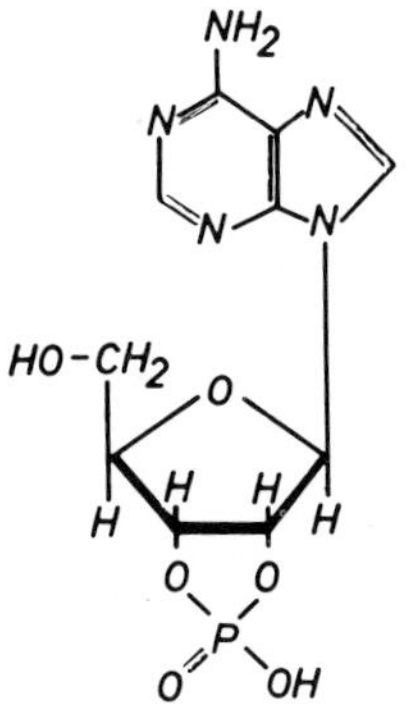

XXV

Adenosine-2′-3′-
cyclic phosphate

tion of RNA. Later these compounds were in fact found in RNA hydrolysates (Markham and Smith, 1952) and were identified by comparison with synthetic compounds (Bockstahler and Kaesberg, 1962).

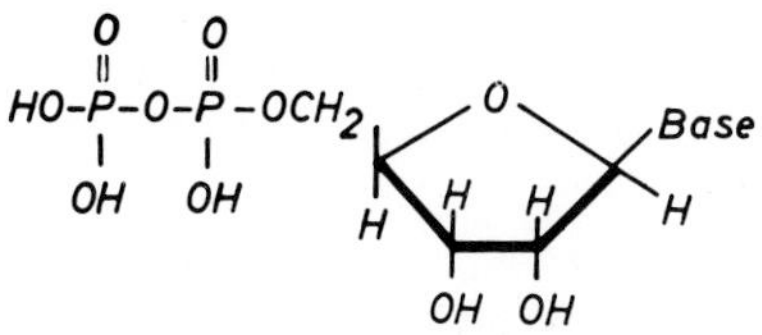

XXX

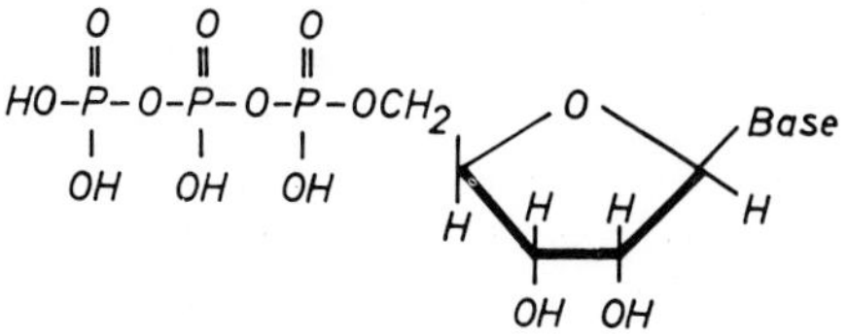

XXXI

The structure of the nucleoside di- and triphosphates (XXX, XXXI), which serve as energy-rich precursors in the biosynthesis of polynucleotides (cf. pp. 168 and 198), follows directly from the potentiometric titration curves, which indicate two or three primary and one secondary phosphorate group. Adenosine triphosphate, in addition, reacts with periodate and increases the conductivity of boric acid solutions. These findings demonstrate the existence of the free *cis*-2′,3′-diol group in the ribose residue.

CHEMICAL SYNTHESIS OF NUCLEOTIDES

In recent years, numerous synthetic procedures have been developed which permit, in principle, the synthesis of any desired nucleoside mono-, di- and triphosphates. A full discussion of the many synthetic methods is beyond the scope of this book; the interested reader is referred to the detailed reviews of Khorana (1961) and Cramer (1966). Moreover, commercially unavailable mononucleosides may be prepared from polynucleotide hydrolysates. To illustrate the steps in the synthesis of a nucleotide, some of the basic reactions will be described, with emphasis on procedures useful for the synthesis of nucleotide analogs or for the incorporation of radiophosphorus.

The synthesis of a nucleotide can take place either by combining a base with a phosphorylated sugar (Tyunosin, 1962) or, more commonly, by phosphorylation of a nucleoside. The conversion of a nucleoside into a nucleotide generally requires three reactions. First, all hydroxyl and amino groups which are not to be phosphorylated must be blocked by treatment with suitable reagents, selected so as to be easily split off after phosphorylation. The next step is phosphorylation with a sufficiently active phosphate. The final step involves the removal of the protective groups of any phosphate esters. In Fig. 10, the synthesis of a nucleoside-5′-phosphate is shown diagrammatically. The isopropylidene derivative of the nucleoside, which is easily obtained according to the procedure of Hampton (1961), is phosphorylated in the 5′-position with dibenzylphosphoryl chloride; the dibenzyl ester groups are removed by catalytic hydrogenation, and the isopropylidene residues are split off by dilute acid.

The second reaction scheme (Fig. 11) shows the preparation of a deoxyribonucleoside-5′-phosphate. The 3′-hydroxyl group is protected by the acetyl residue which is introduced after triphenylmethylation of the 5′-hydroxyl. The acid-sensitive trityl residue is removed by warming in acetic acid after acetylation and phosphorylation have been carried out according to Fig. 10. If deoxycytidine-5′-phosphate is obtained as in

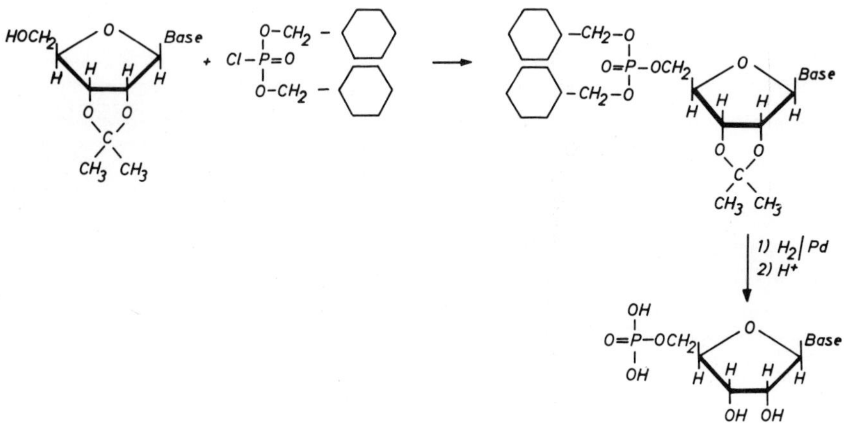

Fig. 10. General scheme for the synthesis of a 5′-ribonucleotide from the corresponding ribonucleoside.

Fig. 11, then both the 3′-hydroxyl group and the amino group in the 6-position of the pyrimidine base become acetylated. This eliminates possible side reactions during phosphorylation. Removal of the N-acetyl residue with alkali presents no difficulties.

The triphenylmethyl residue cannot be used as a protective group for the synthesis of 5′-purine deoxyribotides, because the acid hydrolysis required to remove this residue may also break the glycosidic bond between the purine and the deoxyribose. Khorana and co-workers (Schaller and Khorana, 1962), therefore, have used the mono-, di- or tri-p-methoxy-triphenylmethyl residue in place of the previously used triphenylmethyl residue since the severity of acid treatment needed to remove the pro-

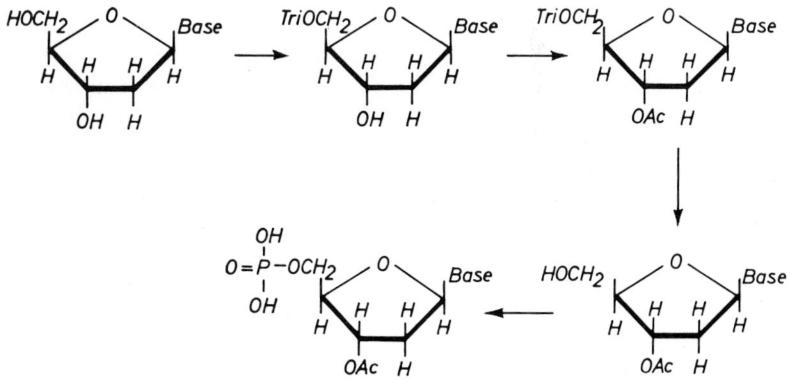

Fig. 11. Synthesis of a 5′-deoxyribonucleotide from the deoxyribonucleoside.

tective group decreases considerably as the number of p-methoxy groups increases.

The use of this modified triphenylmethyl residue may be avoided by peracetylation of the nucleoside, followed by partial hydrolysis with alkali; since the 5'-acetyl residue is split off more readily than the 3'-acetyl residue, one thus obtains the 3'-acetate necessary for subsequent phosphorylation.   3'-Deoxyribonucleotides may be obtained in analogous fashion by reacting the 5'-triphenylmethyl ether (or p-methoxytriphenylmethyl ether) with an active phosphorylating agent and removing the protective group.   The 2'- or 3'-ribonucleotides are prepared most simply in the same way as the 3'-deoxyribonucleotides; the resulting isomeric mixture of 2'- and 3'-phosphates may be separated by chromatography.   To raise the yield of the desired isomers, one can treat the by-product with acid, partially converting it to the desired isomer via the cyclic intermediate stage described on p. 20.

During the last fifteen years, numerous phosphoric acid derivatives have become available as phosphorylating agents, thus enabling all the ribonucleosides and deoxyribonucleosides to be converted to the corresponding ribonucleotides or deoxyribonucleotides.   A list of these compounds with a detailed discussion of their mode of action is included in the review by Khorana (1961).   Only one example from each of the two large groups of phosphorylating agents—the phosphoryl chlorides and the pyrophosphates—will be discussed here.

Figure 10 illustrates the synthesis of a 5'-ribonucleotide by means of dibenzylphosphoryl chloride.   This phosphoryl chloride (XXXII), developed by Todd and co-workers for nucleotide synthesis, acts as an electrophilic reagent and preferentially attacks primary hydroxyl groups with the formation of a phosphate triester.   The two benzyl residues are easily removed by hydrogenation under mild conditions.   With this reagent, numerous 5'-nucleotides have been synthesized.

The second group of phosphorylating agents is the pyrophosphates, these include cyanoethyl phosphate in combination with dicyclohexylcarbodiimide, the so-called Tener reagent (XXXIII; Tener, 1961).   The reaction mechanism probably involves the conversion of cyanoethyl phosphate to a meta- or a polyphosphate, which then reacts with primary and secondary hydroxyl groups via phosphorylation.   The resulting product (type XXXIV) is split by mild alkali treatment into the nucleotide derivative and acrylonitrile.   The reagent generally gives very high yields and is also suitable for the synthesis of 3'-purine deoxyribonucleotides which are difficult to synthesize by other methods (Schaller, personal communication).   Because cyanoethyl phosphate can be easily prepared in a P^{32}-labeled form (Tener, 1961), the Tener reagent is also suitable for the synthesis of P^{32}-containing nucleotides.

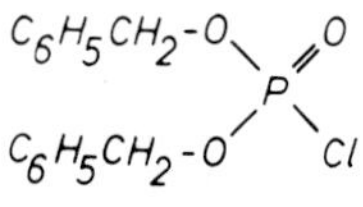

$$C_6H_5CH_2-O \diagdown \diagup O$$
$$P$$
$$C_6H_5CH_2-O \diagup \diagdown Cl$$

XXXII

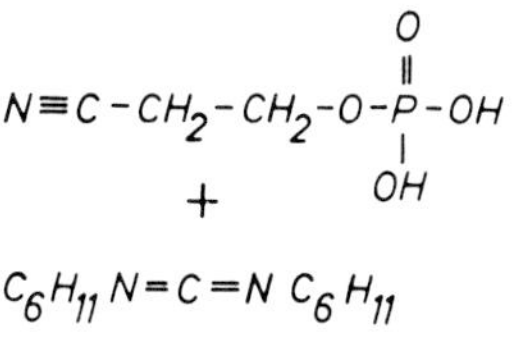

$$N \equiv C - CH_2 - CH_2 - O - \overset{\overset{O}{\|}}{\underset{\underset{OH}{|}}{P}} - OH$$

$+$

$$C_6H_{11} N = C = N\ C_6H_{11}$$

XXXIII

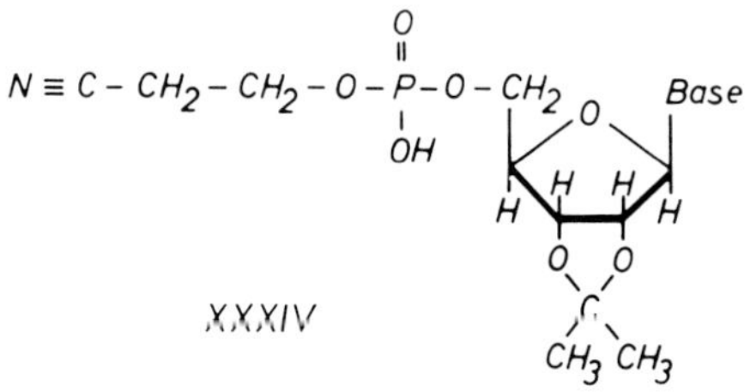

$$N \equiv C - CH_2 - CH_2 - O - \overset{\overset{O}{\|}}{\underset{\underset{OH}{|}}{P}} - O - CH_2 \quad Base$$

XXXIV

$$CH_3 \quad CH_3$$

Ribonucleoside and deoxyribonucleoside di- and triphosphates, used as nucleic acid precursors in enzymatic RNA and DNA syntheses, may be prepared in a one-step reaction from the monophosphates (Smith and Khorana, 1958). For that purpose, the tri-*n*-butylammonium salts of the monophosphates are reacted in pyridine with excess phosphoric acid, with the addition of cyclohexylcarbodiimide. Depending on the reaction conditions, the end products are di- or triphosphates, with a series of 5′-polyphosphates as intermediates.

PROPERTIES OF BASES, NUCLEOSIDES
AND NUCLEOTIDES

The compounds discussed so far are colorless, crystalline substances, all of which melt above 150°C, usually with decomposition. The bases are less soluble in water than the parent aromatics from which they are derived. The ribonucleosides and deoxyribonucleosides dissolve more readily than the pure bases because of their pentose residues; the solubilities of the ribonucleotides vary—perhaps due to the formation of zwitterions.

The chemical properties of the bases and nucleosides are clearly summarized by the scheme shown in Fig. 12 and devised by Pullman and Pullman (1961); the properties and reactions are in substantial agreement with those predicted by molecular orbital theory, even when these compounds exist as polynucleotides.

In contrast to the bases, nucleosides and nucleotides are optically active because of their pentose component. The purine ribonucleosides and deoxyribonucleosides exhibit negative rotation in common with the

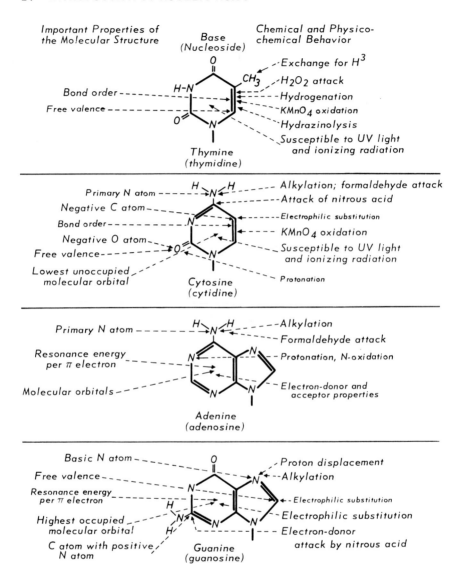

Fig. 12. Charges and chemical properties of DNA bases and nucleosides (according to Pullman and Pullman, 1961).

two sugars in mutarotation equilibrium, while the corresponding pyrimidine nucleosides and nucleotides are dextrorotatory, except for thymidine-5'-phosphate which is weakly levorotatory.

The pK values and other characteristics of bases, nucleosides and nucleotides are shown in Tables 2A and 2B. From these, it can be seen that adenine and cytosine are about as basic as the primary aromatic amines (e.g., aniline: $pK_a' = 4.6$), whereas guanine is much less so. It is not yet known which nitrogen of the free bases is protonated. However, nuclear magnetic resonance measurements have shown that the nitrogen atom in position 1 is probably involved in adenosine and cytidine, and nitrogen 7 in guanosine. It is likely that the same will hold true for the corresponding nucleotides (Jardetzky and Jardetzky, 1960). The sugar moiety in nucleosides lowers the basicity of the amino groups by 0.5–1.5 pK units, while the phosphate residue of nucleotides increases their basicity. The dissociation of the —NH—CO— group in positions 1 and 6 of uracil, thymine and guanine is slight ($pK = 9.5$) and is hardly affected by conversion to the corresponding nucleosides and nucleotides. The hydroxyl groups in the pentose moiety of nucleosides have about the same pK value as glucose ($pK = 12.1$). In nucleotides, this dissociation

TABLE 2A. Properties of Purine Bases, Nucleosides and Nucleotides

Compound	Molecular Weight	Melting Point (°C)	Specific Rotation		pK
			α	c (g/dl)	
Pyrimidine	80.09	22.5			1.30
Uracil	112.09	338 (dec)			9.5, 13
Cytosine	111.10	320–5 (dec)			4.6, 12.2
Thymine	126.11	326 (dec)			9.8, 13
Uridine	244.20	163–165	+9.6	2; H_2O	9.2, 12.5
Cytidine	243.23	215	+34.2	2; H_2O	4.2, 12.3
Thymidine	242.23	186	+32.8	1.04; 1N NaOH	9.8, 13
Deoxycytidine	227.22	215	+82.4	1.31; 1N NaOH	4.3, 13
Uridine-3′-phosphate	324.18	191 (dec)	+22.3	2.0; H_2O	1.02, 5.88, 9.43
Uridine-5′-phosphate	324.18				6.4, 9.5
Cytidine-3′-phosphate	323.21	234 (dec)	+49.4	1.0; H_2O	0.8, 4.28, 6.0
Cytidine-5′-phosphate	323.21	233			4.5, 6.3
Thymidine-3′-phosphate	322.21				
Thymidine-5′-phosphate	322.21		−4.4	0.4; H_2O	1.6, 6.5, 10.0
Deoxycytidine-3′-phosphate	321.24				
Deoxycytidine-5′-phosphate	321.24	187	+35	0.2; H_2O	4.6, 6.6

TABLE 2B. Properties of Purine Bases, Nucleosides and Nucleotides

Compound	Molecular Weight	Melting Point ($^{\circ}C$)	Specific Rotation		pK
			α	c (g/dl)	
Purine	120.11	216–217			2.5, 8.9
Adenine	135.11	360–365			4.15, 9.80
Guanine	151.15	365			3.3, 9.2, 12.3
Adenosine	267.24	234–235	−61.7	0.7; H_2O	3.45, 12.5
Guanosine	283.26	235 (dec)	−72	1.4; 0.1N NaOH	1.6, 9.2, 12.3
Deoxyadenosine	251.24	189–190	−26	1.0; H_2O	
Deoxyguanosine	267.24	>300	−47.7	0.86; 1N NaOH	
Adenosine-5′-phosphate	347.22	200 (dec)	−26	1.0; 10% HCl	3.8, 6.2
Adenosine-3′-phosphate	347.22	196 (dec)	−45.4	0.5; 0.5M Na_2HPO_4	
Deoxyadenosine-5′-phosphate	331.22	142	−38	0.23; H_2O	4.4, 6.4
Deoxyadenosine-3′-phosphate	331.22				
Guanosine-5′-phosphate	363.24	190–200 (dec)			
Guanosine-3′-phosphate	363.24		−57	1.0; 0.5N NaOH	0.7, 2.3, 5.92, 9.38
Deoxyguanosine-5′-phosphate	347.23	>360 (Ba salt)	−31	0.43; H_2O (Ba salt)	2.9, 6.4, 9.7
Deoxyguanosine-3′-phosphate	347.23				

is no longer detectable, while two more dissociation equilibria occur at the phosphate residue due to the ionization of the first (pK = 0.7–1.6) and the second (pK = 5.8–6.6) phosphate hydroxyls. Altogether, therefore, a nucleotide has three or four dissociation constants. Depending on the base, these are, in order of increasing pK values: the first dissociation of the phosphate residue, the dissociation of the NH_3^+ group, the second dissociation of the phosphate, and finally the dissociation of the —NH—CO— group. The differences in pK values of the NH_2^+ or N^+H groups, respectively, groups can be used for the electrophoretic separation of bases, nucleotides and nucleosides.

The bases, especially when they are hydroxylated in the 2- and 6-positions, show little aromatic character. Thus, the carbon-carbon bond in the 4,5-position of uracil has such strong double-bond character that it can be hydrogenated even under relatively mild conditions. Both the amino group and the —NH— groups (in positions 1 and 3) of the bases react with formaldehyde. Lewin (1964) has studied these reactions extensively with adenine and uracil, and has found that in adenine, only the amino group undergoes a reversible reaction, forming either a mono- or a dimethylol compound, —NHCH$_2$OH or —N(CH$_2$OH)$_2$, respectively. The Schiff base —N=CH$_2$ does not participate in the equilibrium. At 20 °C, the stability constant of the monomethylol compound is approximately 22 liters/mole. In uracil, the —NH— group in either the 1- or the 3-position reacts to form a monomethylol compound with much lower stability, the K_{stab} amounting to only about 2 liters/mole at 20 °C. This formaldehyde reaction also occurs with the polynucleotides and can serve to stabilize denatured polynucleotides (cf. p. 84).

Guanine reacts with diazonium salts to form deeply colored azo compounds. However, the nucleoside does not exhibit this reaction. Rather,

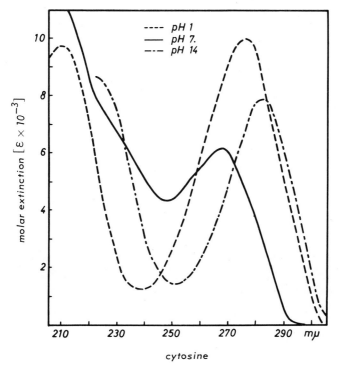

Fig. 13. Ultraviolet spectra of cytosine solution at different pH values. pH 1: NH$_2$ groups protonized. pH 14: C=O groups in enolate form.

guanosine forms a 2-diazoamino compound in about 60% yield (Kossel, 1965). The remainder of the nucleoside undergoes a radical reaction leading to the formation of fluorescent 8-aryl derivatives of guanosine, with the elimination of nitrogen (Hoffman u. Müller, 1966). This reaction is not restricted to guanosine; both adenosine and polynucleotides can undergo it as well.

The ultraviolet absorption of the base, however, is due to the same electronic transition as in benzene; i.e., the absorption band at about 260 mμ results primarily from a $\pi \longrightarrow \pi^*$ transition; in the purine bases, the absorption band consists of two independent excitations of this kind (Mason, 1954). In addition, about 10% of the ultraviolet absorption of

TABLE 3A. Optical Properties of Purine and Pyrimidine Bases and Nucleosides (from Beaven et al., 1955)

Compound	pK_A	λ_{max_1} (mμ)	E_{max_1} (×10⁻³)	λ_{min_1} (mμ)	E_{min_1} (×10⁻³)	λ_{max_2} (mμ)	E_{max_2} (×10⁻³)	λ_{min_2} (mμ)	E_{min_2} (×10⁻³)	E_{260} (×10⁻³)	E_{250}/E_{260}	E_{280}/E_{260}	E_{290}/E_{260}
Adenine	<0	—	—	—	—	—	—	—	—	≤11.5	—	—	—
	4.1	262.5	13.15	229	2.55	—	—	—	—	13.0	0.76	0.375	0.035
	9.8	260.5	13.35	226	2.55	—	—	—	—	13.3	0.76	0.125	0.005
		269	12.3	237	3.35	—	—	—.	—	10.45	0.57	0.60	0.025
Guanine	<0	—	—	—	—	—	—	—	—	≤ 8.8	—	—	—
	3.2	275.5	7.35	267	7.15	248.5	11.4	224	3.55	8.0	1.37	0.84	0.495
	9.6	275.5	8.15	262	7.05	246	10.7	224.5	4.3	7.2	1.42	1.04	0.54
	~12.5	273.5	8.0	255	6.05	246	6.3	234	6.0	6.4	0.985	1.135	0.585
		274.0	9.9	239.5	5.15	—	—	—	—	11.3	0.705	0.19	0.007
Cytosine	4.5	276	10.0	238.5	1.2	210	9.7	—	—	6.0	0.48	1.53	0.78
	12.2	267	6.13	247	4.3	—	—	—	—	5.55	0.78	0.58	0.08
		282	7.86	250.5	1.4	—	—	—	—	2.35	0.595	3.28	2.6
Thymine	~0	—	—	—	—	—	—	—	—	—	—	—	—
	9.9	264.5	7.89	233.5	1.9	207	9.5	—	—	7.4	0.67	0.53	0.09
	>13	291	5.44	244	2.2	—	—	—	—	3.7	0.65	1.31	1.41
		—	—	—	—	—	—	—	—	≤3.6	—	—	—
Uracil	~0.5	260.0	7.80	228.5	1.52	—	—	—	—	7.8	0.795	0.30	0.05
	9.5	259.5	8.20	227.5	1.8	—	—	—	—	8.2	0.84	0.175	0.01
	>13	284	6.15	241.5	2.15	—	—	—	—	4.1	0.71	1.40	1.27
		—	—	—	—	—	—	—	—	4.1	—	—	—
Adenosine	3.4	257	14.6	230	3.5	—	—	—	—	14.3	0.84	0.215	0.03
	13.0	259.5	14.9	227	2.25	—	—	—	—	14.9	0.78	0.144	0.002
		—	—	—	—	—	—	—	—	—	—	—	—
Guanosine	2.2	256.5	12.2	228	2.4	—	—	—	—	11.75	0.94	0.695	0.50
	9.5	252.5	13.65	223.5	2.85	—	—	—	—	11.7	1.15	0.67	0.275
		258–66	11.3	231	4.1	—	—	—	—	11.3	0.89	0.61	0.13
Cytidine	4.1	280	13.4	241.5	1.7	212.5	10.1	—	—	6.4	0.45	2.10	1.55
	~13	271	9.1	250.5	6.5	229.5	8.3	226	8.2	7.55	0.86	0.93	0.28
		273	9.2	251.5	6.03	—	—	—	—	7.0	0.865	1.17	0.56
Thymidine	9.8	267	9.65	235	2.25	207.5	9.55	—	—	8.75	0.65	0.72	0.235
	>13	267	7.38	240.5	4.58	—	—	—	—	6.65	0.75	0.67	0.16
		—	—	—	—	—	—	—	—	≤6.5	—	—	—
Uridine	9.28	262	10.1	230.5	2.05	—	—	—	—	9.95	0.74	0.35	0.03
	>13	262	8.5	236.5	4.48	—	—	—	—	7.35	0.83	0.29	0.02
		—	—	—	—	—	—	—	—	≤7.1	—	—	—

[a]In general, maxima and minima of absorbance values have been measured at pH values at least 1.5–2 pH units away from the nearest pK.

TABLE 3B. Optical Properties of Nucleotides (from Beaven *et al.*, 1953)

Compound	pH	λ_{max} (mμ)	E_{max} ($\times 10^{-3}$)	λ_{min} (mμ)	E_{min} ($\times 10^{-3}$)	λ_{max} (mμ)	E_{max} ($\times 10^{-3}$)	E_{260} ($\times 10^{-3}$)	$\dfrac{E_{250}}{E_{260}}$	$\dfrac{E_{280}}{E_{260}}$	$\dfrac{E_{290}}{E_{260}}$
Adenosine-3'-	2	—	—	—	—	—	—	14.2	0.85	0.22	0.038
phosphate	7	—	—	—	—	—	—	15.0	0.80	0.15	0.009
	12	—	—	—	—	—	—	15.0	0.80	0.15	—
Adenosine-5'-	2	257	15.1	230	3.5	—	—	14.2	0.84	0.22	0.038
phosphate	7	—	—	—	—	—	—	15.0	0.79	0.15	0.009
	12	259	15.4	227	2.6	—	—	15.0	0.79	0.15	—
Guanosine-3'-	1	257	12.2	228	2.6	—	—	11.8	0.93	0.69	0.49
phosphate	7	—	—	—	—	—	—	11.4	1.15	0.68	0.285
	12	256	11.1	230	4.3	—	—	11.2	0.89	0.60	0.11
Guanosine-5'-	1	—	—	—	—	—	—	11.8	0.99	0.70	0.51
phosphate	7	—	—	—	—	—	—	11.4	1.15	0.68	0.285
	12	—	—	—	—	—	—	11.2	0.89	0.60	0.11
Cytidine-3'-	2	279	13.0	240	1.7	—	—	6.8	0.45	2.00	1.43
phosphate	7	—	—	—	—	—	—	6.8	0.86	0.93	0.30
	12	272	8.90	250	6.78	—	—	7.6	0.86	0.93	0.30
Cytidine-5'-	2	281	13.6	—	—	—	—	6.2	0.46	2.10	1.55
phosphate	7	—	—	—	—	—	—	7.4	0.84	0.99	0.33
	12	274	7.3	250	6.6	—	—	7.4	0.84	0.99	0.33
Uridine-3'-	2	262	9.9	230	1.95	—	—	10.0	0.76	0.32	0.03
phosphate	7	—	—	—	—	—	—	10.0	0.73	0.35	0.03
	12	261	7.3	242	5.37	—	—	7.4	0.83	0.28	0.02
Uridine-5'-	2	261	9.7	—	—	—	—	10.0	0.74	0.38	0.03
phosphate	7	—	—	—	—	—	—	10.0	0.73	0.40	0.03
	12	—	—	—	—	—	—	7.4	0.82	0.33	0.03
Thymidine-5'-	2	267	9.6	—	—	—	—	8.4	0.64	0.72	0.23
phosphate	7	—	—	—	—	—	—	8.4	0.65	0.73	0.24
	12	—	—	—	—	—	—	6.7	0.74	0.67	0.17

the bases results from an $n \longrightarrow \pi^*$ transition, due to the excitation of free electrons on the nitrogen atoms of the aromatic ring. The $n \longrightarrow \pi^*$ transition has a minor absorption band at a longer wavelength than the $\pi \longrightarrow \pi^*$ transition and is largely masked by the latter (Mason, 1954). Proof of the two kinds of transition was obtained by means of polarized ultraviolet light with single crystals.

The effect of pH on the position and intensity of the ultraviolet bands of cytosine may be seen in Fig. 13. Ionization of both the —NH$_2$ and the —NH—CO— groups is accompanied by a bathochromic shift and a marked rise in absorption. The spectra of the other bases and of the nucleosides and nucleotides show a similar pH dependence. Therefore, reliable quantitation of these substances on the basis of spectrophotometric data is possible only in a pH range where either the ionic or the neutral form exists alone (Holiday, 1930). Measurements of optical density must be made in a pH range that differs sufficiently from the pK values (1–2 units).   Conversely, the strong pH dependence of the base spectra makes possible the determination of pK values unobtainable by conventional methods (Flexer *et al.*, 1935; Fox and Shugar, 1952).

Volkin and Cohn (1954) have shown that the ratios of extinction co-efficients at given wavelengths and pH values may be used to identify bases, nucleosides and nucleotides; this eliminates the necessity of routinely determining the full spectrum (cf. p. 87).   Tables 3A and 3B list some of these ratios for the absorption at 290, 280, 260 and 250 mμ, the wavelengths of the maxima and minima and the corresponding extinc-

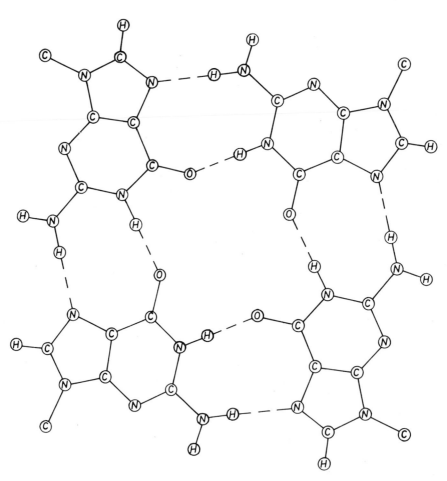

Fig. 14. Hydrogen bonds between four guanine molecules (from Gellert *et al.*, 1962).

tion coefficients. Detailed tables of these values are available (e.g., Volkin and Cohn, 1954; Beaven *et al.*, 1955).

In conclusion, we would like to draw attention to a peculiarity of guanosine-3′- and 5′-phosphate. In 0.2M NaCl solution at pH 5, above a critical concentration, both isomers form helical aggregates (Gellert *et al.*, 1962) which apparently are composed of layers of four nucleotides, connected by eight hydrogen bonds (Fig. 14). In the case of 5′-GMP, the distance between layers is 3.25 Å; strong van der Waals and hydrophobic forces may therefore be responsible for their cohesion. The helical form results because each layer is rotated against its neighbor by $-25°$. In the case of the 3′-isomers, two single layers seem to form a subunit, thereby giving rise to additional periods of 6.73 Å.

CHAPTER III

Chemistry of the Nucleic Acids

PRIMARY STRUCTURE OF THE NUCLEIC ACIDS

Type and Location of the Internucleotide Bond

The primary problem in the structure of nucleic acids is the way in which the individual nucleotides are linked. In principle, these bonds can originate from any available functional group of the mononucleotide, i.e., from the amino and carbonyl groups of the bases, from the hydroxyl groups of the pentoses, and from the phosphates. Titration curves of nucleic acids obtained by Levene and Simms as early as 1925, and also by Hammarsten (1924), showed a characteristic disappearance of the second dissociation of the phosphate residues. This distinguishes them from the mononucleotides and suggests that a phosphodiester bond is involved in the internucleotide linkage. Levene has given the following general scheme (A) for the primary structure of nucleic acids, a scheme subsequently proved to be correct:

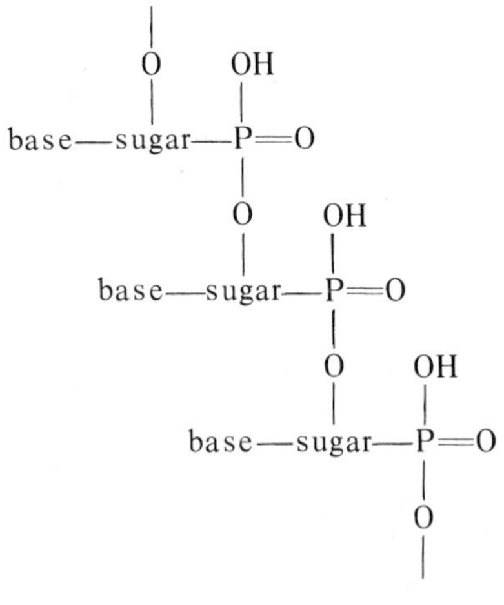

$$
\begin{array}{c}
\mid \\
\mathrm{O} \quad \mathrm{OH} \\
\mid \quad\quad \mid \\
\text{base}-\text{sugar}-\mathrm{P}{=}\mathrm{O} \\
\mid \\
\mathrm{O} \quad \mathrm{OH} \\
\mid \quad\quad \mid \\
\text{base}-\text{sugar}-\mathrm{P}{=}\mathrm{O} \qquad\qquad \text{A}\\
\mid \\
\mathrm{O} \quad \mathrm{OH} \\
\mid \quad\quad \mid \\
\text{base}-\text{sugar}-\mathrm{P}{=}\mathrm{O} \\
\mid \\
\mathrm{O} \\
\mid
\end{array}
$$

However, titration curves did not exclude the possibility of branched phosphotriester bonds occurring in the nucleic acid primary structure (see scheme B).

```
        |                         |
        O                         O
        |                         |
base—sugar—┐          ┌—sugar—base
           |          |
           O          O
           |          |
       O=P—O—sugar—base              B
           |
           O
           |
base—sugar—┘
           |
           O
           |
```

Because Todd and co-workers have shown such phosphotriesters to be very sensitive to acid and alkaline hydrolysis under conditions in which RNA and DNA do not yield detectable amounts of cleavage products, it is unlikely that these branches occur in either RNA or DNA.

With RNA, there is a further possibility of branching (as shown in scheme C) which involves esterification of all three available ribose hydroxyl groups:

```
        |              |
        O          OH  O
        |          |   |
base—sugar—┐    O=P—sugar—base
           |       |
           O       O
           |       |
       O=P—O—sugar—base              C
           |       |
          HO       O
                   |
               O=P—O—sugar—base
                   |
                   OH
```

Such branches would be more resistant to alkali than the usual phosphodiester bonds, since the α-hydroxyl group catalyzing the hydrolytic breakdown is blocked (cf. Fig. 9); furthermore, nucleosides would occur in the RNA alkali hydrolysates. However, no alkali-resistant trinucleotides appear on hydrolysis, and degradation leads quantitatively to nucleotides (Markham and Smith, 1952). Hence it is improbable that RNA contains the type of branching shown in scheme C.

The phosphodiester bonds between the pentose hydroxyl groups (scheme A) could conceivably be linked through the following carbon atoms:

(a) *In RNA*: 3′-5′; 2′-5′; 2′-3′; as well as alternating 2′-2′, 3′-3′ and 5′-5′.

(b) *In DNA*: 3′-5′; or alternating 3′-3′ and 5′-5′.

The observation that RNA, on alkaline hydrolysis, quantitatively yields mononucleotides (2′- and 3′-phosphates and cyclic 2′,3′-phosphates) excludes the 2′-3′ bond as well as the following alternating inter-nucleotide bonds: 2′-2′, 3′-3′; 2′-2′, 5′-5′; 3′-3′, 5′-5′. Either these bonds would not be split at all by alkali (cf. p. 18), or they would yield only the dinucleotide with 5′-5′ bridges and a phosphate residue in the 2′- or 3′-position. By enzymatic hydrolysis of RNA (Fig. 15), it was possible to decide in favor of the 3′-5′ linkage from the two remaining alternatives (2′-5′ or 3′-5′).

A phosphodiesterase from spleen degrades RNA quantitatively to ribonucleoside-3′-phosphates (Volkin and Cohn, 1952; cf. p. 213). There-fore, a phosphodiester bond in RNA must originate from the 3′-hydroxyl group of the ribose residue. Snake venom diesterase (see p. 214) leads to the liberation of 5′-ribonucleotides (Cohn and Volkin, 1952); hence, the second phosphodiester linkage must be attached to the 5′-hydroxyl group of the adjacent nucleotide. These findings may be regarded as verification of the primary structure of RNA shown in Fig. 15.

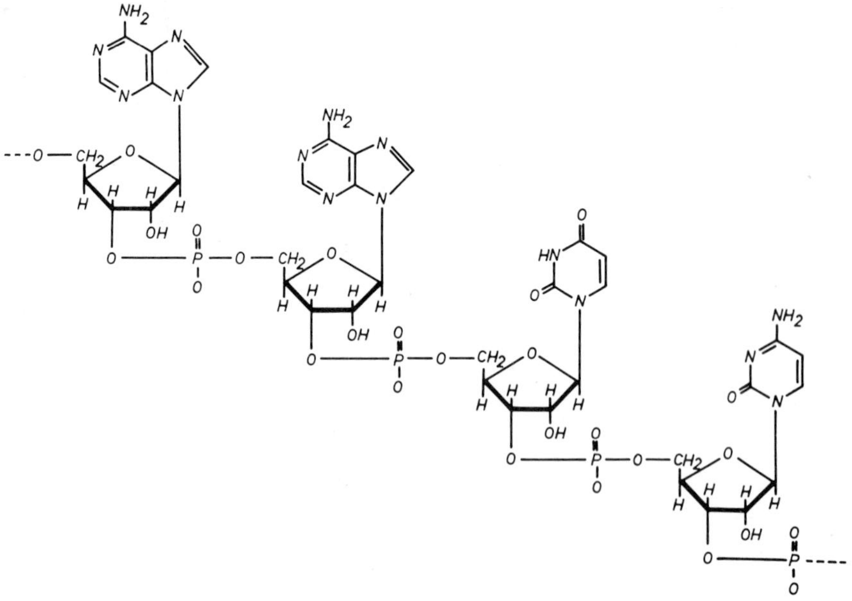

Fig. 15. Primary structure of ribonucleic acid.

The primary structure of DNA was demonstrated by splitting DNA almost quantitatively into 5′-deoxyribonucleotides by sequential treatment with deoxyribonuclease and snake venom phosphodiesterase (Hurst *et al.*, 1951). The alternating linkage (B) was not compatible with this result; thus, there remained only the 3′-5′ linkage as in the RNA primary structure. The primary structure of RNA given in Fig. 15 is therefore valid also for DNA, with thymine replacing uracil and deoxyribose replacing ribose.

Chemical Synthesis of Internucleotide Bonds

It is now possible to synthesize polynucleotides enzymatically with relative ease, but straightforward chemical synthesis is difficult. Whereas practically any desired dinucleotide can be prepared, the stepwise synthesis of a polynucleotide with a particular sequence of bases is still not feasible. Khorana (1961) has critically and comprehensively described the various synthetic methods; a few of them will be presented here to illustrate the major problems.

$$\overline{XXXV}$$

We shall first deal with the synthesis of uridyladenosine (XXXV). To obtain this compound, 3′-UMP can be combined with adenosine in the 5′-position, or a linkage of 5′-AMP with uridine through the 3′-hydroxyl group of the latter is possible. To achieve a good yield, protective groups must be used, in both cases, to block the functional groups which, after suitable activation, would otherwise also react with the phosphate

residue. In the synthesis of this dinucleotide, Khorana and co-workers
started from a 3′-UMP derivative (XXXVI), combined with an adenosine
derivative (XXXVII) by means of dicyclohexylcarbodiimide, with the
formation of the 3′-5′ bond. The synthesis of the UMP derivative (XXXVI)
takes place via the following steps (Fig. 16): 3′-UMP is converted in
dilute solution with dicyclohexylcarbodiimide (DCC) to the cyclic 3′,5′-
phosphate. The free hydroxyl group in the 2′ position is transformed to
the 2′-o-tetrahydropyranyl ether with dihydropyran. Then, the intramolecular phosphate bond is split with barium hydroxide, yielding two
isomers, with an 80% yield of the desired one. Reaction of the latter
with triphenylchloromethane (trityl chloride) produces XXXVI.

The synthesis of the adenosine derivative XXXVII, needed to react
with XXXVI, can be achieved as follows (Fig. 17). The 5′-hydroxyl group

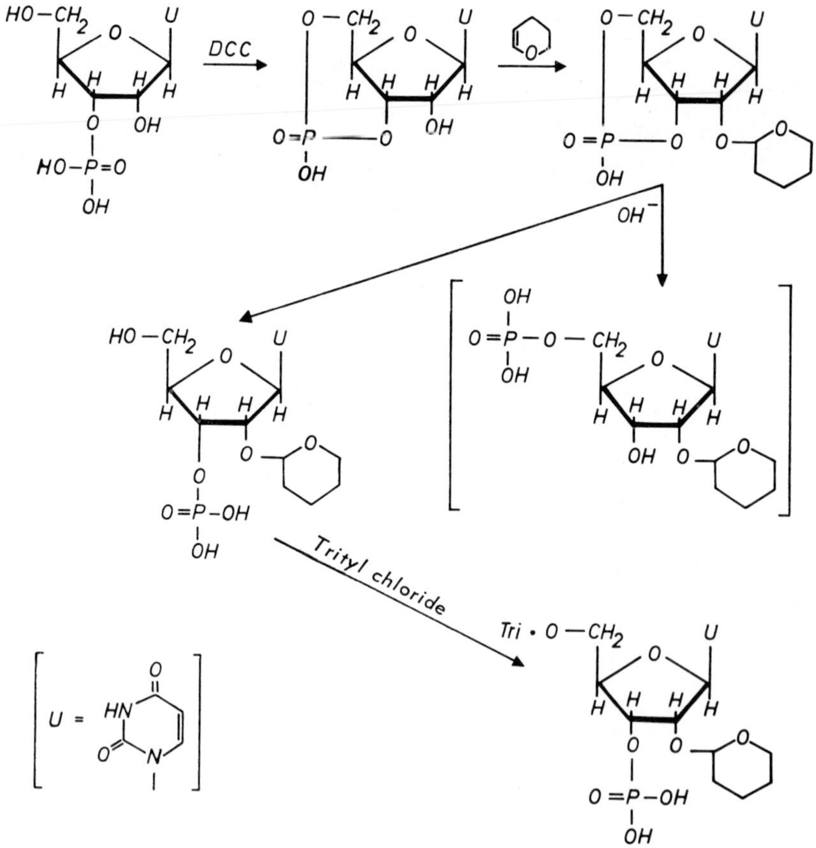

Fig. 16. Stepwise synthesis of XXXVI (for explanation, see text; after
Khorana and co-workers, 1961a and b).

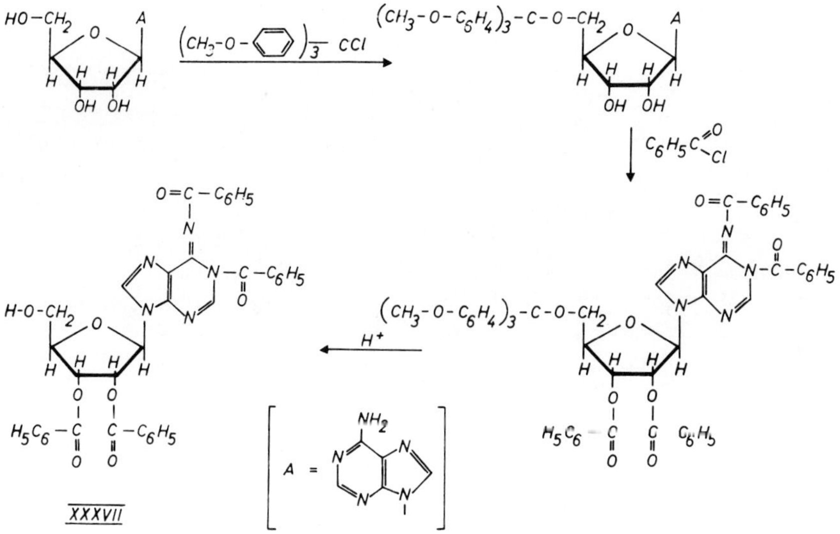

Fig. 17. Stepwise synthesis of XXXVII (for explanation, see text; after Khorana and co-workers, 1961a and b).

of adenosine is blocked by etherification with tri-*p*-methoxytriphenyl-methyl chloride; the adenosine derivative obtained is then converted with benzoyl chloride to a tetrabenzoate, from which the modified "trityl" residue is removed by mild acid treatment, with the formation of XXXVII. The 3′-UMP derivative XXXVI is combined with XXXVII by means of dicyclohexylcarbodiimide (Fig. 18); the dinucleotide derivative produced

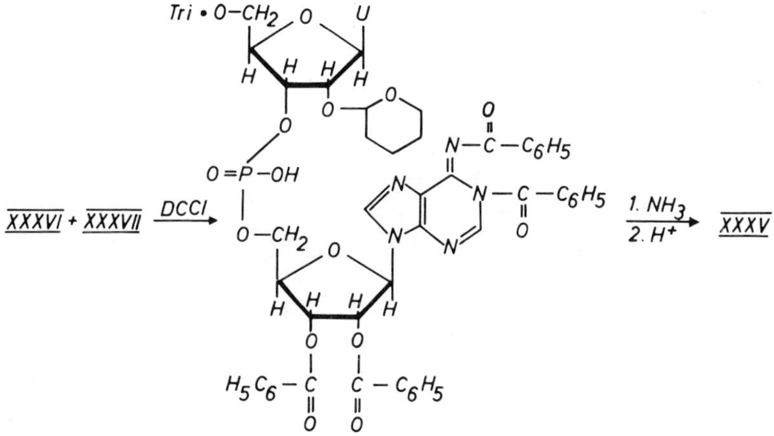

Fig. 18. Masked dinucleotide as an intermediate in the synthesis of XXXV (after Khorana and co-workers, 1961).

is treated with ammonia to remove the benzoyl residues and finally with dilute acid to remove the trityl and tetrahydropyranyl residues. With respect to the amount of compound XXXVI employed, the yield of the AU dinucleotide is almost 100%.

In the synthesis shown schematically in Fig. 18, dicyclohexylcarbodiimide is the condensation agent used to form the 3'-5' internucleotide bond. An analogous activation of the phosphate monoester may be achieved with trichloroacetonitrile, developed by Cramer and co-workers (Cramer and Bàldauf, 1960; Cramer and Weimann, 1961). Carbonyldiimidazole (Staab, 1961) and N-methyl-5-phenylisoxazolium methyl sulfate (Cramer, 1966) have also been utilized as activating agents. Particularly promising reagents are picryl chloride, which Cramer, Wittmann et al. (1963) have used for di- and oligonucleotide syntheses, and the 2,4,6-triisopropylbenzenesulfonyl chloride introduced recently by Lohrmann and Khorana (1966).

If mononucleotides are employed in the coupling reactions outlined above and there are unprotected hydroxyl groups on the pentose residue, then with an adequate supply of dicyclohexylcarbodiimide, the reaction can go beyond the dinucleotide stage and lead to the formation of oligo- or even polynucleotides. This polymerization is favored by a high monomer concentration in the reaction mixture (cyclicphosphates are formed preferentially in dilute solution); therefore, the type of solvent used appears to be crucial. Khorana and co-workers (Tener et al., 1958; Khorana and Viszsolyi, 1961) found pyridine particularly suitable, and since, of all the nucleotides, thymidine-5'-phosphate dissolves most readily in this solvent, its polymerization was studied first. It proved possible to fractionate the resulting polymerization products by ion exchange chromatography on ECTEOLA- and DEAE-cellulose (Peterson and Sober, 1956). One fraction contained polythymidylic acid in linear 3'-5' linkage (XXXVIII), the other contained cyclic products (XXXIX) of varying ring size (Fig. 19).

The formation of cyclic by-products may be largely avoided by the addition of 3'-O-acetylthymidine-5'-phosphate. This monomer derivative is always incorporated terminally, preventing the cyclization reaction (Khorana and Viszsolyi, 1961) and increasing the yield of linear products with $n > 2$ by more than 45%. In place of 3'-O-acetylthymidine, N-3'-O-diacetyldeoxycytidine-5'-phosphate may be used to form terminally modified polynucleotides. It has been shown that these polymerization reactions are not limited to thymidine phosphate as a monomer. If the amino groups of the bases are protected, polymers may also be produced from deoxycytidine and deoxyadenosine-5'-phosphates (Khorana, Turner and Viszsolyi, 1961; Ralph and Khorana, 1961). Despite its low solubility, deoxyguanosine-5'-phosphate can also be polymerized. Tri- and tetra-

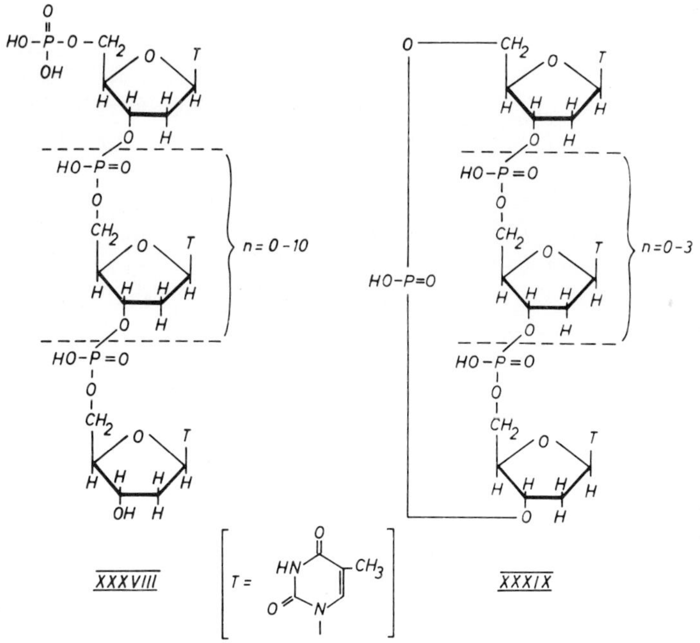

Fig. 19. Polymerization products of thymidine-5′-phosphate (after Khorana and Viszsolyi, 1961).

nucleotides of dGMP are of special interest because they form stable aggregates in aqueous solution (Ralph *et al.*, 1962). Di- and trinucleotides can be polymerized in a manner similar to the masked mononucleotides (Schaller and Khorana, 1963; Weiman *et al.*, 1963). This leads to products which can act as templates in the enzymatic synthesis of DNA and RNA (Kornberg *et al.*, 1964; cf. pp. 168 and 201), although their molecular weights are relatively small.

With polyphosphate esters as condensing agents, Schramm and co-workers (1962) have polymerized mononucleotides to a mean molecular weight of 8×10^4, corresponding to approximately 230 monomer units. It seems unlikely that polymerization occurs only via the 3′-5′ bond since it has been shown that after hydrolysis of the pyro- and polyphosphate bonds, the remaining oligonucleotides contain a maximum of four subunits (Gottlich and Slutsky, 1964; Hayes and Hansbury, 1964).

SECONDARY STRUCTURE OF NUCLEIC ACIDS

X-ray Structural Analyses

The previous section has shown that DNA and RNA are made up of essentially the same components, with only trivial differences in their primary structures. However, the behavior of these substances in salt solutions is so dissimilar that even if allowance is made for variations in the order of magnitude of their average molecular weights, different molecular conformations (secondary structures) must be assumed.

X-ray analysis has contributed considerably to the elucidation of the secondary structure of ordered macromolecules. The application of this method presupposes that the macromolecules can be oriented so as to approximate crystalline conditions. This is possible in the case of nucleic acids when fibers are drawn slowly from a concentrated, gel-like solution, so that the molecules are oriented parallel to the axis of the fiber. Depending on the degree of crystallinity, X-ray fiber patterns are obtained with reflections which, though seldom as sharp as those from single crystal photographs, still provide information on the secondary molecular structure. Although this method was successfully applied to structural studies of DNA, considerable difficulties were encountered with RNA, and only recently have results been obtained for some types of RNA.

Astbury and Bell, in 1938, were the first to obtain useful X-ray fiber patterns of DNA threads. These indicated a regular stacking with a periodicity of 3.34 Å perpendicular to the fiber axis, which in turn is parallel to the axis of the DNA molecule. The authors therefore concluded that the stacking was at the level of individual nucleotides. Furthermore, a longitudinal period of about 27 Å was demonstrated. No attempts were made at that time to derive a detailed secondary structure. Later, Furberg and co-workers (1956) found that the planes of the ribofuranose rings in ribonucleosides and ribonucleotides were nearly perpendicular to those of the bases (cf. Fig. 20 and also p. 14). Since this steric arrangement also applies to deoxyribonucleosides and deoxyribonucleotides, Astbury's conclusions had to be modified since only the bases in DNA, and not the whole nucleotide, are stacked with a spacing of 3.34 Å. Furberg suggested two models of the secondary structure of DNA based on the three-dimensional structure of the mononucleotides, but he was unable to account for the 27-Å periodicity.

Subsequently, Pauling and Corey (1953) described a molecular model of DNA made up of three polynucleotide strands wound around one another, with the phosphate groups on the inside and the bases on the outside. However, this arrangement does not accord with the chemical behavior of DNA whose phosphate groups can be titrated, but whose amino and —NH—CO— groups cannot (cf. also p. 75).

In 1953, Watson and Crick published a model for the secondary structure of DNA which agreed with the then available X-ray data and the behavior of DNA on titration with acids or alkalies and also predicted the

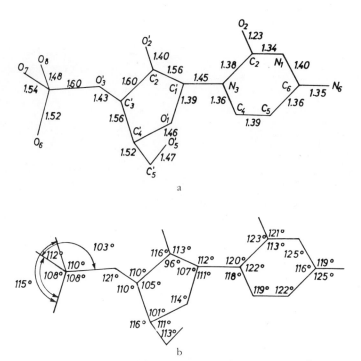

Fig. 20. Bond length (a) and angles (b) of cytidine-3′-mono-phosphate (from Alver and Furberg, 1959).

base ratio relationship observed by Chargaff (1950) and Wyatt (1951) in DNA from a variety of sources. In the Watson-Crick model, two poly-nucleotide strands are wound counterclockwise around an axis, like a double helix (Fig. 21); the two individual strands run opposite each other, i.e., the sequence C(3′)—C(4′)—C(5′)—O—P—O runs up in one strand and down in the other. The bases, arranged perpendicular to the helical axis in the center of the molecule, have a spacing of 3.4 Å. After ten bases, or nucleotides, the helix has made one full turn; thus, it re-peats every 34 Å. In this model, the two polynucleotide strands are held together in what was considered, at the time Watson and Crick presented it, an altogether novel and remarkable manner. They proposed that every base of one strand forms hydrogen bonds with a coplanar base of the other strand. To achieve a constant-diameter helix, a pyrimidine must always be opposite a purine. On the basis of the data of Chargaff and Wyatt, who showed that the molar ratio of adenine to thymine and of guanine to cytosine equals one in nearly all samples of DNA, Watson and Crick proposed specific hydrogen bonds between each pair of bases.

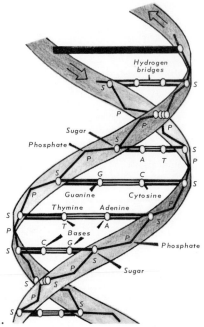

Fig. 21. The Watson-Crick model of DNA.

The nature of the base pairing is shown in Figs. 22 and 23 which also indicate the length of the hydrogen bonds.

The hydrogen bonds in Figs. 22 and 23 are only two of 24 possible combinations. Donohue (1956) has listed all of these, including those for two identical bases, and has discussed their stereochemical implications. He assumed that the NH...O distance was 2.85 Å and the NH...N distance 3.00 Å, and that the maximum permissible distortion of the NH...X bond was 15°. All 24 combinations cannot exist in polynucleotides since the sugar moieties do not always permit the necessary steric arrangements.

Recently, Sobell *et al.* (1963) have demonstrated that the hydrogen bonds required for the guanine-cytosine pair in DNA can also be formed between monomeric bases. For example, 9-ethylguanine and 3-methyl-5-bromocytosine co-crystallize from dimethyl sulfoxide in single crystals. It has been shown by X-ray analysis that guanine forms hydrogen bonds with cytosine of an adjoining unit cell, with the planes of the bases twisted like a propeller with an angle of 6.5°. Within the unit cell, the bases form layers with a spacing of 3.4 Å. An analogous structure has recently been demonstrated for single crystals of 9-ethylguanine and 3-methylcytosine (O'Brien, 1963).

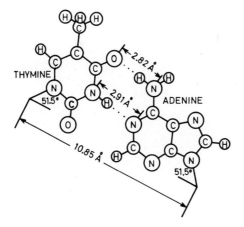

Fig. 22. Hydrogen bonds between thymine and adenine in the DNA helix.

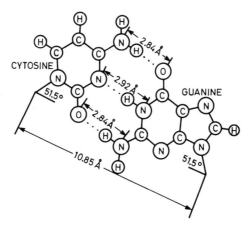

Fig. 23. Hydrogen bonds between guanine and cytosine in the DNA helix.

The kind of base pairing suggested by Watson and Crick in no way limits the base sequence of DNA, except that the sequence of bases in one polynucleotide strand is determined by that of the other, *the two single strands being complementary*. This arrangement is the basis for the reproduction of identical DNA molecules (Watson and Crick, 1953b; cf. p. 178).

Subsequent detailed X-ray analyses by Franklin and Gosling and by Feughelman *et al.* (1955) have shown that the secondary structure of DNA depends to a certain extent on the moisture content of the sample.

At less than 70% relative humidity, the sodium salt of DNA "crystal-lizes" in the so-called *A-form*, characterized by a high degree of order and correspondingly distinct X-ray diffraction diagrams. The monoclinic unit cell ($a = 22.2$ Å, $b = 40.0$ Å, $c = 28.1$ Å, $\beta = 97°$) contains a com-plete helical turn of two adjoining DNA molecules, whose axes run in the *C*-direction. The base pair in this crystal form is inclined at about $70°$ to the helix (Fig. 24). At higher moisture and salt contents (> 90% relative humidity), the crystalline A-form changes to the paracrystalline *B-form*. Although its X-ray diagram shows comparatively diffuse reflec-tions, a 3.4-Å base stacking periodicity and the spiral period of 34.6 Å, which corresponds to ten base pairs per period, can be inferred without difficulty. Moreover, its intensity distribution shows that the phosphate residues are arranged at a distance of 9 Å from the axis of the helix.

The original Watson-Crick model of this form has been refined by Wilkins and his co-workers with the aid of more recent stereochemical data. In this model, the base pairs are perpendicular to the helical axis; the slightly puckered pentose rings are assumed to be planar, requiring only trivial bond angle distortions and are arranged so as to yield maximum inclination of the ring planes with respect to the axis of the helix (Fig. 25). This results in a gradual decrease of electron density toward the periphery of the molecule and accounts for the weakness of the secondary reflections on the equator of the X-ray diffraction diagrams (see below). In contrast to the original Watson-Crick proposal, the helix axis runs be-tween the hydrogen bonds that link the bases. Figure 26 (see frontispiece) shows a space-filling molecular model of DNA-B, with the functional groups and hydrogen atoms in the bases omitted for the sake of clarity.

More recently, Wilkins and co-workers have studied the lithium salt of DNA in detail (Langridge *et al.*, 1960). The advantages of lithium over sodium are its much smaller scatter (lower atomic number) and the fact that the B-form of lithium deoxyribonucleate crystallizes as well as the A-form of the sodium salt.

At a relative humidity below 44%, the lithium B-salt changes its con-formation to the so-called *C*-form, which differs only slightly from the *B-form*. The relatively high ordering of lithium DNA-B leads to very sharp and clear X-ray reflections (cf. Fig. 27), whose intensity can be measured quantitatively. The dimensions of the orthorhombic unit cell of this lithium DNA are $a = 22.5$ Å; $b = 30.9$ Å; $c = 33.7$ Å. Table 4 lists the major crystallographic properties of crystalline and semicrystalline DNA fibers.

Determination of chemical structures by means of X-rays depends on the fact that the distances of individual atoms within a molecule are of the same order of magnitude as the wavelengths of X-rays. A regular molecular arrangement, as in a single crystal, a partially stretched film or a stretched fiber, acts like a diffraction grating for the X-rays; i.e., the scattered X-rays can be reinforced only in certain directions, while

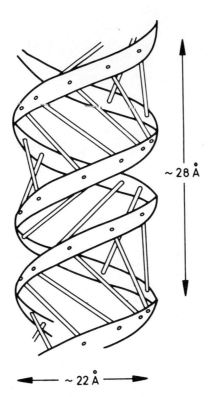

~ 28 Å

~ 22 Å

Fig. 24. Stereochemical relationships in DNA-A (from M. Fuller *et al.*, 1965). (Reproduced with permission of Academic Press Inc., New York)

they are quenched in all other directions. The direction of the diffracted beams, visualized photographically as so-called reflections, is used to measure the distance between the grating lines or lattice planes, as well as their relative positions in the unit crystal. Since a diffracted ray results only from definite geometrical conditions of orientation, the crystal is rotated about an axis so as to achieve the necessary orientation for as many lattice planes as possible ("rotation photograph"). From such

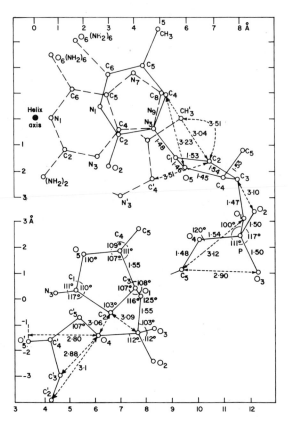

Fig. 25. Relative position of a base pair and a deoxyribose residue to the helical axis in DNA-B (from Langridge *et al.*, 1960). (Reproduced with permission of Academic Press Inc., New York)

diffraction diagrams, it is then possible to determine the dimensions of the unit cell and the angle between lattice planes.

The reflection intensities measured either densitometrically on photographic film or directly by a Geiger-Müller counter give a measure of the

TABLE 4. Molecular Conformation and Crystal Habit of DNA Fibers (from Wilkins, 1963)

Conformation	Nucleotide Pairs per Helix Repeat	Inclination of the Base Pair to the Helical Axis	Cation	Conditions[a]	Crystal Class and Crystallinity	Molecular Positions	Dimensions of Unit Cell a(Å)	b(Å)	c(Å)	β(°)
A	11.0	70°	Na, K, Rb	75% r. h.	Monoclinic Crystalline	$0\ 0\ 0$; $\frac{1}{2}\ \frac{1}{2}\ 0$	22.24	40.62	28.15	97.0
B	10.0	90°	Li	66% r. h. 3% LiCl in the fiber	Orthorhombic Crystalline	$0\ 0\ \frac{1}{6}$; $\frac{1}{2}\ \frac{1}{2}\ -\frac{1}{6}$	22.5	30.9	33.7	—
			Li	75–90% r. h.	Orthorhombic Partially crystalline	$0\ 0\ \frac{1}{8}$; $\frac{1}{2}\ \frac{1}{2}\ -\frac{1}{8}$	24.4	38.5	33.6	—
			Li, Na K, Rb	92% r. h.	Hexagonal Partially crystalline	$0\ 0\ 0$; $\frac{1}{3}\ \frac{2}{3}\ \frac{1}{6}$; $\frac{2}{3}\ \frac{1}{3}\ -\frac{1}{6}$	46	—	34.6	—
B_2	9.9	90° (?)	Na	75% r. h. under tension	Tetragonal Partially crystalline	$0\ 0\ 0$; $\frac{1}{2}\ \frac{1}{2}\ \frac{1}{4}$	27.4	—	33.8	—
C	9.3	95°	Li	44% r. h.	Orthorhombic Partially crystalline	$0\ 0\ \frac{1}{8}$; $\frac{1}{2}\ \frac{1}{2}\ -\frac{1}{8}$	20.1	31.9	30.9	—
			Li	44% r. h.[b]	Hexagonal Partially crystalline	$0\ 0\ 0$ or $\frac{1}{2}$; $\frac{1}{3}\ \frac{2}{3}\ \frac{1}{6}$ or $\frac{1}{3}$; $\frac{2}{3}\ \frac{1}{3}\ -\frac{1}{6}$ or $\frac{1}{3}$	35.0	—	30.9	—

[a] r. h. = relative humidity.
[b] This form occurs only in some samples.

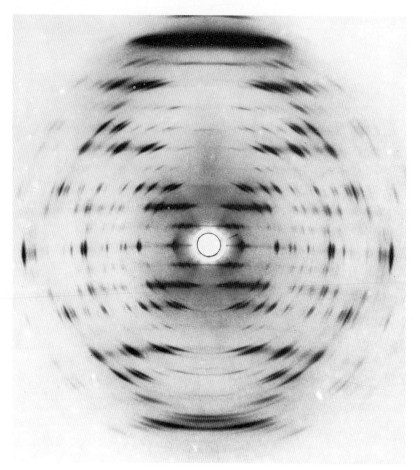

Fig. 27. X-ray fiber diagram of a lithium DNA-B (Wilkins, unpublished).

amplitudes of the scattered X-ray waves. From the amplitudes and the relative phase displacement of the scattered waves, one can calculate the so-called structure factor and, hence, the electron density distribution within the unit cell. This, in turn, provides information on the arrangement of the atoms in the unit cell and the molecular structure of the material under study. The relative phase displacement cannot always be determined by X-ray analysis alone and, therefore, not all structures can be worked out in this way. Frequently, because of symmetry properties, the phase angle assumes only certain values and the refraction is much simplified. For a more detailed consideration of these problems, the interested reader is referred to the specialized literature (Bijvout *et al.*, 1951; Henry *et al.*, 1953; Lipson and Cochran, 1957).

For the biochemist concerned with the secondary structure of polynucleotides, X-ray analysis of stretched fibers is of special interest.
The fibers, which have been drawn from a highly viscous gel, consist
mainly of crystallites embedded in amorphous material. The techniques
of photography and interpretation have much in common with the rotating
crystal method. However, the fibers do not need to be rotated, since a
sufficient number of crystallites are oriented with respect to the fiber
axis in such a way as to be in a reflecting position. If a well-focused
beam of monochromatic X-radiation is concentrated on a polynucleotide
fiber about 30–100 μ in diameter and the X-ray film is placed perpendicular to the incident beam 10–45 mm behind the fiber, reflections are obtained whose sharpness or diffuseness is determined by the degree of
orientation of the crystallites.

If the place where the incident beam strikes the film is designated the
zero point, then the layer lines are arranged perpendicular to the fiber
direction and symmetrically above and below this zero point. The zero
layer line, i.e., that line which passes through the zero point itself, is
called the equator, and the reflections on it are termed equatorial reflections. The line which runs perpendicular to the layer lines through the
zero point is called the meridian, and the reflections lying on it are called
the meridional reflections. The structural periodicity in the direction of
the fiber axis can be deduced directly from the distance of the layer
lines from the zero point by the relationship $\sigma = n\lambda/P$, where n denotes
the layer line number, λ the wavelength of the X-radiation and P the
period of the axis.

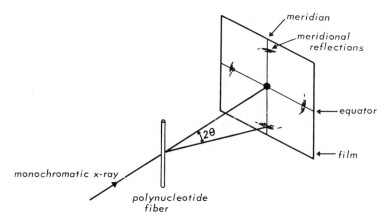

Fig. 28. Schematic drawing showing the technique for taking
X-ray fiber diagrams.

If the polynucleotide molecules, as in DNA, are arranged in a helical secondary structure, the X-ray diagrams of the fibers show certain characteristic features.   Among these, the most obvious is the absence of meridional reflections on the layer lines, due to a mutual extinction of the scattered waves originating from neighboring scattering centers lying on the helix.   The first meridional reflection occurs only on the mth layer line, where m signifies the number of subunits in a helical period; in the present case, m equals the number of nucleotide pairs in one turn of the helix.

Mathematical expressions for calculating the amplitude and phase of the X-ray reflections on the individual layer lines have been developed by Cochran, Crick and Vand (1952) and also by Wilkins *et al.* (1953). From these it follows that the intensity distribution on the nth layer line depends largely on the square of the Bessel function of the nth order (where n is an integer), as well as on the radius of the helix.   Figure 29 shows the calculated intensity distribution about layer lines near two zero points (A and C) for different helix radii.   By considering the first intensity maxima on each layer line and the angle they make with the meridian, the pitch angle of the helix can be established.   This, together with the inclination of the helix, enables one to calculate the diameter of the helix (Stokes, 1955).

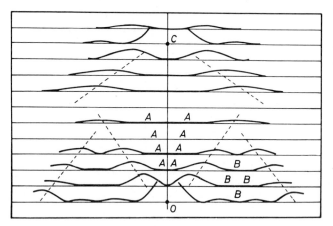

Fig. 29. Diffraction pattern of a system of helices corresponding to the structure of DNA. The squares of Bessel functions are plotted about 0 on the equator and on the first, second, third and fifth layer lines for half of the nucleotide mass at 20-Å diameter, with the remainder distributed along a radius; the mass at a given radius is proportional to the radius. Similar functions are plotted for an outer diameter of 12 Å (from Wilkins *et al.*, 1953).

If a polynucleotide exists not as a simple helix but as a symmetrical double helix, the reflections on the odd-numbered layer lines disappear by interference, and the reflections on the remaining even-numbered layer lines follow an intensity distribution determined by the squares of the Bessel functions of the even-numbered series. In the case of a symmetrical triple helix, one obtains reflections only on every third layer line.

Even though a considerable amount of information can be obtained from an X-ray diffraction pattern, this is generally not sufficient for a detailed determination of the secondary structure of a polynucleotide. With the aid of direct measurements and the available steric information on bond angles and bond lengths, one can devise model structures and construct them with wire or space-filling models. If the reflections calculated for these models by Fourier transforms agree in position and intensity with those measured experimentally, then the suggested secondary structure is most likely correct (cf. Fig. 30).

Until recently, all attempts to obtain some insight into the secondary structure of RNA by means of X-ray diffraction have failed. This would appear, in general, to be attributable to the low degree of organization of the secondary structure of RNA. The behavior of RNA in dilute solution supports this belief. Only fragments of transfer RNA (Spencer et al., 1962), Reo virus RNA (Langridge and Gomatos, 1963) and, as recently shown, of the replicative form of MS2 virus RNA (Langridge et al., 1964) yield fibers sufficiently oriented to give useful X-ray diffraction patterns. The great similarity between these diffraction patterns and those of the A-form of DNA (shown schematically in Fig. 24) suggests a helical structure with similar parameters. In contrast to DNA, transfer RNA fragments are not double stranded. It must, therefore, be assumed that in this case, the single polynucleotide chain is made up of about 90 monomers, parts of which form double helices (cf. p. 101 and Fig. 55). The base pairs are linked by the same hydrogen bonds as DNA (with uracil replacing thymine) and are inclined about 70° to the helical axis.

In addition to naturally occurring nucleic acids, synthetic polynucleotides may be prepared enzymatically; e.g., polynucleotide phosphorylase (see p. 198) can be used to produce RNA-like polynucleotides. The so-called homopolymers, whose behavior in solution indicates the existence of ordered secondary structures (cf. Chapter IV), are of particular interest.

One of these homopolymers, the acid form of polyadenylic acid (polyA) will be discussed first. Below pH 6, polyA yields fibers with high negative birefringence ($\Delta n = -0.10$), which show a strong reflection on the fourth layer line of the X-ray diagram; this corresponds to stacked bases with a spacing of 3.84 Å. The layer line spacing suggests a repeat

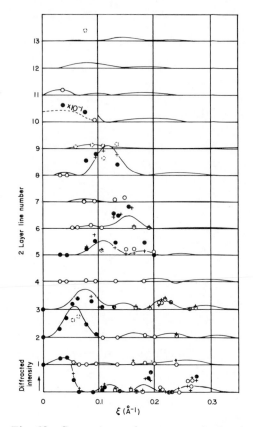

Fig. 30. Comparison between calculated and measured intensity distributions along the layer lines of an X-ray diagram of DNA-B (from Langridge *et al.*, 1960). (Reproduced with permission of Academic Press Inc., New York)

period of 15.2 Å (Rich *et al.*, 1961); the meridional reflections missing up to the fourth layer line indicate a helical structure for polyA with 4 monomer units per turn, on the assumption of a single-stranded helix. In order to yield the observed 3.84-Å layering of the bases, the long axis of each adenine residue would have to be turned 90° with respect to the next base residue. The shortness of the sugar phosphate chain makes this sterically impossible. To overcome these difficulties, Rich and co-workers have suggested a model in which two parallel strands of polyA are twisted to form a symmetrical double helix. The resulting diad symmetry axis requires a helix turn of only 180° per identity period, and the displacement of the base axes need therefore amount to only 45°. In this model, as in DNA, the bases are oriented toward the helical axis and the planes are positioned with respect to one another like propeller blades. The adjacent bases of the two strands are assumed to be linked by hydrogen bonds as in Fig. 31. The X-ray patterns were made from fibers drawn from a gel at pH 5. As indicated by the titration curve (Beers and Steiner, 1957), nearly every base at this pH value is protonated on the nitrogen atom in position 1, so that each positive charge in the helix lies near a phosphate group on the opposite strand. The resulting electrostatic stabilization by formation of an internal salt explains the dense packing of this homopolymer in the helix, as well as its high sta-bility in solution (cf. p. 74).

Like polyadenylic acid, polycytidylic acid (polyC) also forms a stable secondary structure in salt solutions at pH 5. Fibers drawn from a gel of this material are characterized by an unexpectedly high degree of or-ganization (Langridge and Rich, 1963). The strong negative birefringence

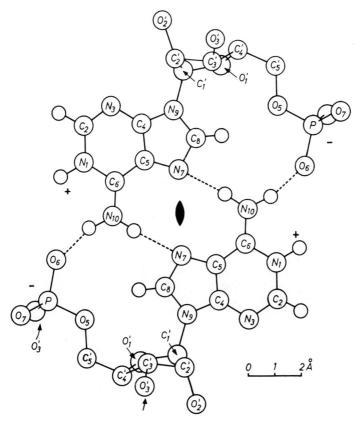

Fig. 31. Hydrogen bonds in polyA (from Rich *et al.*, 1961).

suggests an orientation of the bases perpendicular to the helical axis. In the X-ray pattern, again typical of a helix, the first meridional reflection occurs at the sixth layer line, indicating that there are six bases for a repeat period of 18.65 Å. A base spacing of 3.11 Å can be inferred from the position of the meridional reflection. Since this spacing is significantly smaller than the normal value of 3.34 Å, a slight tilting of the base planes yielding denser packing may be inferred. The nearly meridional reflections on the fourth layer line also suggest such a distortion. To reconcile these data with a single-stranded helical model, it is necessary that the six successive bases of a complete turn be displaced by 60°. Although sterically possible, this is very unusual in polynucleotides. For polyC, Langridge and Rich (1963) suggest a symmetrical double helix with two parallel strands of 13.5-Å diameter in which, because of the twofold symmetry, a repeat period occurs in one-half of the helix repeat proposed for polyA. The six bases would then be displaced by only 30°, an arrangement with no steric difficulties. The suggestion of a symmetrical double helix, moreover, is supported by the following: At pH 4, the degree of organization of the polyC secondary structure is reduced to the point where the symmetry of the double helix is partly destroyed. This should lead to the formation of new layer lines for a longitudinal period of 37.3 Å in the X-ray fiber diagram. These can, in fact, be observed experimentally so that the double-helix model for polyC may be regarded as essentially correct.

The cytosine residues are assumed to be hydrogen bonded (see Fig. 32) in the interior of the molecule, as in DNA. In this scheme, a proton is necessary for the NH...N bond; this explains the marked dependence of the degree of organization of polyC on pH. At pH 4, nearly every cytosine residue is protonated; this interferes with the formation of a

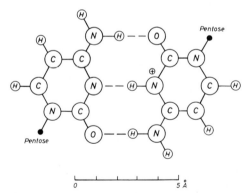

Fig. 32. Hydrogen bonds in polyC (from Langridge and Rich, 1963).

symmetrical double helix. The equatorial reflections indicate a hexagonal crystal structure, with a double helix tightly packed into each rhombohedral unit. Consequently, it is virtually impossible to accommodate additional water molecules. This, in turn, explains the highly crystalline and moisture-independent structure of the fiber.

The secondary structure of polyinosinic acid (polyI) in solution (see Chapter IV) differs in many respects from the types discussed so far. Rich (1958) has suggested a structure for polyI based on X-ray data. These fibers, obtained in the usual way, show a high negative birefringence ($\Delta n = -0.08$), with the X-ray diagram exhibiting a strong meridional reflection at 3.4 Å. The two findings suggest an arrangement of the hypoxanthine residues perpendicular to the fiber axis, with the normal 3.4-Å spacing. The equatorial reflection suggests lattice planes with a 23.8-Å spacing parallel to the fiber axis. The remaining equatorial reflections are higher orders of the latter. In addition to the meridional and equatorial reflections, the diagram shows only two layer lines with 9.8- and 5.2-Å spacings. From these values, it follows that the second cannot be a higher order of the first; the reciprocal lattice spacing from the equator to the first layer line is, however, the same as the reciprocal spacing of the 3.4-Å meridional reflection to the 5.2-Å layer line. This is typical for a helix with a nonintegral number of subunits per turn (Rich, 1958).

An attempt to interpret the findings in terms of a single helix with 3.4-Å spacing yields 2.88 bases per turn and requires that the longitudinal axes of the base be displaced by 125°. This is sterically impossible and so several polyI strands must form a helix of a higher order in which coplanar base linkage occurs by hydrogen bonding. This is possible only for triple or quadruple helices. Since the triple helix would have a diameter of 24 Å, which is in good agreement with the observed equatorial reflection of 23.8 Å, Rich prefers it to the quadruple helix (with a diameter of about 28 Å). The bonding scheme, by which the bases may be linked in the triple helix, is given in Fig. 33. The Fourier transform calculations carried out for the suggested model gave essentially the same diffraction pattern as observed experimentally.

The ability to form helical secondary structures with several polynucleotide strands held together by hydrogen bonds is not limited to homopolymers of the same kind. Whenever steric conditions exist that favor the formation of hydrogen bonds like those in the Watson-Crick model, all types of homopolymers will form helical structures. The oldest example of this kind is the polyA-polyU complex (polyU = polyuridylic acid). If polyA is mixed with polyU in a solution of 0.1 ionic strength, there results a 1:1 complex (Felsonfeld and Rich, 1957) from

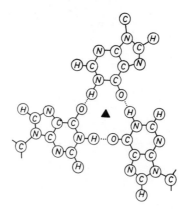

Fig. 33. Scheme suggested by Rich for the hydrogen bonding between the hypoxanthine residues in polyI triple helix (from Rich, 1958).

which strongly negative birefringent fibers may be drawn. The X-ray diffraction pattern of this material is very similar to that of DNA (Rich and Davies, 1956); the relatively greater intensity of the first layer line of the X-ray pattern shows that the helix diameter is somewhat larger than that of DNA (Rich, 1957). In principle, the same secondary structure is assumed for the poly(A + U) complex fibers as for DNA; there is also a similar dependence of structure on moisture content. The hydrogen bonds between the two bases are thought to be the same as those of the thymine-adenine pair in DNA (cf. Fig. 22).

In addition to the 1 : 1 complex, polyA can also form a 1 : 2 complex with polyU, where the structure of the 1 : 1 complex is assumed to be retained, with a second polyU strand placed in the larger "grooves" of the 1 : 1 helix (Felsenfeld et al., 1957). In the third strand, hydrogen bonds are thought to exist between the carbonyl group (carbon 6) of uracil and the amino group of adenine, as well as between nitrogen 1 of uracil and nitrogen 7 of adenine.

In analogous fashion, polyA can also form a 1 : 1 complex with polyI which is stable in solution. Rich (1958) has investigated the strongly negative birefringent fibers of this complex by X-ray analysis. The pattern, obtained at 66% relative humidity, is very similar to that of DNA-B; the strong meridional reflection between 3 and 4 Å indicates the usual base stacking perpendicular to the fiber axis with a 3.4-Å spacing, in agreement with the negative birefringence of the fibers. Moreover, the typical helix diffraction patterns indicate a longitudinal period of 38.8 Å and a base displacement of 31.5°. The equatorial reflections correspond to a hexagonal lattice with $a = 24.4$ Å. Although further study is needed, there is scarcely any doubt that this complex forms a double

helix of the DNA type. Stabilizing hydrogen bonds probably exist between the carbonyl oxygen on carbon 6 of hypoxanthine and the amino
group on carbon 6 of adenine (CO...NH), as well as between nitrogen 1
of hypoxanthine and the nitrogen 1 of adenine (NH...N).

The synthetic polynucleotides discussed so far have been ribonucleotides. With the help of DNA polymerase (cf. p. 168), deoxypolynucleo-
tides, each with two bases, may also be synthesized.

Radding *et al.* (1962) described the enzymatic synthesis of a poly(dC +
dG) complex, made up of one strand each of polydeoxycytidylic and poly-
deoxyguanylic acids; the strands are believed to be linked by hydrogen
bonds in accordance with the Watson-Crick scheme (cf. Fig. 23). The
complex is very stable in solution; no X-ray diffraction studies of the
secondary structure have yet been reported. The DNA polymerase system can also produce a polynucleotide with deoxyadenylic and deoxy-
thymidylic acids arranged alternately within a strand (Schachman *et al.*,
1960; cf. p. 176), forming a double-stranded complex in solution. The
X-ray studies of Davies and Baldwin (1963) indicate that this deoxy-
polynucleotide possesses the same secondary structure as natural DNA.
Replacement of the methyl groups in the thymine bases by bromine atoms
(see p. 171) has no effect on the helix structure.

Nucleic Acids as Macromolecules in Solution

While it is possible to use X-ray analysis to obtain detailed information on the secondary structure of nucleic acids, this information is
limited to the crystalline or paracrystalline state of the dried gel fibers
and cannot be related directly to conditions in aqueous solution. Above
all, X-ray analysis of fibers is not capable of giving information on
molecular weight or molecular weight distribution within the sample investigated. To complete the characterization of macromolecules as to
shape and size, further physical measurements are necessary. In this
connection, factors must be considered which are unimportant if one is
working with substances of low molecular weight. When nucleic acids
are isolated from a tissue or microorganism, there usually results a mixture of macromolecules made up of individual molecules of varying chain
lengths; therefore, one cannot speak of a unique molecular weight but
only of an *average*. Since the method of averaging will depend on the
technique employed, the type of mean value used (weight average molecular weight, number average molecular weight, or a more complex one)
must be specified. In the number average, every molecule, irrespective
of size, counts as one, and the molecular weight therefore is defined by

$$M_n = \sum_i n_i \cdot M_i \Big/ \sum_i n_i \qquad (1)$$

where n_i is the number of molecules of molecular weight M_i per unit volume. In the weight average, size is considered as well as the number of single molecules; therefore, the weight average molecular weight is defined as

$$M_w = \sum_i n_i M_i^2 \Big/ \sum_i n_i M_i = \sum_i c_i M_i \Big/ \sum_i c_i \qquad (2)$$

where c_i represents the weight concentration of component i.

The ratio by number or molecular weight of one molecular species to the sum of the others is a rough index of the distribution of the molecular weight. Systems in which such a distribution exists are termed "polydisperse." In addition to polydispersity, still another property of macromolecules in solution must be considered, one for which there is likewise no parallel in low molecular weight compounds. This is the property, especially characteristic of long-chain polymers, of existing in solution not as compact, spherical particles, but as either random coils or rigid rods. Such molecules aggregate appreciably at concentrations where compact spherical molecules of equal mass scarcely aggregate at all. Therefore, determination of the properties of single molecules requires very much more dilute solutions than those used for low molecular weight substances; in turn, this means analytical methods of far greater than ordinary sensitivity. Three thermodynamic and two hydrodynamic methods that satisfy these demands will now be briefly discussed, along with results that provide some insight into the structure of nucleic acids in solution.

The best known thermodynamic method of molecular weight determination is the measurement of osmotic pressure. Because osmotic pressure is a function of the number of solute molecules per unit volume, their molecular weight can be calculated if the concentration (w/v) and the number of particles are known. With polydisperse systems, this method gives a number average molecular weight. The mathematical expression linking the measured parameters with the molecular weight follows directly from the equation of state for real gases in the form:

$$PV/nRT = 1 + B\,(n/v) + C\,(n/v)^2 + \ldots \qquad (3)$$

In this equation, P denotes the pressure, V the volume, n the number of molecules, R the gas constant and T the absolute temperature; B and C are virial coefficients. If P in Eq. (3) is replaced by the osmotic pressure π, and n/V by the weight concentration divided by the molecular weight, one obtains the analogous relationship:

$$\pi/cRT = 1/M + Bc/M^2 + Cc^2/M^3 + \ldots \qquad (4)$$

Although osmotic pressure has rarely been used to determine the molecular weight of nucleic acids, the method illustrates the general approach of all three thermodynamic methods, as well as a typical manner of interpretation. As stated above, the properties of individual macromolecules in solution can be determined only at extreme, or strictly speaking at infinite, dilution. Hence the parameters, whether osmotic pressure, intensity of scattered light or sedimentation constant, must be determined for a series of dilutions and then extrapolated, if possible, to infinite dilution.

Inasmuch as macromolecules are studied in very dilute solutions, the third and subsequent terms on the right-hand side of Eq. (4) can be neglected, since they are small in comparison with the first two terms. If the experimental values for osmotic pressure are divided by cRT and plotted against c, a straight line will usually result which, when extrapolated to $c = 0$, intersects the ordinate at $1/M$. B, the second virial coefficient, is obtained from the slope of this line. This value, which for gases is the volume of the individual molecules, may, in the case of macromolecules in solution, yield information on the dissociation of complexes with increasing dilution.

The osmotic pressure of a solution is determined experimentally by separating it from the pure solvent by a membrane permeable only to the solvent and then measuring the increase in pressure of the solution resulting from the difference in chemical potential of the solvent on the two sides of the membrane. The method is applicable to particles whose molecular weights range from 10,000 to several hundred thousand; the desired accuracy sets the upper limit since π decreases with increasing molecular weight [cf. Eq. (4)]; the lower limit results from the permeability properties of the membrane.

The second method for determining molecular weight is to measure the intensity of light scattering. Diluted gases decrease the intensity of an incident beam of light by scattering it. Nearly 100 years ago, Raleigh investigated this phenomenon and showed that if a beam of light passes through a scattering medium, the intensity is reduced according to the following equation.

$$I = I_0\, e^{-\tau l} \tag{5}$$

where I_0 is the intensity of the incident light, I the intensity of the beam of light after passing through the scattering medium, l the length of the light path in this medium and τ the turbidity coefficient of the medium.

The decreased intensity of the incident beam may also be characterized by determining the intensity of the scattered light i_θ at some arbitrary angle θ to the incident beam and at a distance r from the scattering center. When θ is $90°$, then the following relationship exists between τ, i_{90} and r:

$$\tau = \frac{16\pi}{3} \cdot \frac{i_{90}\, r^2}{I_0} = \frac{16}{3}\pi R_{90} \qquad (6)$$

R_{90} represents the "reduced scattered light intensity" and is termed the "Raleigh quotient."

Equation (7) can be used to calculate this quotient for an arbitrary angle and for unpolarized light:

$$R_\theta = \frac{2\pi^2\,(n-1)^2}{\lambda_0{}^4} \cdot \frac{1}{\nu}(1 + \cos^2\theta) \qquad (7)$$

where n is the refractive index of the gas, λ_0 the wavelength of the incident light in vacuo and ν the number of molecules per milliliter.

Einstein (1910) and Debye (1947) have shown that Eq. (7) also holds for very dilute solutions if the difference between the refractive index of the solution and the solvent $(n - n_0)$ is substituted for $(n - 1)$. Moreover, if the number of particles ν is expressed as a concentration c (grams per volume), then after rearrangement one obtains

$$R_{90} = \frac{2\pi^2 n_0{}^2\,[(n-n_0)/c]^2}{N\lambda_0{}^4} \cdot cM = KcM \qquad (8)$$

where N is Avogadro's number.

It follows from this expression that the intensity of scatter for a scattering angle of $90°$ increases linearly with the molecular weight. If the system is polydisperse, then this method measures the weighted mean molecular weight. The factor K can best be determined experimentally with the help of a known compound.

Strictly speaking, Eq. (8) holds only for a system of independent scattering centers. The finite dimensions of the individual solute particles, especially large macromolecules, lead, in the case of light scattering measurements, to a concentration-dependent interaction of the individual centers which, according to Einstein (1910), obeys the following relationship:

$$K \cdot \frac{c}{R_{90}} = \frac{1}{M} + 2\frac{B}{M^2} \cdot c + \dots \qquad (9)$$

In this equation, the analogy to the concentration dependence of the osmotic pressure [Eq. (4)] is apparent. Hence, light scattering measurements must also be done at a series of dilutions; when Kc/R_{90} is plotted against c and extrapolated to $c = 0$, the intercept on the ordinate equals the reciprocal of the molecular weight of the unit particle.

The relationships derived up to this point are valid only as long as the maximum size of the scattering particles does not exceed 5% of the wavelength of the incident light. If the particles are larger than 200–300 Å, as is often the case for nucleic acids, the scattered waves which emerge from different regions in the same molecule may interact and the intensity of the scattered light is reduced by interference. Since the magnitude of this effect increases with increasing angle of scatter, it is necessary to make measurements at different angles of scatter and to derive the value for the angle $0°$ by extrapolation; the smallest angle at which reliable measurements may be made with commercial equipment is about $25°$.

It is advisable to make measurements over as large a range as possible (normally between 135 and 30°), since variation of scattered light intensity with the scattering angle reflects the mass distribution within the scattering particles. Some information on the form of the scattering particles may be obtained from this angular dependence. A complete analysis, therefore, requires a determination of both the concentration and the angular dependence of the scattering capacity of a compound in solution.

The angular dependence of the scattered light intensity fits the following general expression:

$$K \cdot \frac{c}{R_\theta} = \frac{1}{MP_\theta} + 2Bc \tag{10}$$

The quantity P_θ is termed the "particle-scattering factor" and is equivalent to the square of the structure factor from X-ray structural analysis (cf. p. 46); it is determined by the shape of the scattering particles. Mathematical relationships between the shape of the particle and P_θ have been formulated for ideal molecular shapes, particularly spheres, rods, random coils of unbranched chain molecules, so-called rigid coils and small disks. For further details, the interested reader should consult the specialized literature (Stacey, 1956; Peterlin, 1959).

The particle scattering factor P_θ may also be represented by an exponential series as a function of the radius of gyration, R_G, and of the scattering angle θ,

$$P_\theta^{-1} = 1 + \frac{R_G^2}{3}\left[\frac{4\pi \sin \frac{\theta}{2}}{\lambda'}\right]^2 + \dots \tag{11}$$

where λ' signifies the wavelength of the light in the test solution. The radius of gyration R_G is defined as the mean distance of the units of a particle from the center of mass,

$$R_G{}^2 = \frac{1}{N} \sum_i n_i \, r_i{}^2 \tag{12}$$

where N is the number of scattering centers and n_i is the number of units at a distance r_i from the center of mass.

In accordance with Eq. (11), R_G may be determined quite accurately from the scattering angle, regardless of the type of particle involved. Once its basic form is known, the dimensions of a particle can be readily inferred with the aid of R_G. For example, in the case of rod-like molecules, the over-all length is $R_G \sqrt{12}$. In a random-coil molecule, the root-mean-square end-to-end distance $\sqrt{<r^2>}$ is $R_G \sqrt{6}$.

The interpretation of light scattering measurements of macromolecules requires, as already noted, a double extrapolation to $c = 0$, $\theta = 0$. This is best done as described by Zimm (1948). One plots Kc/R_θ as a function of $\sin^2 \frac{\theta}{2} + kc$, where k indicates an arbitrary constant chosen so that the experimental values for the different concentrations are distributed widely enough. If one first connects the experimental points for constant angle and different concentrations, and then those for equal concentrations and different angles, extrapolating for $c = 0$, and $\theta = 0$, respectively, a grid-like diagram results which, for small values of c and θ, is bounded by two lines connecting the extrapolated points. Figure 34 shows such a "Zimm plot" for DNA. From the slope of the boundary line, joining the scattering values for $c = 0$, the radius of gyration may be calculated from Eqs. (10) and (11); the corner point on the ordinate gives the reciprocal of the molecular weight.

If the dimensions of the particle under study are so large that the radii of gyration considerably exceed 1000 Å, this method of interpreting light scattering data is no longer valid because the necessary extrapolations cannot be made with sufficient reliability and the numerical values for molecular weight and scattering mass radius become inaccurate. In polydisperse systems, evaluation is even more difficult because P_θ also depends on the distribution of the molecular weight of the sample. If the radius of gyration of the particle is greater than 2000 Å, the limiting value of a plot of $K \frac{c}{R_\theta} \sin^2 \frac{\theta}{2}$ at large values of θ may be useful. One may obtain the reciprocal value of twice the molecular weight from the intersection of the asymptote with the ordinate, and either the radius of gyration (random coil) or the mass per unit length (rod) can be obtained from the slope (Benoit, 1953). An asymptote which cuts

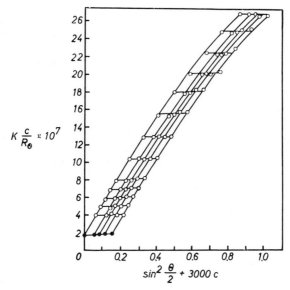

$$K \frac{c}{R_\theta} \times 10^7$$

$$sin^2 \frac{\theta}{2} + 3000\,c$$

Fig. 34. Light scattering values of calf thymus DNA
in 0.2M NaCl solution (from Reichman *et al.*, 1954).

the ordinate in the negative region is characteristic of a so-called zigzag
chain molecule (Hermans and Hermans, 1958; Sadron, 1960).

As seen from Eq. (10), the scattering power of a molecule, in con-
trast to the osmotic pressure, increases with increasing molecular weight.
This permits the investigation of large molecules at very high dilutions.
The use of this method with calf thymus DNA has yielded molecular
weight values between 6×10^6 and 10×10^6 at radii of gyration of 2200 Å
(Hermans, 1959).

If 8×10^6 is taken as an average value, then for a Watson-Crick double
helix, the total length for a completely stretched DNA molecule is about
38,000 Å. If the molecule in solution exists in this rod-like form, the
scattering mass radius should be about 11,000 Å; this, however, is not
the case. If the coiled-chain model is used to interpret the experimental
findings, then on the basis of the radius of gyration one would calculate
an average end-to-end distance of about 7000 Å, a value much too large
for a random coil with a maximum length of 38,000 Å. Hence, the DNA
molecule in solution is best described as a rigid coil made up of a
"worm-like" chain molecule, as suggested by Porod (1949). The Zimm
grid plot (Fig. 34), which fits neither a random coil nor a rod, but an
intermediate form, supports this interpretation.

DNA light scattering data have been critically discussed by Sadron
(1960). According to this author, the large radii of gyration of DNA

particles preclude exact determination of the molecular weight with present techniques. A zigzag molecule is suggested from the intersection of the asymptote for the curve $K \dfrac{c}{R_\theta} \sin^2 \dfrac{\theta}{2}$ at large angles. The mass per unit length of 200–220 Å obtained from the slope is in agreement with the Watson-Crick model.

Zimm and co-workers (1965) have recently developed a new type of cuvette that permits reliable measurements at much smaller angles than 25°. Using these cuvettes, they were able to measure molecular weights of approximately 25×10^6 for T7 bacteriophage DNA, values that are in good agreement with results obtained by other absolute methods.

In contrast to DNA, RNA in solution behaves like a random-coil molecule. Light scattering studies (Boedtker, 1959) of RNA from tobacco mosaic virus (TMV) have yielded a molecular weight of 2×10^6 and a radius of gyration of about 300 Å. In contrast to the intact virus, for which the rod model is appropriate, the free RNA molecule must be very highly coiled by reason of its small R_G value. The results for RNA from *E. coli* are similar (Littauer and Eisenberg, 1959) to those from TMV RNA, but microsomal RNA from calf liver has a molecular weight of about 120,000 according to Hall and Doty (1958).

Of the numerous homo- and copolymers known from X-ray analysis to form double helices, polyA and poly(A + I) complexes have been examined by light scattering (cf. p. 59). Steiner and Beers (1957) have reported a mean molecular weight of 3×10^6 and a radius of gyration of 690 Å for a polyA sample in $0.5M$ potassium chloride solution at pH 6.5. The form of the Zimm grid plot indicates a random-coil molecule for ionic strengths about 0.01; at lower ionic strengths, the molecule lengthens as a result of increasing electrostatic repulsion of the negatively charged phosphate groups, with a corresponding increase in the radius of gyration. Rich (1958a) has shown that on complex formation, the average molecular weight of the individual particles of poly(A + I) increases from 0.8×10^6 to 2.5×10^6.

The third thermodynamic method for determining molecular weight is the measurement of the sedimentation equilibrium. This method, useful for molecular weights ranging from about 10^2–10^6, depends on the fact that solute molecules distribute themselves in a gravitational field in a manner similar to that in which molecules of the atmosphere are distributed in the gravitational field of the earth. This distribution may be described by the formula:

$$\frac{dp}{p} = -\frac{Mg}{RT} \cdot dh \qquad (13)$$

where p is the pressure, h the height, M the molecular mass, and g the acceleration due to gravity. If a solution of any substance is placed in the gravitational field of an ultracentrifuge (which may exceed that of the earth by a factor of up to 3×10^5), the ultimate distribution equilibrium can be calculated by means of Eq. (13) if the concentration c is substituted for the pressure p, if the distance x from the center of rotation is substituted for h, and if the term g is replaced by the term for centrifugal acceleration $\omega^2 x$ (ω = angular velocity). Because of the buoyant effect of the solvent, M, the molecular mass, must be corrected by the factor $(1 - \bar{v}\rho)$, where $\bar{v}\rho$ denotes the ratio of the density of the solvent to that of the solution. Thus, one obtaines from Eq. (13)

$$\frac{dc}{c} = \frac{M(1 - \bar{v}\rho)\omega^2 x \, dx}{RT} \tag{14}$$

Integration of Eq. (14) yields an expression relating the molecular weight of the solute to the concentrations c_1 and c_2 at distances x_1 and x_2 from the rotational center. Like the light scattering method, the sedimentation equilibrium for a polydisperse system gives a well-defined average molecular weight. The great disadvantage of this method— especially with fairly large molecules—is that it takes a long time to achieve equilibrium. A variation of the method, introduced by Archibald (1947), is based on the fact that the condition of equilibrium (namely, that the particle flow through a plane at right angles to the rotor radius is zero) is fulfilled at any time at the bottom and the meniscus of the solution. Consequently, the concentration gradient (dc/dx) at these two points (which is measured directly in the cell of an analytical ultracentrifuge) can be used for the molecular weight determination as follows:

$$M_w = \frac{RT}{(1 - \bar{v}\rho)\omega^2} \cdot \frac{(dc/dx)_m}{x_m c_m} \tag{15}$$

Equation (15) is derived from Eq. (14) by direct transformation; the subscript m refers to measurements at the meniscus; an analogous relationship can be obtained for the cell bottom. The use of this method in ultracentrifuge studies is mainly due to Klainer and Kegeles (1955).

As with other thermodynamic procedures, this technique, applied to finitely dilute solutions, yields molecular weights which—as a function of the size of the virial coefficients—are concentration dependent. To determine the molecular weight of an individual molecule, one must still extrapolate the reciprocal values from a dilution series to $c = 0$. Although in contrast to the sedimentation velocity technique (cf. p. 67) the equilibrium method does not presuppose a knowledge of the diffusion constants, technical problems have prevented its application to polynucleotide studies in all but rare instances. Since the molecular weights involved are frequently greater than 10^6, measurements of these particles by the *Archibald method* require such small gravitational fields to avoid sedimentation that both rotor stability and speed control become problems.   Efforts to overcome these difficulties by using magnetic rotor suspension and stabilization are in progress.   For further study of the variety of techniques and procedures possible with the modern analytical ultracentrifuge, the interested reader is referred to the excellent monograph by Schachman (1929).

In contrast to the thermodynamic methods described so far, hydrodynamic methods do not depend on equilibrium conditions, but deal with particle movement in solution under the influence of external forces.   In a simple diffusion process, one must distinguish between translational and rotational diffusion.   Both processes can be characterized by their respective diffusion coefficients, D for translation and θ for rotation.   If the particle movement is due to external forces, then new phenomena occur. For example, if a particle of sufficiently large mass is subjected to the gravitational field of an ultracentrifuge, it will sediment.   This sedimentation can be characterized by a sedimentation coefficient s.

The definition of the translational diffusion coefficient D is derived from Fick's law. This holds that the number of particles dn passing a plane of unit area per unit time dt in the direction of a negative concentration gradient $(- dc/dx)$ is given by

$$dn = Dq \left(\frac{dc}{dx} \right) dt \qquad (16)$$

where q is the diffusion cross section and D is the diffusion coefficient.

The sedimentation coefficient s is defined as the sedimentation velocity of a particle in a unit gravitational field according to

$$s = \frac{(dx/dt)}{\omega^2 x} [\text{seconds}] \qquad (17)$$

where x denotes the distance of the particle from the center of rotation, t the time and ω the angular velocity.

If a streaming solution contains particles distinctly larger than the solvent molecules, they increase the shear stress needed to maintain the original velocity gradient. This occurs because large particles distort the flow pattern. Since the viscosity η of a streaming medium is defined as the ratio of shear stress to velocity gradient, the presence of solute particles increases the viscosity of the solution beyond that of the pure solvent.

To characterize the viscosity of solutions, one uses the "Staudinger index" $[\eta]$, generally called the "intrinsic viscosity." This is defined as the limit at zero concentration of the specific viscosity η_{sp} divided by the concentration. η_{sp} is the relative viscosity minus one

$$\eta_{sp} = \frac{\eta_{solution}}{\eta_{solvent}} - 1$$

$$[\eta] = \lim_{c \to 0} \frac{\eta_{sp}}{c}$$

The nonequilibrium quantities D, θ and $[\eta]$ always contain a factor which takes into account the presence of a force or pair of forces acting against the hydrodynamic resistance. This resistance is a function of the size and shape of the molecule, and may be represented by the frictional coefficient f, approximately the same in all kinds of motion. By measuring two of the four parameters, it is possible to derive an expression for the molecular weight and to eliminate f. An example is the Svedberg equation [Eq. (19)] in which the sedimentation coefficient has been combined with the translational diffusion coefficient D. The following relationship applies at infinite dilution:

$$D_0 = kT/f; \quad s_0 = M(1 - \bar{v}\rho)/Nf \tag{18}$$

where s_0 and D_0 denote sedimentation and diffusion coefficients at infinite dilution, k is the Boltzmann constant, T the absolute temperature and N Avogadro's number.

Eliminating f, one obtains the Svedberg equation:

$$M = \frac{s_0 RT}{D_0 (1 - \bar{v}\rho)} \tag{19}$$

This expression for the molecular weight depends only on the two measured values s_0 and D, on the ratio of the density of the solvent to that of the solution, and on the absolute temperature.

The use of the Svedberg equation to determine the molecular weight of polynucleotides is not wholly successful. The determination of s_0 for high molecular weight polymers involves measurements at great dilutions, since the large concentration dependence of s makes sufficiently reliable extrapolations to $c = 0$ difficult (Fig. 35); for DNA, this dependence prevents a linear plot of s versus c. There is instead a linear relationship between $1/s$ and c which can be utilized to determine s_0 in the concentration range below 5×10^{-4} g/ml. In the region between 5×10^{-5} and 1×10^{-5} g/ml, there is a nearly linear relationship between s and c; however, the measurement of such low concentrations in the

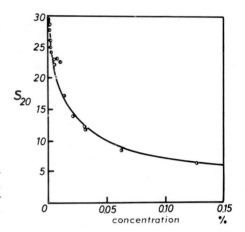

Fig. 35. The sedimentation coefficient of calf thymus DNA plotted
as a function of the DNA concentration (after Shooter and Butler,
1956; also after Oth, 1955).

ultracentrifuge is feasible only with ultraviolet absorption optics or, in some cases, by using an ultracentrifuge cell with a 30-mm light path and Schlieren optics. The s_0 values obtained by the latter method are in good agreement with those from absorption optics (Shooter and Butler, 1956).

In the determination of s or s_0 for large particles, it should be remembered that these parameters depend on the gravitational field in the centrifuge. Therefore, it is necessary to make a series of measurements at different speeds and extrapolate to zero speed of rotation (Eigner *et al.*, 1962). Measurements on concentrations above 5×10^{-4} g/ml are useless, since a so-called infinity gradient is soon formed because of the marked interaction of single DNA molecules through the negative charges on the primary phosphate groups (Coates and Jordan, 1959). The determination of the diffusion coefficient D_0 for polynucleotides requires time-consuming measurements which are subject to large errors; therefore, the value for M obtained from Eq. (19) contains considerable uncertainty. Recently, a method has been reported for measuring diffusion coefficients of viruses, transfer RNA and ribosomal RNA from the form of the sedimentation boundary in the ultracentrifuge (ultraviolet optics). This allows an absolute determination of the molecular weight by means of Eq. (19). The method requires mono- or paucidisperse samples with little or no concentration dependence of s and D. The most serious objection to the use of Eq. (19) for polynucleotides arises because it yields no readily defined mean molecular weight for polydisperse systems.

One may combine the "intrinsic viscosity" $[\eta]$ with the sedimentation coefficient s_0 in a single expression and then calculate the molecular weight of the solute particles. With the model of a semipermeable sphere as a basis, an expression for the so-called *Mandelkern-Flory relation* is obtained for a random-coil molecule (Mandelkern and Flory, 1952):

$$s_0 \, [\eta]^{1/3} M^{-2/3} = \phi^{1/3} P^{-1} (1 - \bar{\nu}\rho)/\eta_0 N \qquad (20)$$

In Eq. (20), ϕ and P are universal constants; η_0 signifies the viscosity of the solvent and N is Avogadro's number. The model known as a semi-permeable sphere refers to particles which do not restrain the movement of solvent molecules at their periphery but do so increasingly toward the center. This model is a rather realistic representation of a random-coil molecule.

Since the random-coil approximation is not generally valid for helical secondary structures in solution, the determination of the molecular weight of these molecules on the basis of $[\eta]$ and s_0 should utilize an expression in which M is reasonably independent of molecular forms.

Mandelkern *et al.* (1952) and Scheraga and Mandelkern (1953) formulated an equation which complies with this requirement to an acceptable degree by using the model of the hydrodynamic equivalent ellipsoid of revolution, an assumed rigid, impermeable ellipsoid of revolution which provides a simple description of the hydrodynamic properties of a real macromolecule. The functional relationship is

$$M^{2/3} = \frac{s_0 \, [\eta]^{1/3}}{\beta} \cdot \frac{\eta_0 N}{(1 - \bar{\nu}\rho)} \qquad (21)$$

The meaning of the symbols here is the same as in Eq. (20).

As can readily be seen, the only difference between Eqs. (20) and (21) is that a parameter β has been introduced in place of the product $\phi^{1/3} P^{-1}$, with Eq. (21) defining β. Thus β is a structural parameter. Whereas $\phi^{1/3} P^{-1}$ is a constant of about 2.5×10^6, β varies with the axial ratio. For a ratio of unity, β equals 2.12×10^6; when the ratio becomes infinite, β assumes a value of 3.50×10^6. This demonstrates that the effect of molecular form on molecular weight as calculated from Eq. (21) is relatively small. For polydisperse systems, Eq. (21) gives a mean molecular weight which corresponds closely to the weight average molecular weight.

The use of Eqs. (20) and (21) to determine the molecular weight of polymers presupposes that the hydrodynamic behavior of the particles under study is well known. Otherwise, one must first determine how closely the behavior of the sample in the chosen solvent may be described by the hydrodynamic models on which Eqs. (20) and (21) are based. If Eq. (20) is to be used for calculating molecular weights, then the product $\phi^{1/3} P^{-1}$ for different members of a homologous series with known molecular size should not depend on the molecular weight. The constancy of this product, therefore, is a sensitive criterion for random coils. To be able to use Eq. (21), one must know the variation of β with molecular size; in turn, this must remain within the limits given above. It follows then that the combination of sedimentation velocity and viscosity is not an absolute method for molecular weight determination, but is applicable only after standardization against compounds whose molecular weights have been determined by absolute methods such as light scattering measurements or a combination of sedimentation and diffusion techniques as in Eq. (19). For the reasons given on pp. 63 and 67, the absolute determination of molecular weight of polynucleotides is difficult, particularly if the molecular weights greatly exceed the value of 6×10^6. Since various sizes of natural polynucleotides can easily be biosynthesized in radioactive form, very reliable autoradiographic methods of molecular weight determination have been developed. These provide standards for the hydrodynamic methods outlined above (Rubinstein *et al.*, 1961; Cairns, 1962). It has been demonstrated recently that the density gradient technique (cf. p. 84) may furnish an absolute molecular weight by measuring the width of the ultraviolet-absorbing DNA band in the gradient.

Determination of the molecular weight of polynucleotides according to Eqs. (20) and (21) involves the use of both the sedimentation coefficient s_0 and the intrinsic viscosity $[\eta]$. Since polynucleotide solutions, especially at low ionic strengths, exhibit non-Newtonian behavior, the measured viscosity of the solutions is dependent not only on concentration, but also on shear stress as determined, for example, by the viscosity gradient in a capillary viscosimeter. Until recently, it was necessary, therefore, to measure the viscosity of polynucleotides in different velocity gradients and to extrapolate the reduced viscosity to zero shear. Only then would the extrapolation to zero concentration yield a meaningful value for $[\eta]$. Reduced viscosities can now be measured readily by the rotating cylinder viscosimeter described by Zimm and Crothers (1962), which involves such low shear stresses that even the reduced viscosity of very high molecular weight DNA (130×10^6) is independent of the velocity gradient.

The importance of viscosity measurements made on macromolecules in solution is not limited to the determination of parameters useful in Eqs. (20) and (21). The axial ratio for rigid particles can also be calculated directly from the intrinsic viscosity. On the assumption that the dissipation of frictional energy that occurs in a streaming solution depends on the orientation of the molecules, Simha (1940) developed a mathematical expression in which $[\eta]$ depends only on the axial ratio and increases almost as the square of this asymmetry factor.

For random-coil molecules, the intrinsic viscosity provides an estimate of the space occupied by fixed solvent molecules. This, in turn, is a gross index of the ratio of the immobilized volume to the actual mass of the inactive macromolecule. An equation relating intrinsic viscosity to molecular weight and mean square end-to-end distance of the molecule was proposed by Flory and Fox (1951). From this equation, simplified relationships were derived for the two extreme molecular forms, a rigid rod and a random coil. Thus, the coefficient of a rod-like molecule varies with the molecular weight according to the relationship

$$[\eta] = KM^2 \tag{22}$$

while the corresponding relationship for a random coil is

$$[\eta] = KM^{0.5} \tag{23}$$

On the same theoretical basis, analogous expressions can be derived for the sedimentation coefficient s_0; for the rod molecule, this becomes

$$s_0 \cong KM^{0.20} \tag{24}$$

and for a coil,

$$s_0 = KM^{0.5} \tag{25}$$

The exponent in Eq. (24) was determined by Doty and Nishihara (1958) with collagen fragments obtained after ultrasonic degradation.

In solution, many macromolecules, especially polynucleotides, adopt conformations that lie between these two extremes. It is important, therefore, to know the range in which the exponents of M might vary with change in conformation. For rod-like molecules [Eq. (22)], a range of 1–2 is permissible; for random coils [Eq. (23)], the numerical value lies between 0.5 and 1.0. In Eq. (24), the exponent should not exceed 0.33 if the rod model is applicable; for coils [Eq. (25)], the range of variation for the exponents is between 0.33 and 0.5.

With regard to molecular weight and molecular conformation, hydrodynamic methods have given the following results for natural polynucleotides and for some synthetic ones. When the molecular weight of calf thymus DNA, a widely studied material, was evaluated by sedimentation and viscosity methods [Eq. (21)], with β generally assigned a value of 2.56×10^6, the values obtained ranged from 6×10^6 to 10×10^6. Doty, McGill and Rice (1958) investigated the relationship between β and the size of the DNA molecule; they found that β increases monotonically from 2.56×10^6 for a molecular weight of 3×10^5 to 3.29×10^6 for a molecular weight of 8×10^6. In these studies, light scattering measurements were used to standardize the molecular weights.

Recent work by Crothers and Zimm (1965) has shown that for high molecular weights, the hydrodynamic properties of DNA are those of a random coil and Eq. (21) yields reliable molecular weights if β is assigned the value of 2.27×10^6. With DNA of molecular weights below 8×10^6, the reduced flexibility of "worm-like chain molecules" becomes increasingly important, the molecules take on a preferred shape, and Eq. (21) can be used only if other values are assigned to β. Crothers and Zimm have taken this into account by developing Eq. (21a). Based upon the theoretical work of Hearst and Stockmayer (1962) and of Eizner and Ptitzyn dealing with the hydrodynamic model of a "worm-like molecule," Eq. (21) takes into account the excluded volume and represents a semi-empirical relationship which applies to molecular weights between 3×10^5 and 2.5×10^8:

$$M^{2/3} = \frac{(s_0 - 2.7)\,([\eta] + 5)^{1/3}\,\eta_0\,N}{2.27 \times 10^{19}\,(1 - \bar{\nu}\rho)} \qquad (21a)$$

The expressions that relate the viscosity and the sedimentation coefficient s_0 to the molecular weight are as follows:

$$0.665 \log_{10} M = 2.863 + \log_{10} ([\eta] + 5) \qquad (21b)$$

$$0.445 \log_{10} M = 1.819 + \log_{10} (s_{w,20}^0 - 2.7) \qquad (21c)$$

Equations (21a) through (21c) do not apply to molecular weights below 3×10^5 because these short DNA molecules increasingly resemble rigid rods in their hydrodynamic behavior. No corresponding equations have as yet been developed for this low molecular weight range, probably because standard samples are unavailable.

In general, bacterial DNA has higher molecular weights than mammalian DNA; this may reflect the fact that DNA can be isolated from microorganisms under less severe conditions than from mammalian tissues. For example, a value of 16×10^6 (Eigner et al., 1962) has been reported for DNA from E. coli, and one of 10×10^6 (Marmur and Lane, 1960) for pneumococcal DNA. Recent autoradiographic studies indicate that the DNA from E. coli, if handled gently, should have a distinctly higher molecular weight, around 10^9 (Cairns, 1962). Massie and Zimm (1965) have shown that with sufficiently mild conditions, it is possible to isolate DNA particles from E. coli and B. subtilis with a molecular weight of 250×10^6. In all likelihood, these represent subunits which in the cell were connected by protein bridges.

Wide variations in the molecular weights of DNA from phages have been reported. The ϕX174 phage contains DNA with a molecular weight as low as 1.7×10^6 (Sinsheimer, 1959), but the values for DNA from T2 phages range from 130×10^6 to 160×10^6 (Davison et al., 1961; Rubinstein et al., 1961). In most cases, the total nucleic acid complement of a virus particle exists as a single molecule.

Hydrodynamic investigations (Crothers, 1963) have shown that Eq. (21) also applies to T2 phage DNA (molecular weight, 130×10^6) if a value of 2.5×10^6 is used for β.

The work by Doty, McGill and Rice (1958) suggests that the intrinsic viscosity for DNA preparations with M between

$$3 \times 10^5 \text{ and } 2 \times 10^6$$

can roughly be expressed as:

$$[\eta] = 1.45 \times 10^{-6} \, M^{1.12}$$

From the numerical values of the exponents given earlier, it can be seen that a value of 1.12 lies at the lower limit for rod-like molecules. The corresponding expression for the sedimentation coefficient

$$s_0 = 0.063 \, M^{0.37}$$

contains an exponent which is also in the range for coil and rod forms. These values indicate an intermediate form between coil and rod, in agreement with the results from light scattering measurements and compatible with the picture of a rigid coil (cf. p. 63).

In contrast to DNA, RNA molecules from mammalian tissues show large variations in molecular size. This may be partly due to the difficulty in distinguishing aggregates from single molecules. Hall and Doty (1958) conducted a detailed investigation of RNA from liver microsomes and found that some aggregates which were partly destroyed in the course of preparation have molecular weights of about 10^6. Other authors (Wang, 1963) have obtained similar results. Upon being heated in solution, these aggregates disintegrate completely into stable single particles whose molecular weight, on the basis of viscosity and light scattering measurements, is about 120,000. The viscosity and sedimentation values were calculated from the Mandelkern and Flory relationship [Eq. (20); see p. 69]. The fact that these values agree with those from light scattering measurements indicates that the RNA in solution forms random coils, because Eq. (20) yields correct molecular weights only for these forms. This is further confirmed by the magnitude of the exponents measured by the same workers for Eqs. (23) and (25) (cf. p. 71). The expressions are

$$[\eta] = 6.2 \times 10^{-4} \, M^{0.53}$$

$$s_0 = 2.1 \times 10^{-2} \, M^{0.49}$$

The exponents of M approach optimal values for random-coil molecules.

A molecular weight of 2×10^6 has been given for RNA gently isolated from tobacco mosaic virus (cf. p. 261); application of the Mandelkern-Flory equation after determination of the hydrodynamic parameters has yielded a value of 1.8×10^6. This satisfactory agreement also suggests that pure TMV RNA exhibits a coiled form in solution. Fresco and Doty (1957) have studied the hydrodynamic properties of polyA in alkaline solution. Since the intrinsic viscosity is a function of the molecular weight according to the expression $[\eta] = KM^{0.65}$, it may be concluded on the basis of the numerical value of the exponent that the molecule is coiled. If the pH of the solution is adjusted to below 5, the exponent of M increases to 0.91. This approach toward a numerical region characteristic of rod-like macromolecules is consistent with the formation of the double-helical structure for polyA in weakly acidic solutions (cf. p. 51).

Denaturation of Nucleic Acids

There are many indications that the secondary structure detectable in the fibers of a number of natural and synthetic polynucleotides is also found in aqueous salt solutions. Since, on the basis of X-ray analysis, hydrogen bonds appear to be the important stabilizers of these structures, one would like to know what effect breaking the hydrogen bonds has on the molecular conformation and other properties of polynucleotides in solution. This may be accomplished by varying ionic strength, pH, dielectric constant or temperature of the solution. The effect of these factors on the properties of dissolved nucleic acid molecules can be measured by light scattering, sedimentation and viscosity determinations, as well as by other optical and electrical methods to be discussed below.

In 1947, Gulland and co-workers found that when DNA (Fig. 36) was titrated with either acid or base starting from the neutral point, the resulting pH values could not be reproduced on back titration. The flattening of the neutralization curve for DNA exposed to extreme pH values is clearly shown in Fig. 36; the shape of the curve indicates that weakly acidic or basic groups which obviously were not available in the first titration of native DNA participate in the neutralization process. At the time this work was done, it was not yet possible to interpret the results as a destruction of the Watson-Crick double helix, but now the generally accepted interpretation is that the addition or removal of protons by

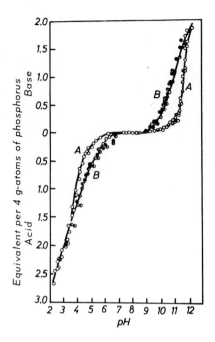

Fig. 36. Titration curve of calf thymus DNA; (A) titration starting at pH 6.9; (B) back titration from pH 12 or 2.5 to neutral point (from Gulland *et al.*, 1947).

acids or bases breaks the hydrogen bonds. When the proportion of broken bonds is high enough, the two DNA strands may be displaced so that the correct base recombination will not take place on neutralization. This, in turn, leads to a relatively high proportion of free bases which, with their —NH_2 and —NH—CO— groups, change the neutralization curve.

Cox and Peacocke (1957) have shown that at 25 °C, protonization of 10% of the amino groups (corresponding to a pH value of 4.2) does not cause irreversible damage to the DNA. This "critical" pH drops as the temperature decreases; at 0.75 °C, the pH of the solution may go down to 2.7 without affecting the titration curves, i.e., no irreversible changes are noted (Peacocke and Preston, 1958). At higher ionic strengths, the "critical" pH also depends on the ionic strength of the solution, the point at which irreversible damage occurs being displaced in the direction of lower pH values (Cavalieri and Rosenberg, 1955). Similar studies have been made of the effect of alkalies on DNA (Ehrlich and Doty, 1958).

This irreversible change in DNA caused by extremes in the hydrogen ion concentration is reflected in drastic changes of the molecular dimensions. Figure 37 shows the viscosity of a DNA solution as a function of pH (Mathieson and Matty, 1957). The steep drop of $[\eta]$ in the pH

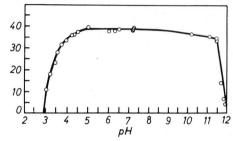

Fig. 37. pH dependence of the intrinsic viscosity $[\eta]$ of calf thymus DNA solution (from Mathieson and Matty, 1957).

range between 3 and 4 and also between 11 and 12 indicates discontinuous changes in the axial ratio of the DNA molecule and, therefore, probably a breakdown of the double helix. The decrease in the DNA radius of gyration (Ehrlich and Doty, 1958; Horn *et al.*, 1952) that occurs when the critical pH is exceeded in either direction, and which was reported almost simultaneously by several groups of workers, supports this interpretation. Today, irreversible structural changes in DNA are generally considered to be a denaturation in the sense of a "helix-coil transition."

In 1954, Thomas observed that the 260-mμ extinction coefficient of DNA in NaCl solutions rose by about 25% when the salt concentration fell below 10^{-3} M. Addition of electrolytes caused the extinction to fall again, but not to the original value. If magnesium sulfate was used as the electrolyte, the extinction began to rise when the salt concentration fell below 10^{-5} M, but could not be restored by the addition of electrolyte, remaining instead at 11% above the starting value. Thomas concluded that there was an irreversible change in the DNA. More detailed investigations by Cavalieri and co-workers (Cavalieri *et al.*, 1956) showed that the extent of the irreversible increase in extinction varied inversely with the DNA concentration; these authors concluded that the concentration of counter ions to the primary phosphate groups was the deciding factor in the alteration of the ultraviolet absorption of DNA. This effect did not occur at pH values above 7.5. With careful addition of acid, the first increase in extinction coefficient occurred at pH 7.0 and a second one at pH 4.5. These changes are accompanied by a fall in viscosity. The unshielded negative charges of the phosphate groups are thought to shift the pK_a value of the amino groups in adenine and cytosine, with the result that the H$^+$ ions of the water protonate the bases, thus breaking the hydrogen bonds (Cavalieri *et al.*, 1956). This rise in extinction resulting from a structural transformation of DNA will be discussed in more detail below.

Organic solvents such as formamide, dimethylformamide or dimethyl sulfoxide also cause irreversible DNA denaturation (Helmkamp and Ts'o,

1961), similar to that resulting from a decrease in electrolyte concentration. In contrast, the structural change which takes place in lower alcohols like methanol, ethanol, or propanol is fully reversible (Herskovitz, 1962). In such studies, optical rotation measurements are frequently employed to follow structural changes. This can be done because the magnitude and the direction of rotation of mononucleotides (see p. 25) change considerably on incorporation into polynucleotides with ordered secondary structures (this is also true for proteins). For calf thymus DNA, this effect is about $+125°$ with sodium light; if the secondary structure is destroyed, the specific rotation falls by this amount. Fresco et al. (1961) found two maxima for the rotatory dispersion of DNA in the near-ultraviolet region. The first is located at 289 mμ and is associated with an $n \longrightarrow \pi^*$ transition, while the second at 257 mμ is due to a $\pi \longrightarrow \pi^*$ transition. On denaturation, the value for the first maximum falls by about 60%, while the second disappears completely.

The changes in the physical parameters of DNA are particularly drastic when solutions of DNA are heated over a definite, fairly narrow temperature range. Cosgrove and Jordan (1950) showed that when DNA was heated in neutral salt solution beyond a certain temperature, its titration curve differed from that of native DNA. Later, Thomas (1954) found that the ultraviolet absorption of DNA above a certain temperature range increased in the same way as it did in solutions below a given ionic strength. In 1953, Zamenhof et al. (1953) showed that at temperatures above 80 °C, the viscosity of a DNA solution falls appreciably and its transforming activity is almost completely lost. Later, Rice and Doty (1957), in the course of light scattering measurements, observed that when a critical temperature was exceeded, the scattering mass radius of the DNA molecule fell from 2800 to 1000 Å (corresponding to about a 15-fold decrease in the mean distance between the ends of the molecule), without in any way leading to a decrease in the molecular weight. Further investigations by Doty, Boedtker et al. (1959) showed that the DNA transition temperature also depends on the ionic strength of the solvent (Fig. 38). Dove and Davidson (1962) were able to show a logarithmic relationship between ionic strength and transition temperature. Figure 39 shows the narrow temperature range in which discontinuous changes take place when calf thymus DNA is dissolved in 0.15M NaCl (+ 0.015M sodium citrate).

This range is similar for nearly all DNA species hitherto investigated; hence, the temperature at which half the maximum increase in extinction is attained may be used as another means for characterizing DNA. This value, termed the T_m value, is comparable to the melting point of low molecular weight compounds. It is presently believed that for a given

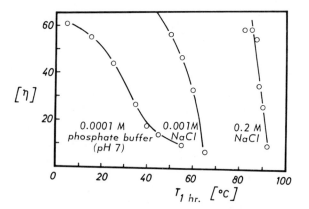

Fig. 38. Change in intrinsic viscosity $[\eta]$ of DNA solutions after heat denaturation at different temperatures in salt solutions of varying ionic strength (from Doty *et al.*, 1959).

ionic strength, the T_m value of a DNA sample depends only on the base ratio.   With DNA from different species of bacteria that exhibit wide variation in gross base composition, it has been shown that increases in T_m with increasing proportions of guanine and cytosine may be described

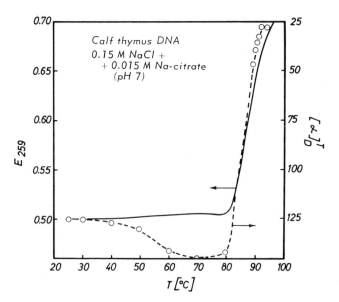

Fig. 39. The effect of thermal denaturation of calf thymus DNA on optical density (260mμ) and optical rotation (from Doty *et al.*, 1959).

by the following relationship (Marmur and Doty, 1962): $T_m = 69.3 + 0.41GC$, where GC denotes the mole per cent of guanine + cytosine in the DNA. Even though this dependence of T_m on base composition indicates that the stability of the DNA helix is largely determined by the number of hydrogen bonds per unit length—the guanosine-cytosine pair is linked by three hydrogen bonds, the adenine-thymine pair by only two—the theory that van der Waals and polarization forces between adjoining base pairs also contribute to the stability of the double helix has become increasingly popular.   A substantial part of the stability of the helix surely also results from effects due to the free energy of the solvent ("hydrophobic forces").

Zimm (1960), using statistical mechanics, developed a theoretical expression for the mathematical treatment of denaturation processes.

Doty, Marmur and Sueoka (1959) have discussed in detail the influence of polydispersity which generally leads to a broadening of the temperature transition range. Heavy metals, diamines, histones (cf. Fig. 40) and other substances that react with DNA cause the T_m value to be displaced considerably and must therefore be removed or masked.

The molecular changes that occur at the point of denaturation will be dealt with in more detail later. Here, however, it may be desirable to discuss why extinction increases when polynucleotides are denatured. In the most general sense, this is due to a destruction of the planar base stacking with its 3.4-Å periodicity (cf. p. 41).   Tinoco (1960) was the first to attempt to explain theoretically the reason for the differences in ultraviolet absorption of DNA molecules of different origin. He suggested that if absorbing groups occur in a polymer in a given sequence, then light-induced electrical dipoles must be arranged correspondingly; this, in turn, raises the possibility of mutual additive interaction of the dipoles.   Because of their short range, quantum mechanical overlapping effects can only involve covalently bonded groups; hence only Coulomb interactions need be considered. Tinoco has developed a mathematical expression to calculate the degree to which the transitional moment, a measure of the amount of light absorbed, is impaired by the sterically possible Coulomb interactions of induced dipoles. From these calculations, it appears that the moment of transition increases when the induced dipoles become linked in a head-to-tail arrangement. If, however, the dipoles are arranged side by side, then both the moment of transition and the absorption decrease. Since the absorption band of the DNA bases at 260 mμ is due to a $\pi \longrightarrow \pi^*$ transition (cf. p. 28), the induced dipoles must lie in the planes of the bases. However, because these are stacked at right angles to the axis of the helix in the Watson-Crick model, the induced dipoles in superposition cause the absorption to decrease.

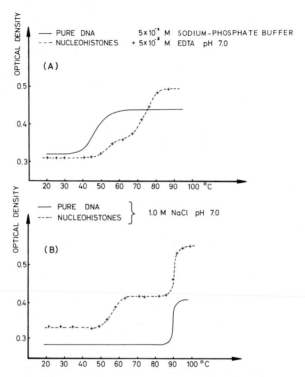

Fig. 40. Thermal denaturation of pure DNA and of nucleohistones isolated from calf thymus: (A) in $5 \times 10^{-5}M$ sodium phosphate buffer, pH 7.0 and (B) in buffered 1.0M NaCl, pH 7.0 (from Harbers and Vogt, 1966). (Reproduced with permission of Taylor & Francis Ltd., London, England).

The decrease in absorption resulting from an ordered secondary structure is termed *hypochromicity*, while the increase in absorption that occurs on destruction of the secondary structure is termed *hyperchromicity*.

The ultraviolet absorption of the heterocyclic bases contains, as already discussed (p. 28), an $n \longrightarrow \pi^*$ transition masked mostly by the intense $\pi \longrightarrow \pi^*$ transition band which is polarized at right angles to the base plane. Therefore, when a helix is formed, an increase in ultraviolet absorption in the region of this band should be detectable, as the induced dipoles undergo a head-to-tail orientation. Using polarized ultraviolet light, Rich and Kasha (1960) were able to locate such an $n \longrightarrow \pi^*$ band with stretched DNA films in the region around 280 mμ; they interpreted the existence of a shoulder in the spectrum of native DNA at this wavelength in terms of a stronger moment of transition that disappeared on denaturation. In the poly(A + U) complex, the $n \longrightarrow \pi^*$ transition band was even sharper. The results therefore agree with the ideas of Tinoco. According to another hypothesis, the hypochromicity of DNA is largely attributable to polarization in the direction of the helix. Further studies are needed to show which of the two views is correct.

If RNA in solution is denatured under conditions similar to those discussed for DNA, then the optical effects are similar but less pronounced and largely reversible; they can be attributed to the formation of intramolecular hydrogen bonds between regions within an RNA strand of complementary or nearly complementary sequence. This also explains the strong coiling at sufficiently high salt concentrations, unusual for single-stranded polyelectrolytes. Figure 41 shows a schematic representation of the secondary structure of RNA in solution based upon these ideas.

Once measurement of the extinction changes in the ultraviolet region proved to be a reliable method for recognizing changes in structure of nearly all natural nucleic acids during denaturation, it was natural to apply this method to investigate the capacity of synthetic polynucleotides to form ordered complexes.

Shortly after Warner (1957) had established that a mixture of polyA and polyU migrated in an electric field as a homogeneous zone (despite the different mobilities of the individual components) and that this zone could not be separated in the analytical ultracentrifuge, Felsenfeld and Rich (1957) tested the ultraviolet extinction of polyA-polyU mixtures as a function of the molar ratio of adenine to uracil. Ratios of $1:1$ in $0.1M$ NaCl solution exhibited a clearly defined minimum; this was interpreted

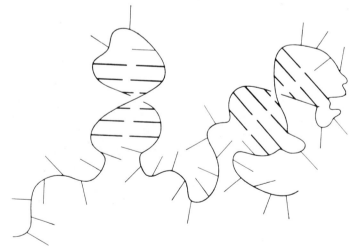

Fig. 41. Possible RNA secondary structures (from Fresco and Straus, 1962).

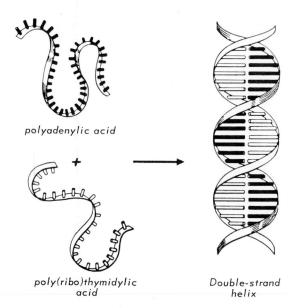

polyadenylic acid

+

poly(ribo)thymidylic
acid

Double-strand
helix

Fig. 42. A double helix formation from coils polyA
and poly(ribo)T (from Rich, 1959).

as due to a double-helix structure, shown diagrammatically in Fig. 42
for polyA + poly(ribo)T. If the two components were mixed in the presence of magnesium chloride, an extinction minimum occurred at an
adenine : uracil molar ratio of 1 : 2; this suggests a poly(A + 2U) complex.
Formation of this complex occurs much more slowly than that of the 1 : 1
complex, but it is greatly accelerated by the presence of Mg^{++}.

Rich (1959) has discussed in detail the many complexes that polyA
and polyI can form with other homopolymers. Later work by Fresco and
Alberts (1960) showed that copolymers of random base sequences can
also interact with homopolymers. This follows from the occurrence of
minima as shown in Fig. 43 for a mixture of poly(A + U) and polyU. It
has, therefore, been suggested that regions in which the bases are not
complementary to each other can be distorted to form loops. Other interesting results occurred during the thermal denaturation of the poly(A + U)
complex. On the one hand, Steiner and Beers (1961) have shown that
the T_m value for molecular weights less than 10^5 depends on the molecular weight of the polymer. On the other hand, in thermal denaturation there occurs a partly reversible but distinct fall in molecular weight.
This indicates a separation into single strands.

The mechanism of the helix-coil transformation of DNA has been the
subject of much study and discussion. Some of the significant findings
will now be discussed.

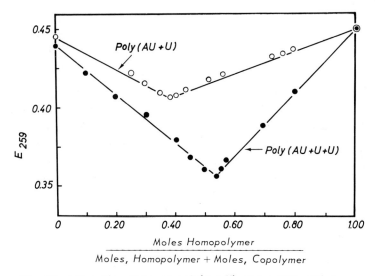

Fig. 43. Interaction between poly(A + U) and polyU with formation of a double and a triple helix (as indicated by measurements of optical density at 259 mμ; from Fresco and Alberts, 1960).

The abrupt changes in molecular dimensions and optical properties of DNA in solution which occur on heat denaturation have left no doubt that the secondary structure is lost when the T_m value is reached. This might simply be due to a breakdown of the double helix leading to the formation of single coiled strands. However, the molecular weight of the particles remains the same after denaturation. At first it was thought that although denaturation by heat modifies the structure, the two single strands remain bonded along those thermally stable regions within the helix which are rich in guanine and cytosine. Light scattering measurements on DNA solutions of temperatures above the T_m value were consistent with this interpretation. Changes in molecular dimensions observed after cooling to below the T_m value, were explained by non-specific base recombinations involving structures with numerous faults. This hypothesis proved to be untenable, however, when thermal denaturation of DNA, previously cross-linked as a result of treatment with alkylating substances such as mustard gas (cf. p. 304), was found to be almost completely reversible if the DNA had only nine cross-links per molecular weight of about 10^7 (Geiduschek, 1961). The work of Doty, Marmur et al. (1960) on the "renaturation" of heat-denatured DNA, determined by the restoration of transforming activity, finally clarified this point. These investigators found that the activity associated with the double-stranded structure was nearly completely lost when samples were quickly cooled after heating. However, if the samples were cooled slowly after thermal denaturation, up to 50% of the activity was recovered (cf. Fig. 134, p. 275). The degree of reversibility of the denaturation process increases with the DNA concentration in solution and shows the expected bimolecular reaction kinetics for the recombination of two single strands to a double strand. Moreover, the reactivation rate increases with the salt concentration in the solution. Doty and co-workers, encouraged by these findings, also studied DNA that had been denatured at great dilution (<40 μg/ml), with an ionic strength of about 0.01; whenever the solution was cooled quickly after denaturation, it contained particles with molecular weights half those of native DNA. Molecular weight was determined with the aid of sedimentation coefficient and intrinsic viscosity measurements after the kinetics of the thermal depolymerizations had been measured. From this, one can conclude that denaturation of DNA leads to strand separation, as evident from the drop in molecular weight, but that this can only be detected if recombination or reaggregation is prevented by sufficiently low DNA concentrations and ionic strengths. This then, is the reason why the expected fall in molecular weight was not observed. Furthermore, the consequences of hybrid formation from N^{14}- and N^{15}-containing DNA strands after denaturation (Meselson et al., 1957; Doty, Marmur et al., 1960) agree with what has just been discussed.

Native and denatured DNA can be separated by density gradient centrifugation (cf. Fig. 44). This method, developed in its present form by

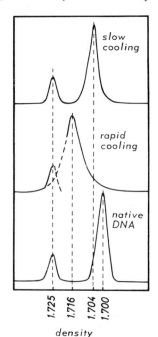

Fig. 44. Effect of thermal denaturation and renaturation on the behavior of pneumococcal DNA in a CsCl density gradient (from Doty et al., 1960).

Meselson *et al.* (1957), is based on the fact that when a concentrated cesium salt solution is centrifuged, a sedimentation equilibrium is established which results in the formation of a density gradient. If the cesium salt solution contains macromolecules, these collect on centrifugation in the region of the gradient corresponding to their own density in solution; this yields sharp zones, particularly with large macromolecules. Since the width of the zones depends essentially on the diffusion velocity of the particles, it is possible to estimate the molecular weight of the particles from the half-width value of the "bands" measured in the analytical ultracentrifuge by means of ultraviolet optics. The resolution of the method is sufficient to separate DNA samples which differ only in their nitrogen mass (N^{14} or N^{15}). With this method, Doty and co-workers were able to show that a mixture of N^{14}- and N^{15}-DNA after denaturation and rapid cooling yielded two well-separated bands, while a renatured hybrid molecule was formed after slow cooling. The extent of strand separation as a function of hyperchromicity has been investigated by Freifelder and Davison (1962) in somewhat similar fashion. When "natural" N^{14}-N^{15} hybrids, in which one strand contains all N^{14} and the other all N^{15} are heated, only N^{14} and N^{15} single strands are liberated, at least if 75% of maximum hyperchromicity is reached. Also, if the extinction increase is complete, unseparated double-stranded DNA is still present in appreciable amounts which "melt out" only if the temperature is raised at least another $5°C$. Double-strand separation has also been detected when DNA was denatured with formamide (Marmur and Ts'o, 1961) or as the result of the action of acids and bases (Schildkraut, Marmur and Doty, 1961). If DNA is denatured in the presence of formaldehyde (1–3% in solution), the addition of methylol groups to the amino groups of the bases will prevent recombination of the strands (cf. p. 275). If the aldehyde is dialyzed out of the solution, then nonspecific aggregation will occur if the DNA concentration is sufficiently high.

Cavalieri and co-workers (1962) have considered the possibility of a four-stranded structure for DNA from cells formed in a cycle of rapid division (e.g., *E. coli* in the logarithmic growth phase). The subunits resulting from denaturation were thought to be of the double-stranded Watson-Crick type, though partially denatured. This suggestion received some support from kinetic studies of the action of nuclease on the DNA of these cells. However, from the investigations of Shooter and Baldwin (1962) on *E. coli* DNA hybrids, which contain bromouracil in place of thymine, it appears that denaturation leads only to single-stranded polynucleotides containing either bromouracil or thymine. These results are incompatible with Cavalieri's hypothesis. Moreover, when T2 and T4 phage DNAs, which on the basis of mass per unit length measurements could only consist of two and not four strands, were denatured under various conditions, they consistently yielded decomposition products with half the molecular weight of the starting material; these products, therefore, must be single stranded (Thomas and Berns, 1961).

Doty *et al.* (1960) also investigated the structure of denatured DNA in solution and found the hydrodynamic behavior of denatured DNA from *E. coli* to obey the following equations:

$$[\eta] = 3.1 \times 10^{-5} M^{0.91}$$

$$s_0 = 0.055 M^{0.36}$$

The exponents were in the border region between random coils and rods, a result that was to be expected from the ionic strength chosen for the polyelectrolytes.

The much stronger coiling of RNA may be due to the formation of additional hydrogen bonds, possibly between the extra 2′-hydroxyl group of the ribose moiety. In contrast to native DNA, denatured DNA shows a hyperchromicity curve characterized by a continuous increase in extinction with increasing temperature (Fig. 45). The DNA of φX174 phage

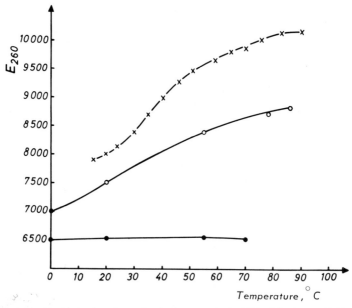

Fig. 45. Optical density of single-stranded DNA (x) from φX174 phage at 260 mμ shown as a function of temperature. Solvent 0.2M NaCl + 10^{-3}M phosphate buffer, pH 7.5. For comparison, the effects on native thymus DNA in 0.1M NaCl (●) and heat-denatured thymus DNA in 0.1M NaCl (o) are also shown (from Sinsheimer, 1959).

which was shown to have a ring-shaped, single-stranded structure gives the same picture (Sinsheimer, 1959).

Reo virus RNA (Gomatos and Tomm, 1963), which shows a sharply defined helix-coil transition in solution because of its secondary structure (cf. p. 263), is formed by two complementary strands. The high T_m value of 99 °C indicates a very stable structure.

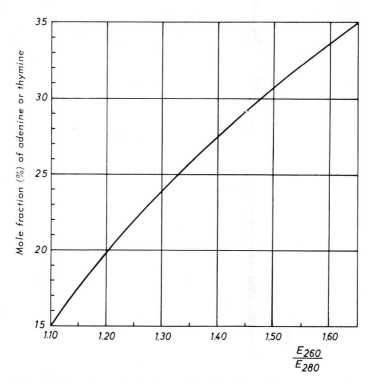

Fig. 46. Per cent of adenine or thymine of total base content in moles plotted as function of the ratio of optical densities at 260 and 280 mμ (from Fredericq *et al.*, 1961).

Fredericq *et al.* (1961) found that the molar ratio of adenine + thymine to guanine + cytosine $(A + T)/(G + C)$ could be derived directly from the extinction coefficients at 260 and 280 mμ (Fig. 46) if the DNA had previously been denatured at pH 3. This procedure constitutes a rapid and convenient method for base determination provided, of course, that the DNA is free from impurities.

CHAPTER IV

Localization of Nucleic Acids

DEOXYRIBONUCLEIC ACID

On the basis of histochemical observations, especially those involving the Feulgen reaction (Feulgen and Rossenbeck, 1924), it has long been known that *DNA occurs in the cell nucleus as a chromosomal constituent.* Appropriate biochemical and cell fractionation procedures leading to the preparation of pure nuclei have since fully confirmed these findings.

Chromosomes are rod-like units that contain dense and thick regions, the chromomeres. Normally, chromosomes can be seen only during nuclear division, when they exist in special form. At telophase they again unwind, and in the succeeding interphase they fill the nucleus without giving rise to any recognizable structure. In interphase nuclei, condensed portions of the chromosomal material (*heterochromatin*) can be distinguished from less dense regions (*euchromatin*) (Fig. 47; cf. p. 89). In the somatic cells of some diptera, the multiple homologous chromosomes occur in thick bundles (*polytene giant chromosomes*).*

The chemical analysis of isolated metaphase chromosomes from Ehrlich ascites tumor cells revealed a content of 13.5% DNA, 13.5% RNA and 68.3% protein (Cantor and Hearst, 1966). The molecular weight of DNA in mammalian cells is generally considered to be around several times 10^7. However, recently, the application of the Cairns technique (see p. 181) to Chinese hamster cells indicated that these contain DNA fibers which are roughly the size of *E. coli* DNA (Huberman and Riggs, 1966). In some rare autoradiographs, DNA fibers as long as 1.6-1.8 mm were found! There was no evidence for circular molecules.

Giant chromosomes remain visible in interphase nuclei. They are polytene bundles of chromosome fibers, each of which is considered to be

---

*In the case of both polyploidy and polyteny, the DNA content is many times that of normal, diploid nuclei, but whereas the number of chromosomes is increased to 2^n in polyploidy, it is the DNA content per chromosome that increases in polyteny, leading in extreme cases to the formation of giant chromosomes. Polytenization, caused by a stepwise doubling of the longitudinal elements, leads to a considerable increase in chromosome thickness. The morphology of giant chromosomes has been reviewed in detail by Beermann (1962).

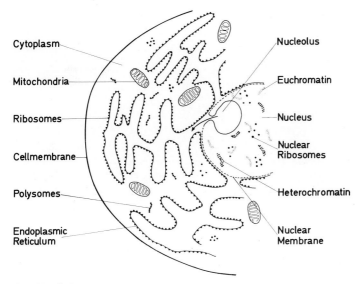

Fig. 47. Schematic representation of a somatic animal cell. For explanation, see text.

equivalent to a single chromosome. The strands are held together in a cable-like manner. At active loci, the structure of the giant chromosomes is modified by "puffing." Puffs are considered to be the sites of messenger RNA formation. In agreement with the phenomenon of puffing, the most intensive RNA synthesis takes place at the euchromatin of normal interphase nuclei (see p. 189).

In histological preparations, euchromatin and heterochromatin differ in their response to acid in the Feulgen reaction; the aldehyde groups in euchromatin are quite reactive, whereas heterochromatin is relatively acid resistant (Sandritter *et al.*, 1965). Some separation by biochemical procedures into fractions that contain predominantly either eu- or heterochromatin has been achieved (Frenster *et al.*, 1963; Harbers. and Vogt, 1966).

DNA does not occur in cell nuclei in the free form but as a DNA-protein complex known as chromatin or, more recently (particularly in the biochemical literature), *nucleohistone*. In addition to DNA and a small quantity of RNA, the nucleohistones also contain basic (histones) and a residual protein (largely acidic).

In sperm nuclei, the DNA has a special structural arrangement. Polarization studies indicate that the individual chromosomes are coiled and are arranged within the sperm head in a specific order (Inoué and Sato, 1962). In the course of spermatogenesis, changes take place in the protein fraction. Thus, fully developed spermatozoa either totally lack non-histone proteins or contain only a very small amount of them; protamines, which are still more basic proteins, have taken the place of the histones (Felix *et al.*, 1956).

The detailed structure of the nucleohistones is not yet known. The protamines are probably located within the small groove of the DNA molecule. Lysine-rich portions of the histones are thought to be localized in the same region, whereas the rest of the histones, which constitute the vast majority, are thought to be localized in the large groove. In most tissues, the mass ratio of total histones to DNA is nearly constant and approximates unity. This ratio remains unaltered when euchromatin and heterochromatin are isolated from nucleohistone fractions. Histochemical studies with giant chromosomes have shown that in the regions of the puffs, where RNA synthesis is most intense, the histone content is about the same as in the remainder of the chromosome (Swift, 1964). Microbial nucleoproteins appear to contain no histones (Wilkins and Zubay, 1959; Zubay and Watson, 1959), and bacteriophages contain neither histones nor protamines.

The molecular weight of histones isolated with mild procedures is generally close to 24×10^3 (Lloyd and Peacocke, 1965). Proline predominates among the N-terminal amino acids in moderately lysine-rich histones, whereas alanine is the predominant N-terminal amino acid of arginine-rich histones. Acetyl groups appear to constitute the N-terminal of very lysine-rich histones (Phillips, 1963); furthermore, methyl groups (Murray, 1964) and phosphoryl groups have been found. The secondary structure of the histones is still largely unknown. Arginine-rich and moderately lysine-rich histones are thought to occur partly as an α-helix, whereas the β-configuration is attributed to the very lysine-rich histones. According to Huang and Bonner (1965), a low molecular weight RNA occurs in nucleo-histones from pea buds. This RNA, which is thought to bridge the histones, has a high content of dihydro-UMP and seems to be made up of about 40 nucleotides. However, these observations could not be confirmed with nucleohistones isolated from mammalian tissues.

When histones are separated from purified nucleohistone preparations, one does not obtain free DNA, but a complex consisting of DNA and a residual (mainly acidic) protein. These acidic proteins contain large amounts of glutamic and aspartic acid. They are very firmly bound to the DNA and are apparently responsible for the gel formation of isolated nucleohistones (Dounce *et al.*, 1966). While Dounce *et al.* (1966) postulated a covalent linkage of DNA to residual proteins, Kirby (1957) discussed the possibility of an interaction via divalent metal ions between the carboxyl groups of the proteins and the purines of DNA. The ratio of acidic protein to DNA, though generally difficult to determine exactly, is smaller than the corresponding ratio between histone and DNA and also seems to be less uniform from tissue to tissue.

Nucleohistones are insoluble in isotonic NaCl (0.12–0.16M). At concentrations above 0.16M, the histones increasingly dissociate from DNA (which occurs in the form of a DNA-protein complex), the dissociation becoming complete at 2.0M (Bayley et al., 1962). The dissociation of lysine-rich histones begins at relatively low ionic strengths. Up to 1M NaCl, most of the histones are lysine rich; as the NaCl concentration increases, the molar ratio of arginine to histone increases in the histones which are no longer bound to DNA. Decreasing the salt concentration to 0.14M leads to significant recombination of DNA and histones, without however restoring the original structural relationships. Hence, the ionic strength of solutions of native nucleohistones must be kept low. For the same reason, the procedure of purifying nucleohistones by alternate precipitation with 0.14M NaCl and re-solution in 1.0M NaCl, a technique commonly in use in the past, should be avoided, particularly as it leads to the progressive loss of histones. Phillips (1962) has extensively reviewed the properties of histones and their subfractions.

Until recently, DNA was thought to occur only in the cell nucleus, but it has now also been found in mitochondria and plastids. The present knowledge of DNA in cytoplasmic organelles has recently been reviewed by Granick and Gibor (1967).

Metzner, in 1952, detected small amounts of DNA in the grana of chloroplasts. This result, which at first seemed doubtful, was confirmed by studies with H^3-thymidine, a specific precursor of DNA (Stocking and Gifford, 1959; cf. p. 137). Subsequently, the chloroplast DNA was separated successfully into two types: DNA which was in all likelihood derived from the grana and had a distinctly higher $(A + T)/(G + C)$ ratio, and nuclear DNA (Sager and Ishida, 1963). Brachet (1959a) reported on extranuclear DNA in *Acetabularia*; Plaut and Sagan (1958) observed similar incorporation of H^3-thymidine into the cytoplasmic material of *Amoeba proteus*. Because of its response to deoxyribonuclease, this material was regarded as DNA.

From the beginning, even the more modern methods of cell fractionation yielded small quantities of DNA which were associated with the cytoplasmic fraction of mammalian tissue extracts. However, these were initially regarded as contaminants resulting from nuclear debris. Only recently has it been shown that DNA occurs in the cytoplasm as a mitochondrial component (Luck and Reich, 1964; Nass and Nass, 1964; Neubert, Helge and Bass, 1965). The DNA content of mitochondria is extremely small, the only exception so far being mitochondria isolated from some rat tumors which have a relatively high DNA concentration (Neubert, 1966; Table 5). The molar base ratio (A + T)/(G + C) of mitochondrial DNA generally differs from that of nuclear DNA from the corresponding cell; it may be either higher or lower. Mitochondrial DNA is quite homogeneous and, therefore, renatures very well after thermal denaturation. The DNA from chicken liver mitochondria was found to be circular (van Bruggen *et al.*, 1966). It seems probable that this is generally true for the DNA of all vertebrate mitochondria.

TABLE 5. DNA Content of Mitochondria (from Neubert, 1966a)

Tissue	µg DNA/mg Protein	
Rat liver	0.8	(210 in nuclei)
Pigeon heart	1.8	
Pigeon muscle pectoralis	1.3	
Morris hepatoma 5123	1.8	
Jensen Sarcoma	5.4	
Shay Chloro-leukemia	4.3	
Walker carcinoma	8.2	
DS carcinosarcoma	8.1	

The fact that mitochondria and plastids have their own DNA agrees with the theory that these particles are genetically independent (Gibor and Granick, 1964) and throws new light on the problem of cytoplasmic inheritance. Hybridization experiments have shown that there is in fact no genetic relationship between nuclear and mitochondrial DNA (cf. p. 276). On the other hand, it is apparent that mitochondria are also determined by nuclear information, inasmuch as a substantial part of mitochondrial RNA is coded for by nuclear DNA (Humm and Humm, 1966). It may be that this RNA is ribosomal in nature, since ribosomes are generally considered to originate in the nucleus (cf. p. 196). The amount of DNA in mitochondria is too small for coding all mitochondrial proteins. More recent observations indicate that a considerable part of mitochondrial proteins (mainly the enzymes) is synthesized outside of the particles and then transferred to the mitochondria (cf. Kadenbach, 1966).

Once DNA had been recognized as the genetic material of chromosomes, it was logical to postulate that each cell contained a constant amount of DNA, just as it had a constant number of chromosomes. Studies with isolated nuclei and analyses of individual cells by means of a quantitative Feulgen reaction and by measuring ultraviolet absorption (Swift, 1955; Sandritter, 1958) showed that *the amount of DNA per nucleus is in fact quite constant* and that diploid cells contain twice as much, and tetraploid cells four times as much, DNA as haploid cells (Davidson *et al.*, 1951; Vendrely, 1955). The amount of DNA per diploid nucleus and its molar base ratios $[(A + T)/(G + C)$; see p. 170] are species specific. Individual chromosomes may show different base ratios, since DNA can be separated into fractions whose base ratios vary. It is not yet known how many DNA molecules make up a single chromosome of a multicellular organism. When all of the DNA in *E. coli* was shown to occur as a single molecule, it became conceivable that—with the exception of polytene chromosomes—each chromosome might also contain a single DNA molecule. This hypothesis is supported by the behavior of DNA during replication and by its uniform rates of duplication (Cairns, 1963b; cf. p. 181).

Chemical determination of cellular DNA content requires isolation of the nuclei (in most cases the DNA of the mitochondria can be neglected). Aliquots of the nuclei are then used for DNA analysis and for counting (as in the counting of blood cells). The results yield average values, but do not give information on possible heterogeneity of the cell material due to the occurrence of polyploid nuclei; hence, the average values may be too high. Histochemical methods, on the other hand, provide a pattern of the distribution of nuclear DNA within a cell population (see Fig. 48). The variation in the DNA content of cells is primarily due to variations in the degree of ploidy. Thus, cells will be found whose DNA contents are in the ratio of $2:4:8$, corresponding to diploid, tetraploid and octaploid nuclei. Intermediate values result from the fact that some cells are undergoing DNA replication (cf. p. 182). Somatic cells, however, tend to have a certain variation in chromosome numbers (Timonen and Therman, 1950) and, therefore, somewhat variable DNA contents. In sperm, the range of DNA content is sometimes considerable (Fig. 49), but the extent to which this affects sperm function is not yet clear. Reduction of the mean DNA content in the sperm of sterile men has been found to be associated with an increase in the variation of DNA values (Leuchtenberger *et al.*, 1953).

The fact that the DNA content of a cell is relatively constant and independent of such exogenous factors as hunger, poisoning, etc., has led to the use of DNA as a reference value in analytical studies, in place of fresh or dry weight, nitrogen content, etc. (Thomson *et al.*, 1953). If the average amount of DNA per cell is known for a given tissue, the results may be easily converted to, and reported as, quantity per cell or enzyme activity per cell. Long-term feeding of p-dimethylaminoazobenzene, which induces liver tumors in rats, brings about a reduction of the mean DNA concentration in liver cell nuclei.

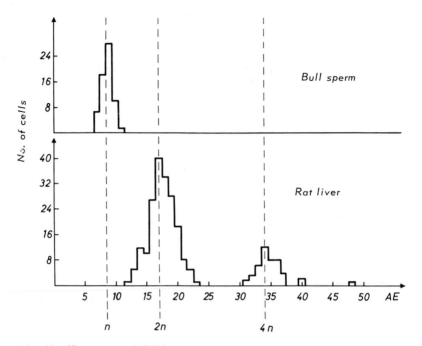

Fig. 48. Histogram of DNA concentration in normal rat liver cells, as measured by a quantitative Feulgen reaction. n, $2n$ and $4n$ refer to the DNA content of haploid, diploid and tetraploid chromosome, respectively, based on bovine sperm as the reference system (from Sandritter, unpublished).

While the DNA in cells and in most DNA-containing viruses (cf. p. 260) normally exists as a double helix, small amounts of a DNA-containing component, which for the most part is present as single-stranded deoxypolyribonucleotide, are found in baker's yeast. In the crab *Cancer borealis*, about 30% of the total DNA in testes and vas deferens corresponds to the dAT polymer (see p. 176). In this naturally occurring AT DNA, adenine and thymine occur alternately along each chain; the molar amount of guanine and cytosine is only 3% of the total base content (Sueoka and Cheng, 1962).

In *E. coli*—and perhaps quite generally in bacteria—all of the genetic material of a cell exists as a single molecule (or under certain conditions as two molecules; cf. p. 181). Recent observations by Massie and Zimm (1965) have raised some doubt as to whether the giant molecules demonstrated by autoradiography (see Fig. 85) consist only of pure DNA. According to these authors, the DNA occurs in *E. coli* and *B. subtilis* as an assembly of subunits with a molecular weight of about 250×10^6, held together by protein.

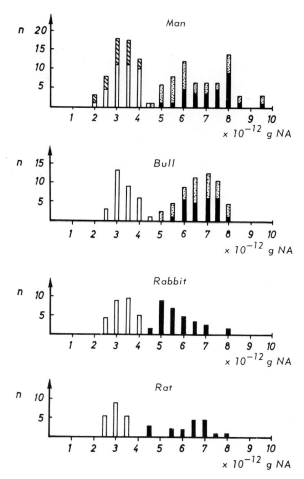

Fig. 49. Interspecies variation in the nucleic acid content of haploid and diploid cells. n = number of cells. Light columns: sperm (cross-hatched part= number of sperm cells after ribonuclease treatment). Black columns: thymus lymphocytes (dotted columns = number of cells after ribonuclease treatment; from Sandritter and co-workers, 1960).

THE VARIOUS RNA OF CELL FRACTIONS

Early cytochemical observations had indicated that RNA is found in the nucleus and the cytoplasm. Biochemical studies have led to the identification of three, perhaps four, types of RNA. A detailed study of cytoplasmic RNA became possible as a result of the introduction of modern cell fractionation techniques.

Cell fractionation requires, as a first step, the gentle disruption of the cell structure, to permit the separation of cytoplasmic from nuclear constituents while avoiding damage to the nuclei. Another important factor is the medium in which the tissue is homogenized. In aqueous solution, some nuclear material, including part of the RNA, is extracted. This can be avoided by the use of nonaqueous media (Behrens, 1956). [A promising new technique has recently been described by Giovanella *et al.* (1966).]   The most common apparatus for homogenizing soft tissues (e.g., liver, kidney, various transplantable tumors of mouse and rat) is the homogenizer described by Potter and Elvehjem (1936; see Fig. 50); other methods, often restricted to the preparation of cell nuclei, have been described by Lang and Siebert (1952), Dounce (1955), and Emanuel and Chaikoff (1957). After homogenization, the homogenate is generally centrifuged in the cold at about 700g for 10 minutes. The nuclei are in the sediment; the cytoplasmic fraction is in the supernatant. For further purification of the nuclei, the homogenization and centrifugation steps should be repeated two more times. It is necessary to constantly check the purity of the isolated nuclei by observation of stained smears under the light microscope or by direct observation with a phase-contrast microscope. The mitochondria, which are the largest particles

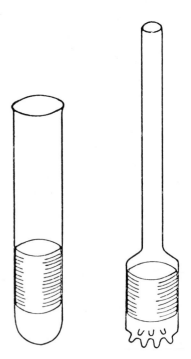

Fig. 50. Potter-Elvehjem homogenizer.

in the cytoplasmic fraction, can be sedimented by centrifugation at about 25,000g. The supernatant contains the microsomes and the particle-free cytoplasmic fraction (soluble cytoplasm). Depending on the medium, 50,000–150,000g is needed to sediment the microsomes quantitatively. An exact description of the conditions of centrifugation should include not only the g values but also the "performance index" (Giebler, 1958). Details of the techniques for the separation of various cytoplasmic fractions and their further purification may be found in several reviews (e.g., Hogeboom and Schneider, 1955; Siebert, 1956). Another method of potential importance for cell fractionation is the technique of carrier-free continuous electrophoresis (Hannig, 1964; Hannig et al., 1964).

Usually 80–90% of the RNA is found as a constituent of particles commonly termed *ribosomes*. Morphologically spherical particles with a diameter of 100–200 Å, ribosomes can occur in the cytoplasm in unbound form, as in microorganisms. Usually, however, they are bound to cytoplasmic structures ("endoplasmic reticulum"; see Fig. 47); in addition, ribosomes can be found in the cell nucleus. Ribosomes from mammalian tissues were originally prepared from the *microsome fraction*, which also contains membrane material from the endoplasmic reticulum. It is possible to separate out the latter by treatment with deoxycholate or by high salt concentrations. Ribosomes so obtained are not homogeneous with respect to particle size. Tissières, Watson et al. (1959) found that the ribosomes from E. coli are composed of well-defined subunits. With increasing Mg^{++} concentration, one 30S and one 50S particle combine into one 70S ribosome; at still higher Mg^{++} concentration, two 70S ribosomes can dimerize to a 100S particle (Fig. 51).

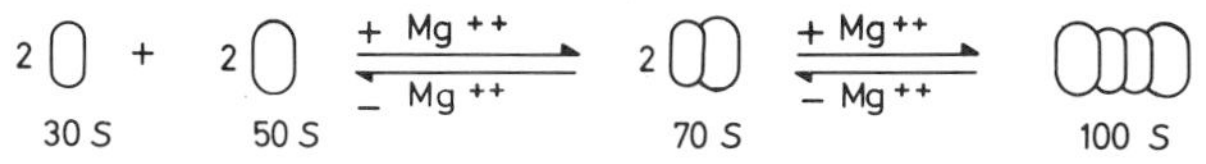

Fig. 51. Subunits (30S and 50S) and aggregates (100S) of 70S ribosomes from *E. coli* (from Watson, 1963). (Reprinted with permission. Copyright 1963 by the American Association for the Advancement of Science).

This subunit structure has been shown to be characteristic of all ribosomes; the concentration of divalent cations determines which particles predominate. Ribosomes from mammalian tissues are somewhat bigger and also more stable than those from bacteria. Their 80S ribosomes (which correspond to the 70S particles of microorganisms) break up into 40S and 60S subunits. Bacterial ribosomes require higher Mg^{++} concentrations for aggregation, and they disaggregate more quickly if the Mg^{++} content falls. The two subunits of the ribosomes also contain two different kinds of RNA (rRNA). RNA of the 30S particle from *E. coli* has a molecular weight of 5.5×10^5 (16S), while the molecular weight of RNA from the 70S particle is 1.1×10^6 (23S). The molar base ratios of *ribosomal RNA* (rRNA) differ in different species. In general, the proportion of guanine is quite high; this is also true for the cytosine content of ribosomes from mammalian cells. rRNA contains a small amount of methylated bases (see p. 101). The proteins of ribosomes are considered to be primarily structural, with similar amino acid composition; they contain high proportions of arginine, glutamic acid and lysine. Although—in addition to the transferase and the translocase (cf. p. 222)—enzymes have been reported in ribosomes, they are probably impurities. The only enzyme that has been found in ribosomes routinely is a ribonuclease bound to the 30S particles in a latent form (see, for example, Tissières and Watson, 1958; Zillig *et al.*, 1959). Detection of enzyme activity requires disintegration of the ribosomes. In addition to rRNA and proteins, ribosomes contain some diamines (Zillig *et al.*, 1959). These, together with Mg^{++} and Ca^{++}, are probably needed to maintain the structural integrity of the ribosome. More recent comparative studies by Stutz and Noll (1967) indicate that at least three classes of ribosomes can be distinguished: (1) 80S ribosomes of animal tissues containing 29/18S rRNA; (2) 80S ribosomes of plant cells (25/16S rRNA); (3) 70S ribosomes (23/26S rRNA) in bacteria and chloroplasts.

At present, little is known about how ribosomes are formed. From kinetic studies with C^{14}-labeled precursors, Britten *et al.* (1962a and b) concluded that ribosomes are formed in a stepwise manner from smaller units, *eosomes* and *neosomes* (Fig. 52). RNA synthesis precedes protein formation, with some analogy to virus formation (see p. 265). Methylation of rRNA takes place after synthesis. It has been suggested that eosomal RNA may serve as the template in the formation of ribosomal proteins. rRNA synthesis is DNA dependent, with the nucleolus considered to be the site of synthesis in mammalian cells (Perry, 1962, 1967).

Ribosomal function requires an additional RNA, termed *messenger RNA* (mRNA), which serves as a template in protein synthesis and provides the correct amino acid sequence (cf. pp. 220 and 229, also Figs. 107 and 110). While, at first, it appeared that a distinction could be made between "heavy" and "light" ribosomes on the basis of their mRNA content, it has now been shown that several ribosomes—frequently five acting as a pentamer—together with mRNA, form a unit cooperating in the template function. The names *ergosome* (Wettstein *et al.*, 1963) and *polysome* (Warner *et al.*, 1962) have been suggested for these rather labile particles, which are held together by mRNA. The number of ribosomes per polysome varies, depending on the length of the mRNA; the pentamers—also demonstrable by electron microscopy (Fig. 53)—have

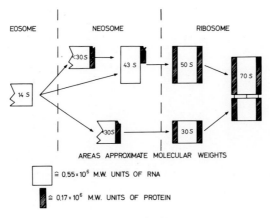

Fig. 52. Biosynthesis of ribosomes in *E. coli*. The open and shaded areas are proportional to weights of RNA and protein, respectively (from Britten *et al.*, 1967). (Reprinted with permission of The Rockefeller University Press).

a molecular weight of about 20×10^6. Without mRNA, e.g., after the action of actinomycin, the particles disintegrate into single ribosomes (Staehelin *et al.*, 1963; cf. p. 294). The size of the mRNA suggests a ratio of approximately 90 nucleotides per ribosome (Staehelin *et al.*, 1964; but cf. p. 225). mRNA is not a homogeneous fraction. Sedimentation constants ranging from 6–8S up to a maximum of 23–30S have been reported for mRNA from *E. coli* (Monier *et al.*, 1962). Because the molecular weights of the proteins vary, and mRNA serves as a template for protein synthesis (see p. 220), this variation in the molecular weight of mRNA is not surprising. In some cases, the mRNA appears to code the amino acid sequences of several proteins (polycistronic messenger).

Membranes of the smooth endoplasmic reticulum of rat liver apparently contain small amounts (0.05–0.1%) of RNA which may be related to the stable cytoplasmic mRNA (Shapot and Pitot, 1966; cf. p. 225).

While it was not clear at first whether the small amounts of RNA found in mitochondria resulted from contamination with microsomes, more careful preparations have proved that these particles indeed contain RNA. The amount of RNA in mammalian mitochondria ranges from 5–40 μg/mg protein (Neubert, 1966a) and includes rRNA as part of the ribosomes in the mitochondria, mRNA and also tRNA (Barnett and Brown, 1967). It may be assumed that mRNA is synthesized in the mitochondria, since they contain both DNA and their own DNA-dependent RNA polymerase (Neubert, Helge and Merker, 1965; cf. pp. 167 and 226). Moreover, inhibition of RNA synthesis also leads to inhibition of protein biosynthesis.

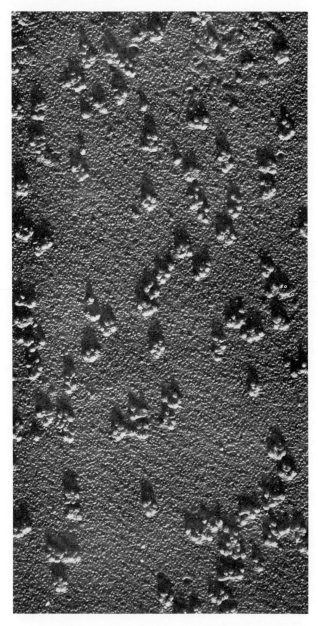

Fig. 53. Electron micrograph of reticulocyte polysomes
(= ergosomes; from Warner, Rich and Hall, (1960).

Another type of RNA is called *soluble RNA* (sRNA; since it was first observed in the soluble cytoplasm), or *transfer RNA* (tRNA) because of its function in protein synthesis. Since all of the material in the sRNA fraction may not participate in the transfer function, these two kinds should be distinguished from each other and not simply equated. For each of the twenty amino acids there exists one or several strictly specific tRNAs (cf. p. 232) which have only recently been characterized adequately. tRNA has a relatively low molecular weight, approximately 25,000. It is single stranded and always has the nucleotide sequence— CCA at the 3′-terminal; in most cases, the 5′-terminal is guanosine. tRNA contains large amounts of rare bases. In addition to pseudouridine, 1-methyluracil, 1-methylcytosine (Hall, 1963a), and thymine dihydrouracil, 1-methylhypoxanthine, 1-methylguanine and N,N-dimethylguanine have been found (Fig. 54). More recently, 2- and 6-thiouracil have been observed in some cases; in addition, 1-methyladenine, 5-methylcytosine, N,N-dimethyladenine (Madison *et al.*, 1966), N-acetylcytosine and N^6-(γ,γ-dimethylallyl)adenine (Biemann *et al.*, 1966) have been found. In addition methyl groups may be attached to the 2′-position of ribose, thus

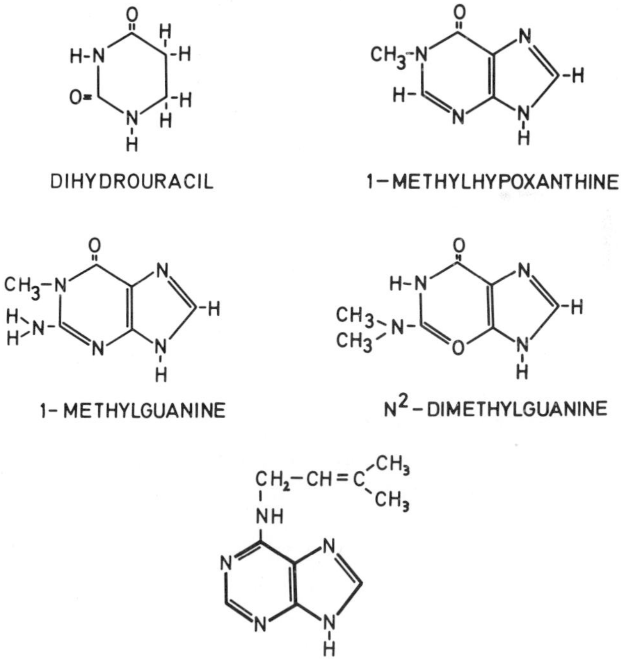

DIHYDROURACIL

1—METHYLHYPOXANTHINE

1- METHYLGUANINE

N^2 – DIMETHYLGUANINE

N^6–(γ,γ –DIMETHYLALLYL)–ADENINE

Fig. 54. Structure of some rare bases found in tRNA.

giving rise to 2^1-O-methylribosides (Hall 1963 b). The presence of rare
bases considerably facilitates the analysis of nucleotide sequences in
tRNA.   Holley and co-workers (1965) were the first to determine the
complete nucleotide sequence of a specific tRNA [alanine tRNA from
baker's yeast; Fig. 55 (A)]. Somewhat later, Zachau et al. (1966) published the complete nucleotide sequence of two serine tRNAs from
brewer's yeast [Fig. 55 (C)]; Madison et al. (1966) reported the sequence
of tyrosine tRNA I from baker's yeast [Fig. 55 (B)], and finally,
Rajbhandary et al. (1967) in Khorana's group determined the sequence of
phenylalanine tRNA.   Earlier observations had already indicated that
numerous bases of the tRNA molecule are hydrogen bonded to each other
and thus form helical structures of the Watson-Crick type. In Figs. 55(A)
and (B), secondary structures for alanine and tyrosine I tRNAs have
been proposed. The basis for these structures is an attempt to achieve
the maximum number of hydrogen-bonded base pairs. In turn, this results
in very similar structures that contain the probable anticodons (cf. p.
230) IGC and GΨA in the same loop. The two serine tRNAs differ only in
three bases [underlined in Fig. 55(C)]. These differences may be due to
transitions (see p. 241) in the tRNA cistrons. The probable anticodon
sequence IGA is identical in serine tRNA I and II; the most probable
anticodon of phenylalanine tRNA is 2′-OMeGAA (Rajbhandary et al.,
1967). The total number of nucleotides is 76 for phenylalanine tRNA, 77
for alanine tRNA, 78 for tyrosine tRNA I and 84 for both types of serine
tRNA. Thus, the molecular weight of tRNAs is not uniform.

Alanine tRNA and serine tRNA differ considerably in their primary
structure. However, the distances between I, diMeG and one GUU, as
well as between T and the —CCA terminal, are the same. The GTΨCG
sequence, apparently common to many tRNAs (Zamir et al., 1965), has
been found in three of the analyzed tRNAs; in serine tRNA I, the last G
of this sequence is replaced by A [Fig. 55(C)]. Contrary to earlier observations, the 5′-terminal of the tyrosine tRNA is pC instead of the
more common pG. Models of serine tRNA, also constructed to achieve
maximum base pairing, show that about 60% of the nucleotides can be
arranged in five helical areas; in these models, the rare bases appear
mainly in the loops and in regions without base pairing (Zachau et al.,
1966). Thus far, nothing is known about the tertiary structure of tRNAs
which would appear to be essential to the specificity of their interactions with activating enzymes, ribosomes and the mRNA template (cf. p.
218). An analysis of small-angle X-ray studies (Lake and Beeman, 1967)
permitted the elimination of various models for the tertiary structure of
tRNA; at the same time, the cloverleaf model, as suggested by Holley
et al. (1965), in which three arms are folded up tightly together while the

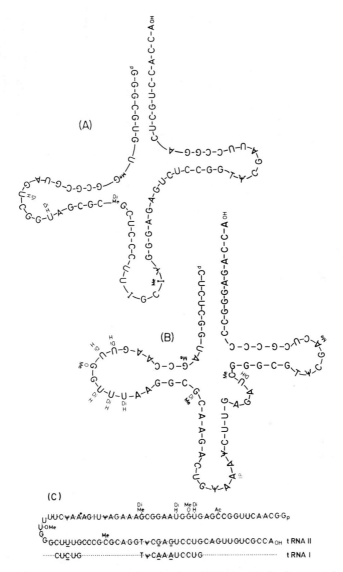

Fig. 55. (A) Proposed structure for alanine tRNA from baker's yeast (according to Holley *et al.*, 1965). (B) Proposed structure for tyrosine tRNA I from baker's yeast (from Madison *et al.*, 1966).   C) Complete nucleotide sequences of two tRNAs for serine from brewer's yeast (from Zachau *et al.*, 1966). Abbreviations for the rare bases: A* = N6-(γ,γ-dimethylallyl)-adenosine; AcC = N6-acetyl-cytidine;   DiMeA = N2-dimethyladenosine;   DiMeG = N2-dimethylguanosine; HDiU = 4,5-dihydrouridine;   I = inosine;   MeA = 1-methyladenosine;   MeC = 5-methylcytidine;   MeG = 1-methylguanosine;   MeI = 1-methylinosine;   OMeG = 2′-O-methylguanosine; OMeU = 2′-O-methyluridine; Ψ = pseudouridine; U* = mixture of U and HDiU; T = ribothymidine. (A) and (B) reprinted with permission. Copyright 1965 and 1966 by the American Association for the Advancement of Science. (C) reprinted with permission of Verlag Chemie.

fourth arm is extended in the opposite direction, received some support. In some cases where thiopyrimidines have been found in tRNA, the possibility of disulfide bridges—which could contribute to the formation of a specific tertiary structure as is the case in proteins—has been discussed (Lipset, 1965). According to observations by Barnett and Brown (1967), mitochondria may have some specific tRNAs.

RNA occurs in various regions of the nucleus, especially in the nucleolus. Small amounts are also found in the chromatin or in the chromosomes. Chromatin or nucleohistones constitute the principal site of RNA synthesis which is highest in the euchromatin (or at the puffs of the giant chromosomes; cf. pp. 88 and 189). It is believed that all RNA, with the exception of some mitochondrial RNA and the RNA in the plastids of plant cells, is synthesized at the nuclear DNA. This view derives support from the fact that all three types of RNA—mRNA, rRNA (as part of the nuclear ribosomes) and tRNA—have been isolated from nuclear material. Moreover, Huang and Bonner (1965) have reported the occurrence of another low molecular weight RNA in pea bud nucleohistones, which is said to bond the histones in bridge-like fashion.

The nucleolus apparently has several functions. One is to concentrate the RNA formed at the chromosomes and to direct it into the cytoplasm (see p. 196). Another involves the methylation of tRNA. Autoradiography has revealed a zone of intensive protein synthesis within or around the nucleoli, which can now be isolated by biochemical techniques. The functional role of the nucleolus has been discussed in several reviews (Sirlin, 1962; Busch *et al.*, 1963).

CHAPTER V

Nucleotide Metabolism

BIOSYNTHESIS OF PURINES

Historical Development

The question of whether purine biosynthesis occurs in the animal organism was answered in the affirmative long ago. Balance studies showed that suckling animals synthesized purines even though they ingested only milk which is low in purines. During incubation, the purine content of hen's eggs was found to increase without exogenous supply; the same was true of sperm formation in starved salmon. On the basis of the chemical structures of purines it had long been assumed that purines were formed by the condensation of two molecules of urea with a three-carbon compound. Experimental support seemed to come from experiments in which birds, after having been fed urea and C_3 compounds, excreted increased amounts of urate. These results proved to be invalid since birds excrete nitrogen mainly in the form of uric acid. In 1886, Minkowski localized the site of purine biosynthesis in geese by tying off the blood vessels leading to the liver and also by hepatectomies. The animals, able to survive for a short time only, excreted ammonium salts rather than uric acid in the urine.

The first successful *in vitro* experiments on purine biosynthesis were done by Krebs in the middle thirties (Edson *et al.*, 1936). He demonstrated hypoxanthine synthesis in pigeon liver slices; addition of oxaloacetate or glutamate led to a distinct increase in purine synthesis. Avian liver proved to be an extremely fortunate choice for experiments of this kind, because it contains no xanthine oxidase to oxidize the newly formed hypoxanthine to uric acid. Working with tissue slices has the disadvantage, however, of liberating large amounts of purines from the unavoidable breakdown of tissue.

With the aid of isotopes, Schoenheimer, as early as 1943-1944, showed that most of the nitrogen in purines and purine nucleotides is derived from ammonium ions (Barnes and Schoenheimer, 1943) and that urea, histidine and arginine are not direct precursors in purine synthesis. Even some of the natural purines and pyrimidines, such as uric acid, guanine, xanthine, hypoxanthine, uracil and thymine, either are not used at all in the biosynthesis of purine bases or are used to only a small extent.

The first aim in the elucidation of the synthetic pathway of purine formation was to determine the origin of the individual carbon atoms, accomplished initially with the aid of the stable carbon isotope C^{13}. The ureido carbon atoms (carbons 2 and 8) (Sonne *et al.*, 1948) were found to be derived from formate, while CO_2 contributed only carbon 6 of the purine ring (Sonne *et al.*, 1946). Hence urea can be excluded as a purine precursor. Glycine, which donates both of its carbons and the amino nitrogen to the purine ring, proved to be the most active precursor (Buchanan *et al.*, 1948). From the relative positions of the three glycine-derived atoms, it became apparent that this amino acid is incorporated *in toto*, without prior splitting. It took longer to determine the origin of the remaining three nitrogen atoms (nitrogens 1, 3 and 9) because many other nitrogen-containing intermediates are also converted to ammonium ions and other amino compounds. Moreover, in the chemical degradation, it was initially impossible to distinguish between the nitrogen atoms in positions 1 and 3 until, with the help of Brandenberger's degradation (cf. p. 123), nitrogen 1 was shown to be formed from aspartate and nitrogens 3 and 9 from the amide nitrogen of glutamine (Levenberg *et al.*, 1956). The origin of the atoms of the purine skeleton is summarized in Fig. 56.

Later the focus of research shifted to the purification of enzymes for the individual steps and the isolation of intermediates. Here, ion exchange chromatography became an important tool. In 1951, Greenberg achieved the first purine synthesis in a cell-free pigeon liver extract, having previously observed hypoxanthine synthesis from radioactive precursors in homogenates. With the aid of soluble enzyme systems, it became possible to minimize nonspecific side reactions of the added

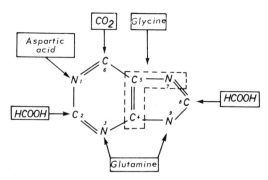

Fig. 56. Biological precursors in the biosynthesis of the purine molecule.

amino acids, and ultimately the synthesis of purine from low molecular weight compounds shown in Fig. 56 was fully confirmed.

Further important additions to our knowledge of purine biosynthesis came from microbiology. By the early 1940's, it had been found that cultures of *E. coli*, when grown in the presence of sulfonamides, accumulated in the medium a diazotizable amine (Stetten and Fox, 1945), whose structure was shown to be 4-amino-5-imidazolecarboxamide (AICA; Shive *et al.*, 1947). Glycine added to the culture medium increased the amount of AICA synthesis. As may be seen from the structural formula XL, the insertion of a carbon atom between the two amino groups would

$$
\begin{array}{c}
O \\
\parallel \\
H_2N-C \\
\diagdown C \diagup N \diagdown \\
\parallel \qquad \qquad CH \\
H_2N-C \diagdown N \diagup \\
\qquad \quad \underset{H}{|}
\end{array}
$$

AICA

XL

convert the molecule into hypoxanthine. At first, it seemed surprising that *E. coli* and other bacteria cannot utilize AICA. Further investigations, however, led to the discovery of mutants which grew just as well with AICA as when the complete purine was added (Bergmann *et al.*, 1952). In addition to sulfonamides, the folic acid antagonist aminopterine (cf. pp. 154 and 308) may also cause AICA accumulation; it seemed reasonable, therefore, to infer that folic acid participates in purine biosynthesis.

C^{14}-labeled AICA is converted by the pigeon mainly to C^{14}-uric acid (Schulman *et al.*, 1950), while in the rat and mouse and in yeast it is incorporated in the purine fraction of nucleic acids (Williams and Buchanan, 1953). These results appeared to confirm the fact that AICA constituted an intermediate product in purine biosynthesis. However, this hypothesis was not compatible with the following observations: Whereas hypoxanthine is synthesized from C^{14}-glycine in pigeon liver homogenates, the isotope is not incorporated in unlabeled AICA added to the same preparation; moreover there is no isotope dilution in the hypoxanthine fraction (Schulman and Buchanan, 1952). C^{14}-formate experiments also were not compatible with AICA being a natural intermediate in purine biosynthesis; C^{14}-labeled formic acid was transformed almost completely into inosinic acid, inosine and hypoxanthine (Greenberg, 1951). From the specific activities and from parallel experiments with C^{14}-glycine, it became clear that inosinic acid is the first purine derivative formed, while the two other compounds represent secondary transformation products. The observations were also the first indication that all of purine biosynthesis took place at the ribonucleotide level, so that AICA probably also occurs in the cell as a ribonucleotide.

How then does C^{14}-AICA become incorporated in purines? The discovery of pentose derivatives suggested the following reactions: Catalyzed by a phosphorylase, AICA reacts with ribose-1-phosphate and the resulting ribonucleoside is then phosphorylated as follows:

$$\text{AICA} + \text{R-1-P} \overset{\text{phosphorylase}}{\rightleftharpoons} \text{H}_3\text{PO}_4 + \text{AICA ribonucleoside}$$

$$+ \text{ATP} \Big\downarrow \text{kinase}$$

$$\text{AICA ribonucleotide} + \text{ADP}$$

Alternatively, AICA could react with 5-phosphoribosyl-1-pyrophosphate (PRPP), (see p. 150), with the aid of a pyrophosphorylase, to form the AICA ribonucleotide directly:

$$\text{AICA} + \text{PRPP} \overset{\text{pyrophosphorylase}}{\rightleftharpoons} \text{AICAR} + \text{PP}$$

Experimental evidence has been adduced for both of these reactions (Greenberg, 1956; Flaks *et al.*, 1957a).

AICA ribonucleoside inhibits the enzyme adenosine deaminase (cf. p. 114) and acts as an inducer for two enzymes involved in the formation of guanosine monophosphate (cf. p. 113; Kuramitsu *et al.*, 1964).

Information on the low molecular weight precursors of the AICA ribonucleotide came from experiments with pigeon liver extracts which will be discussed below. The series of reactions leading from ribose-5-phosphate, ATP, glutamine, glycine, CO_2 and formate to inosinic acid is known as the *de novo synthesis of the purine ribonucleotides*. This should be distinguished from the *utilization of preformed purine bases*, dealt with in a later section (see p. 118).

De Novo Synthesis of Inosinic Acid

It has been known for some time that added ribose-5-phosphate (R-5-P) stimulates nucleotide formation in appropriate systems. However, the R-5-P synthesized in the pentose phosphate cycle (cf. p. 143) must first be activated. This is brought about by a 5-phosphoribose pyrophosphokinase* which has been found in various systems; the enzyme catalyzes the transfer of a pyrophosphate residue from ATP to the aldehyde group

*EC 2.7.6.1   ATP : D-ribose-5-phosphate pyrophosphotransferase.

of R-5-P (Khorana *et al.*, 1958). Moreover, the adenine nucleoside decoyinine and cordycepin triphosphate inhibit the formation of 5-phosphribosyl-1-pyrophosphate (Bloch and Nichol, 1964; Overgaard-Hansen, 1964; cf. p. 303).

The resulting 5-phosphoribosyl-1-pyrophosphate (see Fig. 57) now combines with the amide nitrogen of L-glutamine, with the elimination of pyrophosphate. The stoichiometry of this reaction leading to phosphoribosylamine (PRA) is now completely understood, and the enzyme 5-phosphoribosylpyrophosphate amidotransferase* has been highly purified from hen and pigeon livers (Hartman and Buchanan, 1958a). As yet it has not been possible to isolate the labile reaction product PRA from the enzyme mixture. However, synthetic PRA is readily utilized in the enzymatic reaction (Goldthwait, 1956). 6-Diazo-5-oxo-L-norleucine (DON), azaserine and cordycepin phosphate (cf. pp. 303 and 323) inhibit the reaction leading to PRA (Hartman, 1963). Adenine inhibits the amidotransferase by a feedback mechanism (Nierlich and McFall, 1963), whereas the ribonucleotides of various purine analogs act by a pseudo-feedback mechanism (McCollister *et al.*, 1964). A second pathway for the biosynthesis of 5-phosphoribosylamine has been found in tumor cells, wheat germ, and *E. coli*: The intermediate is formed by an enzymatic reaction of ribose-5-phosphate with ammonia (LeGal *et al.*, 1967).

In the next step, three atoms of the purine ring are incorporated simultaneously. PRA is combined with glycine with the aid of an enzyme called glycinamide ribonucleotide kinosynthetase.** The energy of the Mg^{++}-dependent reaction is supplied by ATP. This reaction, in which glycinamide ribonucleotide (GAR) is formed, has been studied in detail with the aid of purified enzymes (Hartman and Buchanan, 1958b; Nierlich and Magasanik, 1965b). An alternative pathway leading to the formation of GAR has been reported for *E. coli* and hen's liver. R-5-P, found to react with ATP and NH_3 (which leads to better yields than glutamine) without prior formation of PRPP, gives a product which, with glycine, yields GAR directly (Nierlich and Magasanik, 1961). As yet, no functional significance can be attributed to the existence of two GAR pathways in the same cell. In plants, there apparently exists still another kind of GAR synthesis, because L-asparagine has been shown to be the best nitrogen donor in wheat seedlings (Kapoor and Waygood, 1962). Formylglycinamide ribonucleotide (FGAR) is formed through the participation of a derivative of tetrahydrofolate (see p. 153).

If pigeon liver extracts are incubated with R-5-P, ATP, glutamine,

---

*EC 2.4.2.14 Ribosylamine-5-phosphate : pyrophosphatephosphoribosyltransferase.
**EC 6.3.1.3 5′-Phosphoribosylamine : glycine ligase (ADP).

glycine and formate, then GAR and FGAR accumulate in these systems. It has been possible to achieve 60-fold purification of the enzyme glycinamide ribonucleotide transformylase,* which is very sensitive to copper ions (Warren and Buchanan, 1957). When impure enzymes are used, many FH_4 derivatives are active C_1 donors, but only N^5, N^{10}-methenyl-FH_4 (cf. p. 153) reacts with the purified enzyme (Hartman and Buchanan, 1959).

Before the five-membered ring is closed, nitrogen 3 of the purine skeleton is attached to FGAR; ATP again supplies the necessary energy, and glutamine is the nitrogen donor** (Melnick and Buchanan, 1957). Formylglycinamidine ribonucleotide (FGAM) is the product. Azaserine and DON (see p. 323) inhibit this otherwise freely reversible reaction. The enzyme catalyzing the final ring closure of FGAM, phosphoribosyl-aminoimidazole synthetase,[†] leads to the formation of 5-aminoimidazole ribonucleotide (AIR); this enzyme is quite labile and can be purified from hen's liver only with great difficulty. It needs potassium ions for activation and stabilization. The ring closure requires 1 mole of ATP and is freely reversible. The labile product AIR was isolated by exchange chromatography and precipitated as the barium salt. Its aglycone, 5-aminoimidazole, is formed by *Clostridium cylindrosporum* in an enzymatic degradation of purine (Rabinowitz, 1956; cf. p. 123); the corresponding ribonucleoside may be isolated from the filtered culture medium of a purine-dependent *E. coli* mutant (Love and Levenberg, 1959).

Finally the imidazole derivative AIR is carboxylated to form the 5-amino-4-imidazolecarboxylic acid ribonucleotide (AICR). This freely reversible reaction seems unusual since no ATP is necessary for its activation (Lukens and Buchanan, 1959). However, with a purified enzyme preparation, it was found that biotin is involved in the carboxylation (Ahmed and Moat, 1963). The carboxylation[††] proceeds only at high bicarbonate concentration or by removal of the product. AICR has a specific absorption maximum at 249 mμ suitable for quantitative analysis.

In the next step, AICR is converted to its acylamide, 5-amino-4-imidazolecarboxamide ribonucleotide (AICAR), with the amino group of aspartate incorporated in a two-step reaction. (1) One mole of aspartate is added with the aid of phosphoribosyl-aminoimidazole-succinocarbox-

*EC 2.1.2.2 5′-Phosphoribosyl-N-formylglycineamide: tetrahydrofolate-5,10-formyltransferase.
**EC 6.3.5.3 5′-Phosphoribosyl-formylglycineamide: L-glutamine   amido-ligase (ADP).
†EC 6.3.3.1 5′-Phosphoribosyl-formylglycineamidine cyclo-ligase (ADP).
††EC 4.1.1.21 5′-Phosphoribosyl-5-amino-4-imidazolecarboxylate carboxylase.

amide synthetase* (Miller and Buchanan, 1962a). The product of this ATP-dependent reaction is a 5-amino-4-imidazole-N-succinocarboxamide ribonucleotide; it can be analyzed quantitatively by diazotization at $0°C$ and subsequent coupling to give a colored compound. The condensation reaction is ATP specific; UTP, CTP and GTP required in an analogous aspartate addition (see p. 114) do not react here. (2) To obtain the acylamide AICAR, succinate must be split off; the enzyme which accomplishes this, prepared from avian liver, mammalian liver, yeast and various microorganisms (Miller *et al.*, 1959), is probably identical with adenylosuccinase** described on p. 114. In contrast to its precursor, AICAR is stable at $100°C$; it shows a specific absorption maximum at 267 mμ.

Completion of the purine ring requires only the incorporation of a carbon atom in position 2; it is attached to the nitrogen (position 3) by a transformylase;[†] in contrast to the transformylation previously described, N^{10}-formyl-FH$_4$ serves as C$_1$ donor for this purpose (Flaks *et al.*, 1957b).

The reaction product formylated onto the nitrogen in position 3 does not accumulate in the reaction mixture, because an inosicase,[††] which has not yet been isolated, effects the subsequent ring closure to form a purine. It has been suggested that all this is accomplished by a single enzyme with dual activity, since the ratio of the two activities remains constant when the enzyme is purified from hen's liver. Synthetic 5-form-amido-4-imidazolecarboxamide ribonucleotide is readily attacked by inosicase; inosine monophosphate (IMP) is formed by the release of H_2O. The inosicase equilibrium at $1:16,000$ lies well on the purine side (Warren *et al.*, 1957). The enzyme reacts only with the phosphoribosyl derivative; neither the ribonucleoside nor the aglycone is attacked.

Figure 57 summarizes the entire reaction sequence in the *de novo* synthesis of purines up to the formation of IMP.

Reactions Modifying the Base Components of Purine Ribonucleotides

IMP, formed by the *de novo* synthesis dealt with above, is a fundamental purine nucleotide whose base (hypoxanthine), however, does not generally occur as a nucleic acid constituent.

The conversion of IMP to guanosine monophosphate (GMP) takes place

---

*EC 6.3.2.6   5′-Phosphoribosyl-4-carboxy-5-aminoimidazole: L-aspartate ligase (ADP).

**EC 4.3.2.2   Adenylosuccinate AMP-lyase.

[†]EC 2.1.2.3   5′-Phosphoribosyl-5-formamido-4-imidazolecarboxamide : tetrahydrofolate 10-formyltransferase.

[††]EC 3.5.4.10   IMP 1,2-hydrolase (decyclizing).

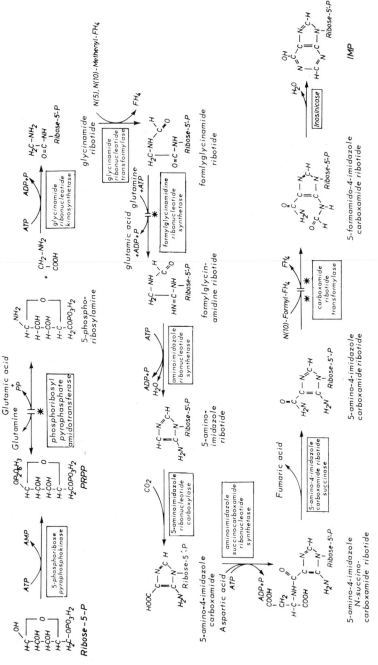

Fig. 57. *De novo* synthesis of inosine-5'-phosphate. ↓‖→ = reaction inhibited by azaserine and 6-diazo-5-oxo-L-norleu-
cine. ↓‖→ = reaction inhibited by folic acid analogs. *
**

in two steps.   The first is dehydrogenation with the aid of inosine-5′-phosphate dehydrogenase,* purified from microorganisms (Magasanik *et al.*, 1957) and animal tissues (Abrams and Bentley, 1955; Lagerkvist, 1958a).   This potassium-dependent, NAD-specific enzyme is inhibited by purine analogs (Hampton, 1963) and is activated by glutathione and cyanide; the reaction product is xanthosine-5′-phosphate (XMP).   Subsequent amination of XMP to GMP proceeds by different routes in bacteria and animal tissues.   The bacterial enzyme, which has been purified 300-fold (Moyed and Magasanik, 1957), requires ATP and NH_3.   The animal enzyme,** on the other hand, shows significantly greater activity with L-glutamine, whose amido nitrogen is transferred to the purine, than with NH_3 (Lagerkvist, 1958b; Abrams and Bentley, 1959).   The formation of an IMP dehydrogenase and of an XMP aminase was induced in an *E. coli* mutant by the addition of AICA ribonucleoside (cf. p. 108).

Isotope experiments have suggested a direct pathway to GMP, bypassing IMP; if the δ-aminomethylene group ($-CH_2-NH_2$) of δ-aminolevulinic acid was incorporated into AICAR, GMP could result by direct ring closure (Nemeth *et al.*, 1957).

In the transformation of IMP to adenosine monophosphate (AMP), a molecule of aspartate is added in the 6-position of the purine nucleus;

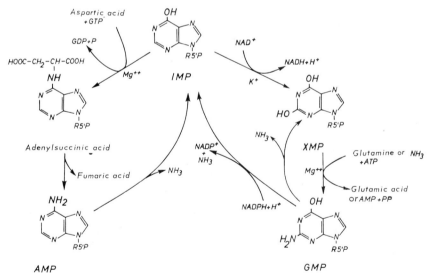

Fig. 58.   Enzymatic transformations of purine ribonucleotides.

*EC 1.2.1.14 IMP : NAD oxidoreductase.
**EC 6.3.5.2 Xanthosine-5′-phosphate : L-glutamine amido-ligase (AMP).

in this condensation, 1 mole of GTP is converted to GDP and inorganic phosphate. GTP cannot be replaced by any other nucleoside triphosphate (Lieberman, 1956b; Davey, 1961). AMP and GMP inhibit this synthetase* (Wyngaarden and Greenland, 1963). Adenylosuccinase, referred to on p. 111, splits off fumarate from adenylosuccinate, with the liberation of AMP (Miller and Buchanan, 1962b).

Deamination reactions lead back to either hypoxanthine or xanthine derivatives. Thus, GMP, guanosine and guanine can undergo hydrolytic deamination, with the formation of XMP, xanthosine and xanthine respectively. A special reductive deamination of GMP has been observed in extracts of *E. coli* (Mager and Magasanik, 1960); IMP and NH_3 are formed with the concomitant oxidation of NADPH.

A hydrolytic deamination of AMP leading to IMP has been reported for muscle tissue, liver and tumors; the deaminase,** which has been crystallized (Lee, 1957; Kizer *et al.*, 1963), may have increased activity during carcinogenesis (Kizer *et al.*, 1964). dAMP is also deaminated by this enzyme, but adenine, adenosine, ADP and ATP are not. An adenosine and deoxyadenosine deaminase*** has been found in intestinal mucosa and bacteria (Zittle, 1946; Koch and Vallee, 1959). The deamination of dAMP, dGMP and their appropriate nucleosides has been investigated in embryonic tissues (Roth and Buccino, 1963).

Recently, several methyl-substituted purines have been isolated from different biological materials. These "rare purine bases" are:

1-Methylhypoxanthine	Found in yeast tRNA (Hall, 1963c) and human urine (Weissmann *et al.*, 1957).
1-Methyladenine	In plant and mammalian RNAs (Dunn, 1961).
1-Methylguanine	In yeast RNA (Adler *et al.*, 1958); in livers (Dunn, 1959; Smith and Dunn, 1959; Bergquist and Matthews. 1962); in mouse tumor RNA (Bergquist and Matthews, 1962); in normal human urine (Weissmann *et al.*, 1957).
2-Methyladenine	In various RNAs (Littlefield and Dunn, 1958) and a vitamin B_{12} derivative (Brown *et al.*, 1955).
3-Methyladenine	In rabbit lung (Axelrod and Daly, 1962).

*EC 6.3.4.4 IMP : L-aspartate ligase (GDP).
**EC 3.5.5.6 AMP aminohydrolase.
***EC 3.5.4.4 Adenosine aminohydrolase.

6-Methylaminopurine	In RNAs of various origins (Little-field and Dunn, 1958; Dunn, 1959; Dunn, Smith and Spahr, 1960; Berg-quist and Matthews, 1962) and in bacterial DNA (Theil and Zamen-hof, 1963).
6-Dimethylaminopurine	In various RNAs (Dunn, 1959; Berg-quist and Matthews, 1962).
3-Methylguanine	In yeast and mouse tumor RNAs (Dunn, 1961).
2-Methylamino-6-hydroxypurine	In liver (Dunn, 1959; Smith and Dunn, 1959a; Bergquist and Matthews, 1962) and in urine (Weissmann et al., 1957).
2-Dimethylamino-6-hydroxypurine	In liver (Dunn, 1959; Bergquist and Matthews, 1962).

It seems possible that the occurrence of these rare bases outside the tRNA fraction represents an artifact (Bergquist and Matthews, 1962). The enzymatic mechanism for the formation of some of these methyl-purines has been investigated and found to involve the transfer to the base of the methyl group from S-adenosylmethionine (Biswas *et al.*, 1961) or methylcobamide (Walerich *et al.*, 1966). Resting cells of *E. coli* B can methylate free adenine and demethylate 6-methylaminopurine (Remy, 1961).

Transformations of Purine Bases into Their Nucleosides and Nucleotides

The free purine bases adenine, guanine, hypoxanthine, xanthine, 2,6-diaminopurine and uric acid (Laster and Blair, 1963) can react with R-1-P to form nucleosides; these reactions are catalyzed by more or less specific nucleoside phosphorylases* and yield inorganic phosphate (Fig. 59). Conversely, purine nucleosides can undergo phosphorolytic scission. Guanine and hypoxanthine react with dR-1-P to form the corresponding deoxyribonucleosides. Purine nucleosides can also be cleaved by hydrolytic nucleosidases** (e.g., Soldo *et al.*, 1966). It is interesting that adenine is excluded from this reaction with dR-1-P; this may be the reason why DNA synthesis and enzyme induction can be inhibited by deoxyadenosine (Maley and Maley, 1960; Welch, 1965; cf. p. 187).

Nucleosides are phosphorylated by kinases, with ATP as the phos-phorus donor, to form the corresponding 5'-phosphates. Kinases have

*EC 2.4.2.1 Purine nucleoside : orthophosphate ribosyltransferase.
**EC 3.2.2.1 N-Ribosyl-purine ribohydrolase.

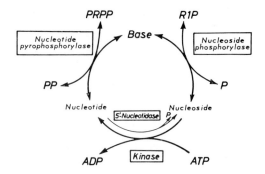

Fig. 59. General scheme of reversible enzymatic conversion of bases to nucleosides
and nucleotides.

been shown to be involved in the reaction of adenosine and xanthosine.
Scarano (1953) has found an alternate path for AMP synthesis in pigeon
liver where ribose-1,5-diphosphate reacts directly with adenine to form
AMP and free orthophosphate.

Ribonucleotides can also be formed from the free bases by the action
of pyrophosphorylases; this involves a condensation with PRPP. Up to
now, two different pyrophosphorylases have been described; one enzyme,* prepared from ox liver, acts on PRPP with adenine and AICA
(Flaks et al., 1957a). The other** is specific for guanine and hypoxanthine; in addition, it reacts with 6-mercaptopurine (Kornberg et al., 1955;
cf. p. 319). The pyrophosphorylase reaction is reversible and, starting
from nucleotides and pyrophosphate, gives rise to the free bases and
PRPP.

Nucleosides can be formed from the corresponding nucleotides by the
action of 5′-nucleotidases, which generally have a low substrate specificity (Reis, 1937; Heppel and Hilmoe, 1951).

A 3′-nucleotidase, which can split the 3′-phosphate esters of adenosine, inosine, guanosine and some pyrimidines, occurs in certain leaves
and plant seeds; the enzyme has been partially purified (Shuster and
Kaplan, 1953).

The nucleoside triphosphates are precursors in the biosynthesis of
DNA and RNA. They are formed by the action of nucleoside mono- and
diphosphokinases, which are enzymes of varying substrate specificity.
In general, the phosphate groups of ATP are used in these kinase reactions; in some cases, however, other triphosphates seem to be necessary
for such transphosphorylations (Krebs and Hems, 1955; Ratliff et al.,

*EC 2.4.2.7 AMP : pyrophosphate phosphoribosyltransferase.
**EC 2.4.2.8 IMP : pyrophosphate phosphoribosyltransferase.

1964). The kinases necessary for the formation of deoxyribonucleoside triphosphates (e.g., Klenow and Lichtler, 1957; Sugino *et al.*, 1966) are formed in large amounts shortly after the T-phage infection of *E. coli* (Bello *et al.*, 1961a; Bessman *et al.*, 1961, 1964; 1965). In mung bean extracts, an enzyme has been found which forms GTP, starting from GDP and phytin (Biswas and Biswas, 1965).

The polyphosphates can be dephosphorylated by phosphatases of greater or lesser substrate specificity. The mechanical work of muscle is largely dependent on the energy liberated in the following reaction:

$$ATP + H_2O \longrightarrow ADP + P_i$$

This ATPase* activity is associated with L-myosin; other nucleoside triphosphates can also be split, depending on the presence of Mg^{++} or Ca^{++} ions (Weber, 1957). The structurally fixed ATPases A and B, located in the cell nucleus, hydrolyze all five ribonucleoside triphosphates and all four deoxyribonucleoside triphosphates normally present. The two enzymes have been separated from kidney cell nuclei (Siebert, 1966a). The ATPase activity with respect to ATP, ITP, CTP and UTP has been increased 300-fold by enzyme purification; the enzyme responsible for GTPase activity has been separated and purified a thousandfold. Nuclear ATPases may well be involved in the regulation of nucleic acid synthesis.

Several different ATPases have been found in bull seminal plasma (Heppel and Hilmoe, 1953); one of them** splits ATP to give AMP and pyrophosphate. A very high nucleoside triphosphatase activity was found in the polyribosome fraction (Raacke and Fiala, 1964); this action seems to be related to protein synthesis. In the case of guanosine-5′-triphosphatase, it was shown that a soluble enzyme is activated by the presence of particles (Chan and McCorquodale, 1965). An unusual kind of dGTP cleavage has been found in *E. coli* extracts† (Kornberg *et al.*, 1958):

$$dGTP + H_2O \longrightarrow deoxyguanosine + P—P—P$$

Nucleoside diphosphates can be converted to monophosphates by more or less specific phosphatases†† (Gibson *et al.*, 1955).

Deoxyribonucleosides of purine bases can be formed from the free bases and deoxyribose-1-phosphate by the reaction catalyzed by nucleoside phosphorylase††† (cf. p. 198). Enzymatic reactions which allow

*EC 3.6.1.3 and 3.6.1.4 ATP phosphohydrolase.
**EC 3.6.1.8 ATP pyrophosphohydrolase.
†EC 3.1.5.1 Deoxy-GTP triphosphohydrolase.
††EC 3.6.1.6 Nucleosidediphosphate phosphohydrolase.
†††EC 2.4.2.1 Purine nucleoside : orthophosphate ribosyltransferase.

purine riboside diphosphates to be converted to the corresponding derivatives of deoxyribose have been studied by Reichard's group in fowl
embryo extracts and *E. coli* (Reichard, 1960, 1961; Bertani *et al.*, 1963;
Larsson, 1963). It seems as if the same reducing enzyme system is acting upon purine ribonucleotides as well as pyrimidine ribonucleotides
(cf. p. 134). The substrate specificity is controlled by an allosteric
mechanism (Larsson and Reichard, 1966b; cf. pp. 134 and 157). In addition, vitamin B_{12} appears to play a part in this reduction, at least in certain bacteria (Manson, 1960; Blakley and Barker, 1964; Downing *et al.*,
1965). A deoxyribosyltransferase,* first found in *Lactobacilli* in 1952
and subsequently purified (Roush and Betz, 1958; Beck and Levin, 1963),
catalyzes a transfer of the deoxyribose molecule from purine and pyrimidine deoxyribosides to the free bases. Later studies have revealed deoxyribosyltransferase activities in various animal organs (De Verdier and
Potter, 1960; Zimmerman, 1962, 1963). Further attempts at purification
have led to the suggestion that the reaction is brought about by a dual
activity of the enzyme thymidine phosphorylase.

Utilization of Preformed Purines

In the discussion of the reactions of purines and their derivatives, we
have indicated the possibility that an organism could utilize preformed
purines like those which arise in the degradation of foodstuffs. Some
tissues, e.g., bone marrow (Lajtha and Vane, 1958), leucocytes (Lowy
et al., 1961), reticulocytes (Cook and Vibert, 1966) and many tumors,
contain only small quantities of enzymes for a *de novo* synthesis and
are therefore forced to utilize preformed purines.

As indicated above, free adenine can rapidly be incorporated into free
nucleotides and nucleic acids by a variety of metabolic pathways. Following the addition of labeled adenine, the tracer is found in the adenine
and guanine moieties of nucleic acids. Conversely, experiments with
labeled guanine have shown that it is converted in various mammalian
tissues to guanosine or deoxyguanosine moieties, but apparently not to
adenine; this might be due to the high guanase content of the tissues in
question. In unicellular organisms and, to some extent, in Ehrlich ascites
tumor cells, guanine is incorporated in both the guanine and the adenine
moieties of the nucleic acids (Balis *et al.*, 1951; Kerr *et al.*, 1951;
Harbers and Müller, 1962). *Tetrahymena* requires guanine for growth and
can utilize adenine only poorly (Flavin and Graff, 1951). Hypoxanthine
and xanthine, depending on the species, are not generally efficient nucleic acid precursors. In rabbit bone marrow, however, hypoxanthine is

*EC 2.4.2.6 Nucleoside : purine (pyrimidine) deoxyribosyltransferase.

very actively incorporated into the adenine and guanine moieties of both nucleic acids, probably because xanthine oxidase is absent.

Uric acid can react with PRPP, in the presence of a bacterial enzyme, to produce uric acid-9-ribosyl-5′-phosphate (Hatfield et al., 1964). A pyrimidine pyrophosphorylase purified from beef erythrocytes catalyzes the reaction of uric acid and PRPP, to form uric acid-3-ribosyl-5′-phosphate, whose phosphate group may be split off in a secondary reaction (Hatfield and Wyngaarden, 1964a and b).

Even purine derivatives which do not occur naturally may be transformed by the purine nucleotide pyrophosphorylases. Thus, 2,6-diaminopurine can be actively incorporated into the guanine moieties of RNA and DNA (Bendich et al., 1950b). Other synthetic purine derivatives have chemotherapeutic importance (cf. p. 319), e.g., 6-mercaptopurine, which will become 6-mercaptopurine ribonucleotide or 6-thioxanthosine-5′-phosphate, and which is a competitive inhibitor in the turnover of guanine (Atkinson and Murray, 1965a and b; Allan et al., 1966).

Adenosine can be used as a nucleic acid precursor by the rat, but not as well as free adenine (Lowy et al., 1952). This may result from differences in the ability of these compounds to penetrate the cell membrane, as well as differences in the degree of deamination preceding incorporation. In rats, guanosine is said to be incorporated in RNA only in trace amounts, whereas GMP is a better nucleic acid precursor (Roll and Weliky, 1951; Lowy et al., 1952). Labeled inosine, but not the deoxyriboside of hypoxanthine, is incorporated in both nucleic acids (Lowy et al., 1952).

Degradation of Purine Bases

Within the great variety of species in biology there exist several different pathways for the degradation of purine bases. The following chapter will deal separately with purine degradations in higher animals and in microorganisms, and degradation by chemical methods will be considered.

Throughout nature, purines constitute important building blocks of nucleic acids, but their metabolic and degradative pathways differ in various organisms and, at least in highly developed ones, the extent of breakdown seems to be a function of their evolutionary level. In the more primitive organisms, purines are completely degraded to low molecular weight compounds, but the higher animals utilize a variety of excretory mechanisms to dispose of these compounds in the form of larger breakdown products.

On the basis of its structure, it has long been suspected that uric acid is involved in nucleic acid metabolism. When uric acid is ingested by healthy subjects, much of it is metabolized. Most of the N^{15}-uric acid is excreted in the urine in the form of N^{15}-urea. This is probably due to uric acid breakdown by the intestinal flora (Green et al., 1950). Intravenously injected uric acid appears in the urine predominantly as uric acid (Wyngaarden and Stetten, 1953); in humans, a significant proportion is also excreted into the gastrointestinal tract, where it is decomposed by the gut flora (Sorensen, 1959).

Xanthine is the first common degradation product of all naturally occurring purine derivatives (nucleotides, nucleosides and free bases; Fig. 60). Most of the enzymes hydrolyze both ribose and deoxyribose derivatives; the reason for the existence of many parallel pathways is not clear. In animal tissues, there is very little adenine decomposition by adenase;* rather, the adenine $\longrightarrow$ adenosine $\longrightarrow$ inosine $\longrightarrow$ hypoxanthine pathway is probably involved. The xanthine oxidase-catalyzed degradation to 8-hydroxyadenine and then 2,8-dihydroxyadenine is normally not very important (Wyngaarden and Dunn, 1957). When large doses of adenine are administered to rats, they develop an ''adenine kidney,'' i.e., deposits of crystalline 2,8-dihydroxyadenine in the renal parenchyma (Bendich et al., 1950a). Guanine is attacked directly by the enzyme guanase,** with the formation of xanthine and NH_3. Neither guanosine nor GMP is degraded by guanase (Kalckar, 1947). In all animals studied, xanthine is further degraded to uric acid. The reaction is catalyzed by the relatively nonspecific xanthine oxidase,† which oxidizes a large number of aldehydes in addition to xanthine, hypoxanthine, adenine, 8-hydroxyadenine and NADH.

In many respects, xanthine oxidase is an interesting and, therefore, widely studied enzyme (De Renzo, 1956; Greenlee and Handler, 1964). The amount of enzyme in a given tissue depends on its function. Regenerating rat liver and hepatomas in which nucleic acid synthesis is intensive contain no xanthine oxidase (De Lamirande et al., 1958). This observation led to attempts to inhibit tumor growth by administering this enzyme (Haddow et al., 1958). Xanthine oxidase can be blocked by hydroxypyrazolopyrimidine, a compound which may prove useful in the treatment of uric acid deposits in the kidney (Rundle et al., 1966; Krenitsky et al., 1967).

Uric acid is the end product of purine degradation in humans, the higher apes, reptiles and birds, all of whom excrete large amounts of uric

*EC   3.5.4.2   Adenine aminohydrolase.
**EC   3.5.4.3   Guanine aminohydrolase.
†EC   1.2.3.2   Xanthine : oxygen oxidoreductase.

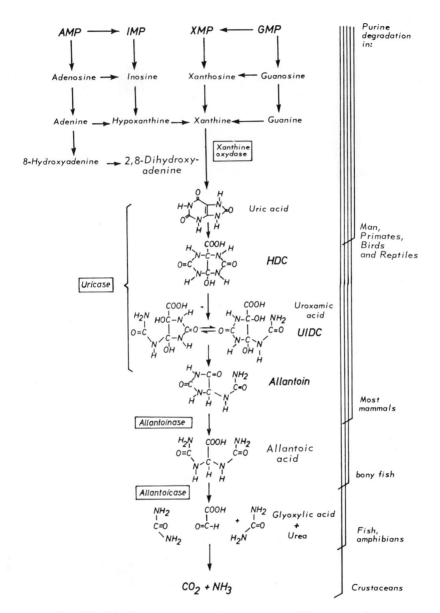

Fig. 60. Comparison of purine degradation in different species.

acid. Pathologic production of uric acid can cause gout in humans and leads to a similar deposition of crystalline urates in the tissues and joints of birds.

With the exception of the species just mentioned, most other animals possess uricase,* an uric acid-decomposing enzyme system (Roush and Shieh, 1963). The product of its reaction is allantoin, a urine constituent discovered in 1790. Allantoin formation requires several steps. First, the pyrimidine ring is opened between nitrogen 1 and carbon 6, then carbon 6 is oxidized to a carboxyl group and a bond is formed between nitrogen 1 and carbon 5. The resulting structure is the equivalent of two condensed imidazolidone rings. Hydroxyacetylenediureinecarboxylic acid (HDC) is formed by the introduction of a hydroxyl group at carbon 4. The existence of ''X,'' a product intermediate between uric acid and HDC, has been postulated on the basis of spectrophotometric findings (Bentley and Neuberger, 1952). HDC is a symmetrical molecule in which nitrogens 1 and 3, and nitrogens 7 and 9, can no longer be distinguished. Hence, subsequent splitting of the imidazole rings proceeds at random. The resulting product is uroxamate which is converted to allantoin by decarboxylation. This is the end of purine degradation in most mammals.

A study of the uricase reaction in different buffers has yielded interesting information about its stereochemistry (Canellakis and Cohen, 1955). In phosphate buffer, the reaction goes according to the scheme shown in Fig. 60. However, if the reaction is allowed to proceed in borate buffer at pH 7.2, a uroxamate-borate complex is formed which in phosphate buffer is decomposed into alloxanic acid and urea. After oxidation of $2\text{-}C^{14}$-uric acid, all of the radioactivity is found in the alloxanic acid, but when $8\text{-}C^{14}$-uric acid is used, all of the C^{14} is found in urea. This demonstrates a certain stereospecificity, at least for the first part of the uricase reaction. In rodent hepatomas, the amount of uricase, as well as xanthine oxidase, was found to be lower than that in the normal liver (De Lamirande et al., 1958).

In bony fish, purine degradation proceeds one step beyond allantoin. The second five-membered ring is split by the enzyme allantoinase** with the formation of a diureidoacetic acid, termed allantoic acid (Stransky, 1933). Most fish and amphibia degrade allantoic acid still further. The enzyme allantoicase*** splits allantoic acid into the two excretion products, urea and glyoxylate (Brunel, 1937). Crustacea and other lower organisms decompose these compounds still further, with CO_2 and ammonia as the final products (Florkin and Duchateau, 1942).

*EC 1.7.3.3 Urate:oxygen oxidoreductase.
**EC 3.5.2.5 Allantoin amidohydrolase.
***EC 3.5.3.4 Allantoate amidinohydrolase.

The pathways of purine degradation found in multicellular organisms are also encountered in some microorganisms.   Various purines are broken down to allantoin by *Pseudomonas* and then further decomposed to glyoxylate and urea (Bachrach, 1957).   Allantoin-decomposing enzymes have also been detected in *E. coli*, *Aerobacter aerogenes* and *Proteus vulgaris* (Young and Hawkins, 1944; Valentine *et al.*, 1962). *Clostridium* species, on the other hand, which have been grown on purines as the sole source of carbon and nitrogen, have evolved a special pathway, shown in Fig. 61 (Rabinowitz and Barker, 1956).   In these

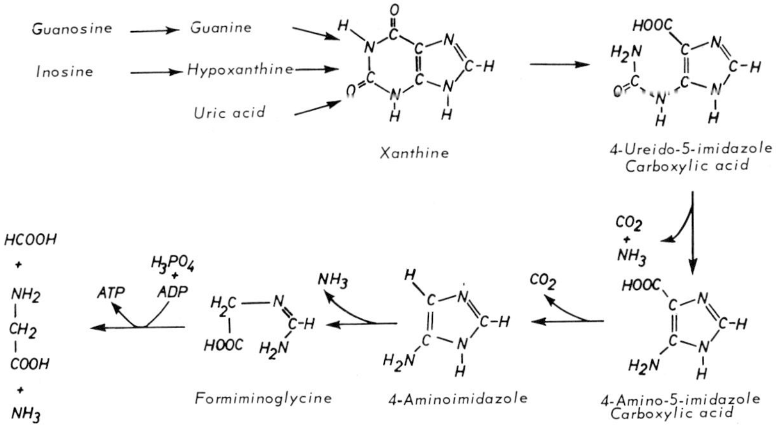

Fig. 61. Purine degradation in *Clostridium*.

organisms, all the purines, including uric acid, are first transformed into xanthine.   The degradation then begins with the opening of the pyrimidine ring.   From the 5-ureido-4-imidazolecarboxylic acid so formed, CO_2 and NH_3 are split off by a Mn^{++}-dependent enzyme.   The resulting 5-amino-4-imidazolecarboxylic acid is decarboxylated and 4-aminoimidazole is then oxidatively deaminated to imidazolone by an Fe^{++}-dependent enzyme.   Then the heterocyclic five-membered ring is split and the formimino group produced is transferred to tetrahydrofolate. Once bound to folate, this fragment can undergo several further transformations (see Fig. 76).

Certain micrococci use still another pathway for purine degradation, with part of the purine converted to pyrimidine derivatives, each of which in turn undergoes its own, relatively long degradation (Whitely, 1951, 1952).

Of the various methods for chemical degradation, Brandenberger's (1954a and b) deserves special mention because it has greatly facili-

tated the localization of tagged purine atoms (cf. p. 106). The Brandenberger method makes possible the stepwise isolation of single ring
atoms and the analysis of their isotope content. Depending on the
severity of the initial treatment, the breakdown will proceed to varying
degrees, but it will always yield defined fragments. Thus, severe hydrolysis with hydrochloric acid yields glycine from the carbon atoms in
positions 4 and 5 and the nitrogen atom in position 7. Treatment with
dilute nitric acid cleaves the atoms in positions 7 (nitrogen), 8 (carbon)
and 9 (nitrogen) from the five-membered ring in the form of urea. A longer
series of reactions, beginning with an oxidation by alkaline hydrogen
peroxide, gradually splits off carbon 6, nitrogen 1, nitrogen 7 and carbon 5.
The analytical data obtained with the Brandenberger method require
careful mathematical treatment, the details of which may be obtained
from the original papers.

BIOSYNTHESIS OF PYRIMIDINES

Like the bicyclic purines, the pyrimidines are also synthesized from
smaller compounds. One of the first results obtained from the biochemical application of isotope techniques was the discovery that in mammals
and birds, N^{15}-labeled ammonium ions were largely incorporated in the
pyrimidine ring (Barnes and Schoenheimer, 1943; Lagerkvist, 1953).
Some years later, CO_2 (Heinrich and Wilson, 1950) and L-aspartate
(Reichard and Lagerkvist, 1953) were recognized as effective pyrimidine
precursors. Investigations of bacterial growth factors provided valuable
additional evidence on the intermediates in pyrimidine biosynthesis.
In 1905, orotic acid, a pyrimidine carboxylic acid, was isolated from
milk by the Italian scientists Biscaro and Belloni. Large quantities of
this acid accumulate in the nutrient medium of pyrimidine-requiring
Neurospora and *Serratia* mutants, apparently as the result of an impaired
pyrimidine biosynthesis (Mitchell *et al.*, 1948; Michelson *et al.*, 1951;
Belser, 1961). Furthermore, orotate, like carbamylaspartate (see Fig. 62
for the structural formula), proved to be an active growth factor for
Lactobacillus bulgaricus 09 (Mitchell and Houlahan, 1947; Wright *et al.*,
1951); if either of the two substances was labeled, the tracer was found
in the pyrimidine moiety of nucleic acids. Similarly, after administration of N^{15}-labeled orotate to rats, the nitrogen isotope was found in the
pyrimidines of the liver nucleic acids (Aroidson *et al.*, 1949).

Certain cases of megaloblastic anemia appear to be due to a disturbance of pyrimidine metabolism, with large quantities of orotate excreted in the urine (Huguley *et al.*, 1959). A direct relationship between
pyrimidine synthesis and orotic acid became more apparent when Korn-

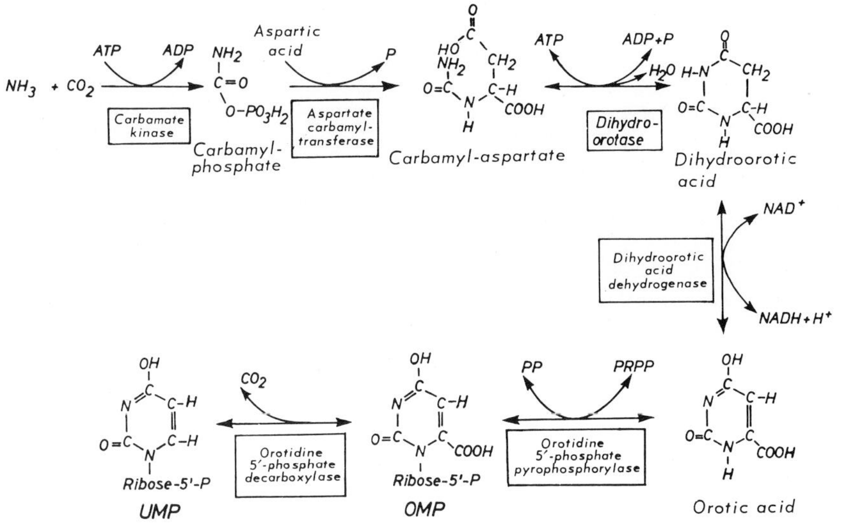

Fig. 62. *De novo* synthesis of uridine-5´-phosphate.

berg succeeded in adapting a soil bacterium to orotic acid by enrichment cultures (Lieberman and Kornberg, 1954). Decomposition of orotate by this microorganism gave rise to carbamylaspartate; the enzymes taking part in the degradation normally catalyze the reverse reaction, whose existence was later demonstrated in mammalian liver (Wu and Wilson, 1956).

This early work was the basis for the subsequent elucidation of pyrimidine biosynthesis. The formation of pyrimidines from aspartate, ammonia and carbon dioxide is known as *de novo* synthesis; its pathways are very similar in microorganisms and in numerous mammalian tissues (including neoplasms). Most of the reactions discussed below have been studied in cell-free systems.

De novo Synthesis of Pyrimidine Nucleotides

The first step in the synthesis of a pyrimidine is the reaction of carbon dioxide and ammonia to form an energy-rich, activated molecule— carbamyl phosphate (Jones *et al.*, 1955, 1960). Carbamyl phosphate, also required for arginine formation, can be synthesized by two different pathways. Many bacteria contain a *carbamate kinase,** which can be crystallized from *Streptococcus* D_{10} (Mokrasch, Caravaca and Grisolia,

*EC 2.7.2.2 ATP: carbamate phosphotransferase.

1966) and which catalyzes the following freely reversible reaction:

$$HCO_3^- + NH_4^+ + ATP \rightleftharpoons H_2N\!-\!CO\!-\!OPO_3H_2 + ADP$$

Carbamyl phosphate synthetase,* found in the livers of urea-forming animals, causes the following irreversible reaction:

$$HCO_3^- + NH_4^+ + 2ATP \longrightarrow H_2N\!-\!CO\!-\!OPO_3H_2 + 2ADP + P_i \cdot$$

This enzyme requires N-acetyl-L-glutamate as its cofactor (Hall *et al.*, 1958); the synthetase has been isolated from frog liver and its mode of action has been studied in detail (Metzenberg *et al.*, 1958; Marshall *et al.*, 1958, 1961). Antibodies produced with the purified enzyme react with the carbamyl phosphate synthetase from many animal species (Marshall and Cohen, 1961). Studies with tadpoles suggest that thyroxine induces the neosynthesis of carbamyl phosphate synthetase (Metzenberg *et al.*, 1961), an enzyme found in a variety of animal tissues (Jones *et al.*, 1961).

A bacterial carbamyl phosphate synthetase has recently been purified which catalyzes the reaction of bicarbonate with either NH_3 or glutamine, a reversible reaction which does not require N-acetylglutamate (Kalman *et al.*, 1966).

Carbamyl phosphate is transferred to L-aspartate by aspartate carbamyl transferase** in a practically irreversible reaction. This transferase has been obtained in crystalline form from *E. coli* (Sheperdson and Pardee, 1960); it has also been detected in animal tissues (Lowenstein and Cohen, 1956). Regenerating rat liver and hepatomas contain much greater amounts of transferase than normal liver (Calva *et al.*, 1959). Aspartate carbamyl transferase is the first in the chain of enzymes involved in pyrimidine synthesis; it has special importance because it controls pyrimidine formation by allosteric inhibition (cf. p. 155). (This carbamyl transferase is not identical with the enzyme ornithine carbamyl transferase). In a pyrimidine-less *Neurospora* mutant, the formation of the aspartate carbamyl transferase can be induced by CO_2 (Nazario and Reissig, 1964); estradiol increases the transferase activity in the uterus of immature rats (Tremblay and Thayer, 1964).

Carbamylaspartate, also called ureidosuccinate, is ultimately converted by closure to a six-membered ring compound. The enzyme needed for this reaction, dihydroorotase,† has been found in bacteria and mammalian liver (Yates and Pardee, 1956a); its reaction is freely reversible.

---

*EC 2.7.2.5 ATP : carbamate phosphotransferase.
**EC 2.1.3.2 Carbamoylphosphate : L-aspartate carbamoyltransferase.
†EC 3.5.2.3 4,5-L-Dihydroorotate amidohydrolase.

The ring compound formed in the preceding reaction is aromatized to the pyrimidine derivative by the enzyme dihydroorotate dehydrogenase.* This enzyme has been isolated in crystalline form from adapted bacteria (Friedmann and Vennesland, 1960). From certain microorganisms, an NAD-dependent dihydroorotate dehydrogenase, and from others an NADP-dependent enzyme, have been purified (Krakow and Vennesland, 1961; Ukada and Vennesland, 1962). As with dihydroorotase, the dehydrogenase was initially isolated from bacteria which decompose orotate. Although it was first considered merely an adaptive microbial enzyme, this dehydrogenase was later also found in rat liver (Cooper et al., 1950; Wu and Wilson, 1956). After feeding of orotate to rats, the biosynthesis of purine ribonucleotides has been found to be greatly increased (Windmueller and Spaeth, 1965).

Whereas purine biosynthesis proceeds by stepwise enlargement of a ribonucleoside-5′-phosphate molecule (cf. p. 109), in pyrimidine nucleotide biosynthesis the pentose phosphate portion is added only to the completed ring. The substrate for this N-glycosidation (Lieberman et al., 1955; Blair et al., 1960) is 5-phosphoribosyl-1-pyrophosphate, which, together with orotate, leads to the formation of the nucleotide orotidine-5′-phosphate (OMP), also called orotidylic acid. The enzyme catalyzing this reaction is orotidine-5′-phosphate pyrophosphorylase;** inhibition by various orotate analogs has been observed with an enzyme purified from yeast (Holmes, 1956; cf. p. 315). The glycosidation reaction is freely reversible; thus from OMP, free orotate and PRPP can be produced by pyrophosphorolysis.

OMP is converted to uridine-5′-monophosphate (UMP) by decarboxylation. The decarboxylase[†] has been concentrated from yeast and rat liver (Creasey and Handschumacher, 1961); the enzymes from these two sources seem to differ in their mode of action. The reaction product UMP inhibits decarboxylation (Blair and Potter, 1961); this inhibition is even greater with some synthetic UMP analogs (Handschumacher, 1958; Smith and Visser, 1965; cf. p. 316). Compared to that in normal liver, the enzyme system leading from orotate to UMP in regenerating rat liver shows considerably enhanced activity (Sköld, 1960b). In a pyrimidine-dependent Neurospora mutant, lack of the decarboxylase leads to the excretion of large amounts of orotidine into the medium (Michelson et al., 1951).

*EC 1.3.3.1 4,5-L-Dihydroorotate: oxygen oxidoreductase.
**EC 2.4.2.10 Orotidine-5′-phosphate: pyrophosphate   phosphoribosyltransferase.
†EC 4.1.1.23 Orotidine-5′-phosphate carboxy-lyase.

Transformations of UMP into Other Pyrimidine Bases or Their Derivatives

The *de novo* synthesis of all pyrimidine-containing nucleotides (Fig. 62) takes place via UMP, which therefore constitutes the key compound in the formation of many of the pyrimidine nucleotides and nucleosides. Reactions similar to those shown in Fig. 59 for purines also apply in principle to the transformations of UMP. Hence, uracil, uridine and UMP are in a constant metabolic equilibrium. Whereas in microorganisms uracil can be converted to both uridine (by a nucleoside phosphorylase*) and UMP (via a nucleotide pyrophosphorylase**), in the normal tissues of higher animals and in tumors only a synthesis via uridine is possible. Hence, when uracil is utilized for nucleotide synthesis in the preformed pathway, this can only occur with uridine as an intermediate (cf. p. 136). Uridine and deoxyuridine kinases are widely distributed (Sköld, 1960a and b; Rada and Gregusova, 1964). The specificity of these enzymes for the various bases differs according to their origin. Deoxyuridine is formed either from uracil and dR-1-P by a thymidine phosphorylase[†] (Friedkin and Roberts, 1954a) or from uracil and thymidine by a trans-N-glycosidation (cf. p. 118). Polyphosphate derivatives of uridine can be formed or decomposed according to the following reactions:

(a) UMP + ATP $\rightleftharpoons$ UDP + ADP $\rightleftharpoons$ UTP + AMP   (Canellakis, 1957b)

(b)   UDP + ATP $\rightleftharpoons$ UTP + ADP[††]

(c)   UDP $\rightarrow$ UMP + P$_i$ (Gibson *et al.*, 1955)

(d)   UTP + UMP $\rightleftharpoons$ 2UDP[‡]

(e)   UDP + GTP $\rightleftharpoons$ UTP + GDP   (Gibson *et al.*, 1956)

(f)   UDP + ITP $\rightleftharpoons$ UTP + IDP

Reaction (c) is catalyzed by a Mg^{++}-dependent nucleoside diphosphate phosphatase, which splits GDP and IDP as well as UDP. The uridylate kinase reaction (d) causes a myokinase-like reaction; nevertheless, this enzyme is clearly different from adenylate kinase. The reaction sequence leading from uracil to UTP is strongly inhibited by alloxan (Younathan *et al.*, 1964).

Recently, some substituted UMP derivatives have been found as minor

---

*EC   2.4.2.3   Uridine : orthophosphate ribosyltransferase.
**EC   2.4.2.9   UMP : pyrophosphate phosphoribosyltransferase.
[†]EC   2.4.2.4   Thymidine : orthophosphate deoxyribosyltransferase.
[††]EC   2.7.4.6   ATP : nucleosidediphosphate phosphotransferase.
[‡]EC   2.7.4.4   ATP : nucleosidemonophosphate phosphotransferase.

constituents in various sRNA preparations: 2-thio-UMP (Carbon *et al.*, 1965), 4-thio-UMP (Lipsett, 1965), and 5,6-dihydro-UMP (Carr and Grisolia, 1964; Madison and Holley, 1965). Various animal tissues reduce UMP with NADH to form 5,6-dihydro-UMP (Mokrasch and Grisolia, 1959).

The 5-methyl-substituted uracil is one of the most important DNA bases, namely *thymine*. Although at first it seemed as if thymine occurred only in combination with deoxyribose, thymine ribosides have now also been found in tRNA preparations (cf. p. 101) and rRNA. The methylation of RNA occurs at the high molecular weight level (cf. p. 210), and therefore, thymine ribotides can be considered as breakdown products of special RNAs. However, a 5-methyl-UMP synthetase* has been described in rat embryo extracts (Maley, 1960, 1962); its activity appears to be a function of the rate of growth. Free thymine can be converted to its deoxyribonucleoside by a phosphorylase (Friedkin and Roberts, 1954a).

In the synthesis of deoxyribose-containing thymidine, the introduction of the methyl groups takes place at the monomer stage, in contrast with what happens in the formation of thymine ribotides. Inhibitor studies have suggested that folic acid derivatives participate in these reactions as coenzymes; vitamin B_{12} also seems to be involved in thymine formation, but its role is not yet completely clear (Downing and Schweigert, 1956; Dinning *et al.*, 1958; Norton *et al.*, 1962).

Deoxyuridine was found, some time ago, to be an active acceptor of the C_1 groups of serine, formaldehyde or formate (Friedkin and Roberts, 1956; Blakley, 1957). The crucial step in the introduction of a C_1 unit involves the enzyme thymidylate synthetase.* The reaction mechanism of this enzyme, which can be obtained in highly purified form from various materials, has recently been studied in great detail (e.g., Blakley *et al.*, 1963; Jenny *et al.*, 1963). dUMP and 5,10-methylenetetrahydrofolate are needed for dTMP biosynthesis (see Fig. 63); dihydrofolate, a product formed in this reaction, has a pronounced absorption maximum at 338 mμ, which has been utilized in kinetic studies (Wahba and Friedkin, 1961). After phage infection, the activity of thymidylate synthetase in *E. coli* increases to nine times the normal value (Bello *et al.*, 1961b; cf. p. 265). Some dUMP analogs have proved to be very effective inhibitors of thymidylate synthetase (Hartmann and Heidelberger, 1961; Blakley, 1963; Mathews and Cohen, 1963a; cf. p. 314).

There are several biosynthetic pathways leading to the formation of

*EC not yet assigned.

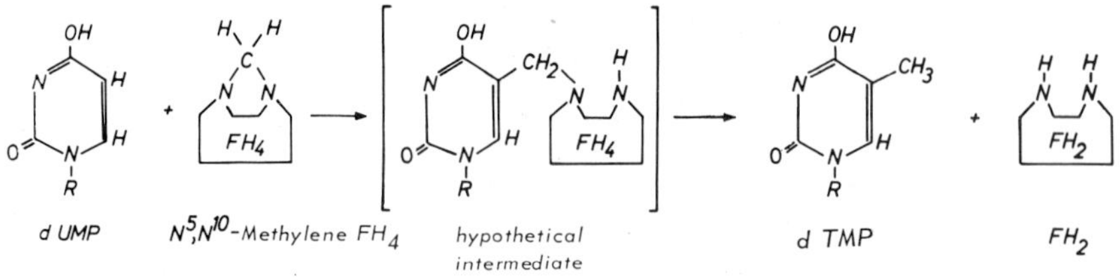

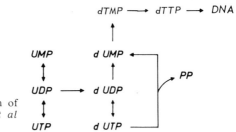

Fig. 63. dTMP formation by methylation of dUMP.   R = 2´-deoxyribose-5´-phosphate.

the important dTMP precursor, dUMP. An ATP-dependent deoxyuridine kinase* (Sköld, 1960b) probably plays only a subordinate role. In the diphosphate stage, uridine phosphates can certainly be reduced enzymatically while attached to the sugar moiety, with the resultant formation of deoxyuridine derivatives. These pentose reductions will be considered in greater detail on p. 134. A third possibility for dUMP synthesis involves the enzymatic deamination of dCMP (cf. p. 134); a corresponding deamination of 5-methyl-dCMP yields dTMP directly (Cohen and Barner, 1957; Scarano, 1960). Potter *et al.* (1960) have discussed the various possible routes for dTMP synthesis and their functional significance. The reactions involved should insure adequate production of the precursors needed for DNA synthesis and should keep false bases like deoxyuridine nucleotides (Fig. 64) from being incorporated into DNA (cf. also Fig. 78).

$$dTMP \longrightarrow dTTP \longrightarrow DNA$$

$$UMP \qquad d\,UMP \longleftarrow$$

$$PP$$

$$UDP \longrightarrow d\,UDP$$

$$UTP \qquad d\,UTP$$

Fig. 64. Blockade of incorporation of dUTP into DNA (after Bertani *et al* 1961).

Oxidative changes have also been described for the thymine methyl group, leading to the formation of 5-hydroxymethyluracil, 5-formyluracil and also, apparently, pyrimidines that are unsubstituted in the 5-position (Fink and Fink, 1962a; Abbott *et al.*, 1964).

The thymidine phosphorylating enzyme system, which is found in many biological systems, is active in the regulation of DNA synthesis and growth (cf. p. 186; Weissman *et al.*, 1960; Bianchi *et al.*, 1961;

*EC not yet assigned.

Potter and Nygaard, 1963; Okazaki and Kornberg, 1964b and c). Although kinases have been found for all other deoxyribonucleosides, normal resting cells contain practically no thymidine kinase* (Hiatt and Bojarski, 1960; Kit et al., 1965). After injection of thymidine and especially during regeneration following partial hepatectomy (Weissman et al., 1960), kinase activity was observed for the reactions: thymidine* $\longrightarrow$ dTMP** $\longrightarrow$ dTDP† $\longrightarrow$ dTTP, whereas the activity of thymidine-decomposing enzymes and of thymine deoxyribonucleotide phosphatases had diminished (Canellakis et al., 1959; Eker, 1965). This control can be attributed to a thymidine kinase inhibitor system, since it is possible to isolate a substance from normal liver which specifically inhibits thymidine kinases (Gray et al., 1960; Bianchi et al., 1962). In some tissues, the enzyme required for the phosphorylation of dTDP appears to be fully active during all phases in the life of the cell (Bianchi et al., 1964).

A thymidine kinase from E. coli has been purified 1200-fold, and its reaction and control mechanisms have been studied in detail. With this enzyme, dATP is more effective as a phosphate donor than ATP; various polyphosphates of deoxyribonucleotides act as activators (Okazaki and Kornberg, 1964b and c). By means of ammonium sulfate precipitation, a thymidine kinase from plant seedlings could be divided into two inactive fractions, referred to as P and T. Recombination of the two fractions restored the active enzyme. While P was found in all plant tissues, the T component was found only in actively growing tissues (Wanka et al., 1964). From B. subtilis and from regenerating liver, an enzyme has been obtained which phosphorylates 3´-dTDP to give 3´-dTTP; the biological significance of this reaction still remains to be established (Cannelakis et al., 1965; Coutsogeorgopoulos et al., 1966).

We will discuss later how the biosynthesis of the DNA precursor dTTP is regulated by negative feedback control. Bacteriophage infection of E. coli leads to an increase in the activity of the existing thymidylate kinase system or to a de novo synthesis of the appropriate enzymes (Bello and Bessman, 1963; cf. p. 265).

Some time ago, an additional uracil derivative, obtained after alkaline or enzymatic hydrolysis, was found in yeast RNA (Davis and Allen, 1957). In 1960, Cohn succeeded in identifying this "fifth nucleotide" as an isomer of UMP. The nucleoside of this RNA building block, usually known as pseudouridine (cf. p. 14), differs from all other known nucleosides in that the sugar is attached not N-glycosidically, but

*EC 2.7.1.21 ATP : thymidine 5´-phosphotransferase.
**EC 2.7.4.9   ATP : thymidinemonophosphate phosphotransferase.
†EC not yet assigned.

through a C—C bond between carbon 1 of ribose and carbon 5 of uracil; hence, it is a 5-ribosyluracil. The nucleotide pseudouridine phosphate (ΨUMP) is found most abundantly in tRNA (Goldberg and Rabinowitz, 1961; see p. 103). The nucleoside pseudouridine is excreted in fairly large amounts in the urine of patients with leukemia or goiter (Adler and Gutman, 1959), but it also occurs in the urine of normal persons (Dlugajczyk and Eiler, 1963).

The synthetic pathway leading to pseudouridine has not yet been fully elucidated. Labeled cytidine was found to be utilized in preference to uridine for pseudouridine formation by a *Neurospora* mutant (Hall and Allen, 1960; Fink, 1963). In contrast, uridine proved to be an active precursor in *E. coli* (Pollak and Arnstein, 1962), ascites tumor cells (Lis and Lis, 1962) and yeast (Robbins and Hammond, 1962). In the case of *E. coli*, a 3,5-diribosyluracil was postulated as an intermediate (Lis and Lis, 1963). Other experiments have suggested an intramolecular rearrangement of uridine (Fink, 1963; Robbins and Kinsey, 1963). An enzyme has been found in *Tetrahymena pyriformis* which condenses uracil with an as yet unknown ribose derivative to give ΨUMP (Heinrikson and Goldwasser, 1963); uridine phosphokinase or OMP pyrophosphorylase from yeast could not effect this reaction. An enzyme system that phosphorylates ΨUMP to ΨUTP has been reported to occur in yeast. This ΨUTP, like the ribonucleoside triphosphates, can be used as a precursor for RNA biosynthesis (Goldberg and Rabinowitz, 1961, 1963b); it is not yet clear, however, how the incorporation of pseudouridine phosphate is always effected at a definite position within the polyribonucleotide chains.

Replacement of the —OH group on carbon 6 of uracil by —NH_2 gives rise to cytosine, whose ribotide is called cytidine. The amination of uracil proceeds at the triphosphate level (UTP $\longrightarrow$ CTP). Only the ribose, and not the deoxyribose, derivatives of uracil are transformed in this manner. Like the amination reactions in the purine series, the bacterial enzymes* require NH_3 and ATP (Lieberman, 1956a; Hurlbert, 1962) for amination, while the enzymes of animal tissues** transfer the amide nitrogen of glutamine (Salzman *et al.*, 1958; Hurlbert and Kammen, 1960). The glutamine-consuming reaction is enhanced by the presence of guanosine monophosphate, whereas the glutamine analog DON (cf. p. 323) inhibits this amination. Recent investigations suggest that in the *E. coli* system, too, glutamine and GMP are the principal compounds used for CTP synthesis (Chakraborsy and Hurlbert, 1961).

*EC 6.3.4.2 UTP : ammonia ligase (ADP).
**EC not yet assigned.

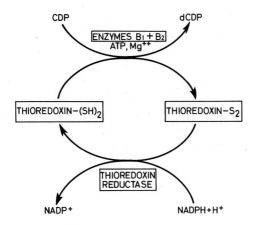

Fig. 65. dCDP formation by an enzyme system purified from *E. coli* (according to Holmgren *et al.*, 1965). (Reproduced with permission of National Academy of Sciences)

The general scheme of reactions shown in Fig. 59 has only limited application for cytosine and its derivatives. Pyrophosphorylases or phosphorylases causing the reaction of cytosine with PRPP or R-1-P, respectively, have not yet been detected, and even the existence of a cytidine kinase has hitherto been inferred only indirectly from the ready utilization of cytidine (cf. Fig. 66). Deoxycytidine can be converted to dCMP by a phosphotransferase* according to the following reaction (Maley and Maley, 1963):

$$dCR + dTMP \longrightarrow dCMP + dTR$$

dCMP can be phosphorylated by a kinase** widely distributed in animal tissues (Mendicino, 1962) and bacteria (Maley, 1958). In the above-mentioned amination of UTP, the RNA building block CTP is formed directly; in addition, CTP can also be formed by phosphorylation of CDP (Ratliff *et al.*, 1964). The activity of the kinases for all deoxyribonu-

*EC not yet assigned.
**EC 2.7.4.5 ATP : deoxyCMP phosphotransferase.

cleotides is greatly increased after phage infection (Bessman *et al.*, 1965).

In general, the DNA precursor dCTP is synthesized in a special way. With the aid of isotope experiments, it was first demonstrated in rats, in 1953, that cytidine can be converted to deoxycytidine without a rupture of the glycosidic bond. This finding was confirmed with extracts from *E. coli* (Larsson, 1966) and from fowl embryos (Reichard, 1961). In the embryo extracts, in addition to CDP, the pentose parts of UDP and of the purine riboside diphosphates were reduced. Detailed studies of the CDP reductase system* isolated from *E. coli* (Laurent *et al.*, 1964; Moore *et al.*, 1964a; Holmgren *et al.*, 1965) and from rat tumors (Moore *et al.*, 1964b) revealed the participation of four protein fractions in addition to ATP and Mg^{++}. Thioredoxin, a new flavoprotein, and its NADP-dependent reductase have been isolated in pure form. ATP, which is not required in stoichiometric amounts, seems to activate the enzymes from *E. coli* B and increase their affinity for CDP. In the absence of ATP, some other nucleotides (dGTP, dTTP) can activate the system. In the presence of ATP, however, dGTP and dTTP inhibit the reaction. Various authors have discussed the question of whether the specificity of the ribonucleotide reduction system toward various substrates is controlled by a feedback mechanism (Goulian and Beck, 1966a and b; Larsson and Reichard, 1966a and b).

Whereas all the systems mentioned so far reduce the pentose moiety of a nucleoside diphosphate, in *L. leichmannii* the formation of the cytidine deoxyribotides occurs at the triphosphate level (Abrams, 1965).

In T6r$^+$ phage-infected *E. coli,* the ribonucleotide units of the phage-induced RNA are transformed into deoxyribonucleotides; the reductase system shows a marked rise in activity, which, however, is detected "late"—about 20 minutes after initiation of the infection (Cohen *et al.*, 1961, 1962).

One pathway leading to the formation of dTTP precursors, particularly important in mammalian organisms, involves the deamination of deoxy-cytidylic acid. A very specific dCMP deaminase,* first found in extracts of sea urchin eggs and embryos (Scarano *et al.*, 1960), was later also found in regenerating (Maley and Maley, 1960) and normal liver (Maley and Maley, 1961; Scarano *et al.*, 1962), in hepatomas (Scarano and Talarico, 1961), in fowl embryos (Maley and Maley, 1964) and in bacteria (Siedler and Holtz, 1963). The deaminase, not detectable in normal *E. coli* cells, appears a few minutes after infection with T-even phages (Keck *et al.*, 1960; Hall and Tessman, 1966; Maley and Maley,

*EC not yet assigned.

1966a; cf. p. 265). Apparently the enzyme can be reversibly inactivated by a microsomal inhibitor (Fiala and Fiala, 1961) and is inhibited by all thymidine phosphates, especially dTDP and dTTP, by means of a feedback mechanism (see p. 157) in which dCTP activates the deaminase. Detailed studies of this control mechanism were made possible by purified dCMP deaminases (Maley and Maley, 1964; Fleming and Bessman, 1965).

In contrast to the deoxycytidylate deaminases isolated from animal tissues, the enzyme isolated from T2r$^+$ phage-infected *E. coli* will not deaminate 5-hydroxymethyl-dCMP, thus preventing the depletion of the precursor essential for the multiplication of the phage (Maley and Maley, (1966a).

As with uracil, compounds can be derived from cytosine with alkyl substituents in the 5-position: 5-hydroxymethylcytosine (5-HMC), 5-methyl-cytosine (5-MC) and "pseudoisocytosine," a compound not yet found *in vivo* but obtained only by enzymatic amination of pseudouridine (Lis and Allen, 1962). 5-HMC and 5-MC play a role only as deoxyribose derivatives. The incorporation of the one-carbon unit into cytosine takes place at the high molecular level, with 5-adenosylmethionine as the methyl donor (Gold and Hurwitz, 1964a and b).

5-MC nucleotides, belonging to the so-called rare bases, are normal constituents of some RNAs and of animal, plant and bacterial DNAs (Amos and Korn, 1958; Littlefield *et al.*, 1962). 5-HMC has so far been found only in the DNA of T-even phages (cf. p. 260), which after infection of the host cells induce the synthesis of numerous enzymes,* including those required for the hydroxymethylation of dCMP (Pizer and Cohen, 1962; DeWaard, 1964; Wiberg and Buchanan, 1964) and for the subsequent phosphorylation to the 5-hydroxymethyl-dCMP (Somerville *et al.*, 1956; Bello and Bessman, 1963). At the same time, a specific phosphatase** is formed to remove dCTP, the "false" precursor of the virus (Koerner *et al.*, 1960; Zimmerman and Kornberg, 1961); dCTP is hydrolyzed to dCMP which can then be converted by hydroxymethylation to a precursor specific for the phage DNA.

By 1954, several groups of workers had shown that the 5-HMC residues present in the DNA of even-numbered phages contained glucose molecules attached glycosidically to the hydroxymethyl groups. In the years following, it was found that glycosidation takes place only after incorporation of the bases in the DNA chain, that UDPG serves to transfer the glucose, that the sugar is bound in α- or β-glycosidic linkage depending on the type of phage (Kornberg *et al.*, 1961; Josse and Korn-

---

*EC not yet assigned.
**EC 3.6.1.12 DeoxyCTP nucleotidohydrolase.

berg, 1962), and that T2 and T6 phages also contain two glucoses per 5-HMC residue as the disaccharide gentiobiose (Kuno and Lehman, 1962; cf. p. 261).

The Utilization of Complete Pyrimidines

Feeding experiments performed in the 1940's with N^{15}-labeled uracil, thymine and cytosine showed that these bases were not incorporated in nucleic acids of mammalian systems. Later experiments with C^{14}-uracil and C^{14}-thymine also showed only slight incorporation of the isotope in the nucleic acid fractions; this utilization was very low compared with that of equimolar quantities of orotate or pyrimidine ribosides. Canellakis (1957b) then demonstrated that the use of uracil in nucleic acid synthesis is dependent on the amount supplied. Small quantities of uracil are broken down practically completely, particularly in the liver; the administration of larger doses of uracil can, however, result in the free base being as acceptable a nucleic acid precursor as its riboside.

The capacity for degradation of uracil is markedly diminished in tumor cells (cf. p. 282). In ascites tumor cells there is extensive utilization of uracil for nucleic acid metabolism via the reactions discussed on p. 116 (Krenitsky et al., 1964). These properties of tumor cells are used to advantage in chemotherapy with uracil analogs (see p. 312). Among the normal tissues, intestinal mucosa is also capable of utilizing uracil (Heidelberger et al., 1957).

Apparently neither cytosine nor thymine can be used effectively by mammals, although a relatively high thymidine phosphorylase activity is present in different organs, particularly in the liver and the mucosa of the small intestine (Friedkin and Roberts, 1954a and b). Consequently, analogs of these bases (e.g., 5-bromouracil; cf. p. 310) are not chemotherapeutically active in mammalian tissues, unless given in the form of nucleosides. Early experiments with *Neurospora* had shown that they utilized the pyrimidine ribosides, uridine and cytidine (Loring and Pierce, 1944).

In vivo experiments on rats have shown that cytidine is incorporated into nucleic acids much better than uridine (Hammarsten et al., 1950). When labeled cytidine was administered orally, the cytosine and uracil portions of the RNA were labeled as readily as cytosine and thymine in DNA. Figure 66 shows once again the universal utilization of the nucleoside cytidine. Administration of biosynthetically obtained C^{14}-deoxycytidine generally does not lead to the labeling of bases in RNA; this also confirms, the irreversibility of ribose reduction (Reichard and Estborn, 1951). In leukemia cells, deoxycytidine causes an increased formation of thymine deoxyribotides (Delamore and Prusoff, 1964). In tumor cells, a more

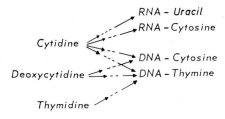

Fig. 66. The importance of pyrimidine nucleosides as precursors for nucleic acid synthesis (after Schulman 1961).

varied utilization of deoxycytidine often appears to be possible (Schneider and Rotherham, 1961). In the perfused cat brain, it was shown that a constant supply of cytidine and uridine is necessary for the tissue to survive (Geiger and Yamasaki, 1957). Thymidine can be used for DNA synthesis by nearly all biological systems. Radioactively labeled thymidine (as the H^3-labeled nucleoside) is frequently used as an indicator of DNA metabolism, especially in autoradiographic studies. The widely held opinion that labeled deoxyribonucleosides represent very specific DNA precursors is not quite correct and may lead to false conclusions. The breakdown products formed from the deoxyribonucleosides, especially when there is an abundant supply, can also give rise to labeled RNA (e.g., Harbers and Heidelberger, 1959a: Fink and Fink, 1962b).

Pyrimidine ribonucleotides, like purine ribonucleotides (cf. p. 118), apparently are taken up and used for nucleic acid production by whole cells only after dephosphorylation: only a comparatively small proportion of radioactive phosphate was recovered in the nucleic acids after doubly labeled nucleotides (C^{14} and P^{32}) had been supplied (Roll et al., 1956).

Degradation of Pyrimidine Bases

The biological breakdown of pyrimidines to small molecules involves either a reductive or an oxidative pathway. Reductive degradation can occur in all higher organisms, as well as in some bacteria.

The first experimental studies of pyrimidine degradation go back to the beginning of this century. Large quantities of uracil and thymine were fed to dogs and led to an increased excretion of urea (Steudel, 1901). Subsequent isotope experiments provided direct proof for what seemed an obvious conclusion. In rats, administration of N^{15}-labeled pyrimidines led to the excretion of N^{15}-urea. Feeding of DNA or thymine in large amounts also resulted in the appearance in the urine of a new amino acid, β-amino-isobutyric acid (Fink et al., 1952). The excretion of this amino acid, never detected in the urine of control animals, was increased even more after a dose of dihydrothymine; β-aminoisobutyrate excretion in the urine was also observed after surgery (Levey et al., 1963).

Surprisingly, administration of 5-methylcytosine (XLI), from which the structure of β-aminoisobutyric acid (XLII) might theoretically be obtained after ring cleavage, did not lead to the excretion of β-aminoisobutyrate.

The original hypothesis that urea is formed as a result of simple cleavage of the pyrimidine ring, was contradicted in 1954 by the observation that the ureido carbon is converted to CO_2 by the liver, both *in vivo* and *in vitro*. The reaction sequence of the entire breakdown, as shown in Fig. 67, was finally elucidated by *in vivo* experiments (Fritzson and Pihl, 1957) and by experiments with tissue slices, tissue extracts (Canellakis, 1956), homogenates, and purified animal and microbial enzymes.

The first step in this degradation is the reduction of the bases uracil or thymine to the corresponding dihydro compounds. The dehydrogenase,[*] prepared from liver, uses NADPH (Grisolia and Cardoso, 1957) for this transfer of hydrogen, while the enzyme[**] purified from *Clostridia* requires NADH (Campbell, 1957). The opening of the heterocyclic ring is brought about by a hydrolase[†] which has been purified from calf liver (Wallach and Grisolia, 1957) and microorganisms (Campbell, 1958). From dihydrouracil, β-ureidopropionate is formed; from dihydrothymine, β-ureidoisobutyrate. According to Marsh and Perry (1964), the dihydrothymine-splitting enzyme shows decreased activity in leukemic leukocytes. Both the reduction of the aromatic ring and the subsequent ring opening are freely reversible. Hence, this has been suggested as an alternate route for the biosynthesis of pyrimidines (cf. p. 125). However, a pyrimidine synthesis from the β-amino acid level has been demonstrated experimentally only with a *Neurospora crassa* mutant (Herrmann and Fairley, 1957), whereas in all other species studied, the decarbamylation[††] of ureidocarboxylic acid to CO_2, NH_3 and a β-amino acid is irreversible (Campbell, 1960).

In the case of the synthetic 5-halogenated pyrimidines, important for tumor chemotherapy (cf. p. 310), degradation takes place in a similar

[*]EC 1.3.1.2 4,5-Dihydro-uracil:NADP-oxidoreductase.
[**]EC 1.3.1.1 4,5-Dihydro-uracil:NAD-oxidoreductase.
[†]EC 3.5.2.2 4,5-Dihydropyrimidine amidohydrolase.
[††]EC 3.5.1.6 N-Carbamoyl-β-alanine amidohydrolase.

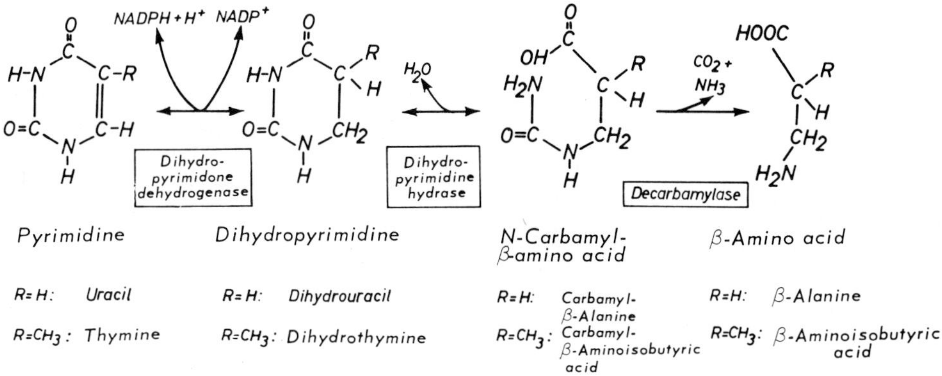

Fig. 67. Reductive pyrimidine degradation in higher organisms and some micro-organisms.

manner. First, the analogs are converted enzymatically to dihydrocompounds; from then on, the paths diverge. As shown in Fig. 68, hydrogen halide is split off from the dihydro derivatives of the thymine analogs 5-bromouracil and 5-iodouracil, and free uracil is produced, which then undergoes decomposition according to the scheme given in Fig. 67 (Prusoff et al., 1960). With 5-fluorouracil, on the other hand, the ring is opened first; α-fluoro-β-ureidopropionate is formed and then further decomposed to α-fluoropropionate and urea (Chaudhuri et al., 1959).

The following pathway of oxidative degradation was consistently found in various microorganisms grown with pyrimidine bases as the sole source of carbon and nitrogen (Hayaishi and Kornberg, 1952: Wang and Lampen, 1952a and b). Four different pyrimidines (uracil, cytosine, 5-methylcytosine and thymine) were investigated. All are broken down according to the scheme shown in Fig. 69. First the NH_2-substituted bases are deaminated, so that uracil or thymine is produced. These are then further oxidized by the same enzyme system, uracil-thymine oxidase.* Thus, from uracil, barbiturate is formed (not isobarbiturate as postulated earlier by other investigators). Barbiturate is decomposed into urea and malonate by a barbiturase** which can be purified from adapted bacteria. 5-Methylbarbiturate, produced from thymine, is broken down to urea and methylmalonate (Biggs and Doumas, 1963).

Various authors have discussed the relationship between the activity of pyrimidine-degrading enzymes and the growth potential of tissues. Uracil and thymine are not broken down in fetal rat liver (Belousova and Grigor'Eva, 1960: Stevens and Stocken, 1960); the degrading enzymes

*EC 1.2.99.1. Uracil:(acceptor) oxidoreductase.
**EC 3.5.2.1 Barbiturate amidohydrolase.

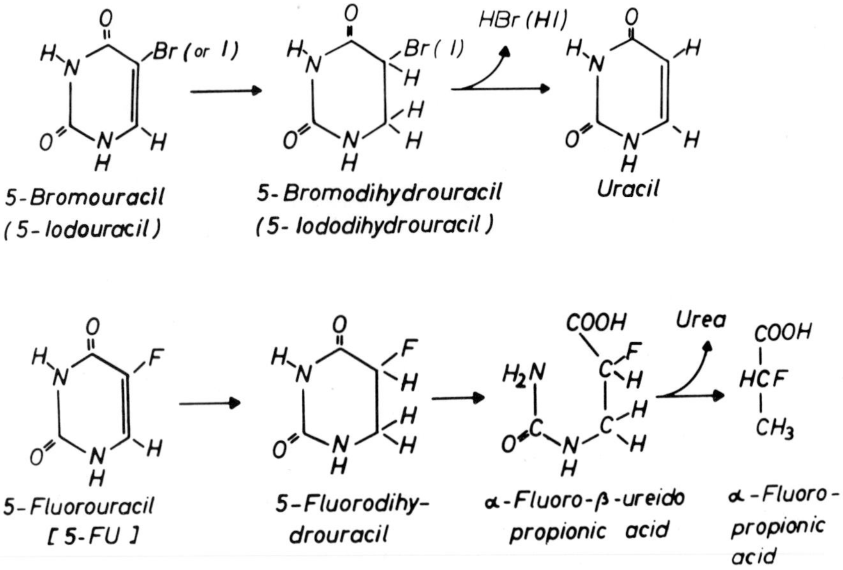

Fig. 68. Enzymatic degradation of halogenated derivatives of uracil.

remain undetectable until the twelfth day after birth. Compared to most normal tissues, tumors show a decreased capacity for decomposing pyrimidine bases (Potter *et al.*, 1958; Reichard and Sköld, 1958; cf. p. 138 and 282). In regenerating rat liver, a striking parallelism has been

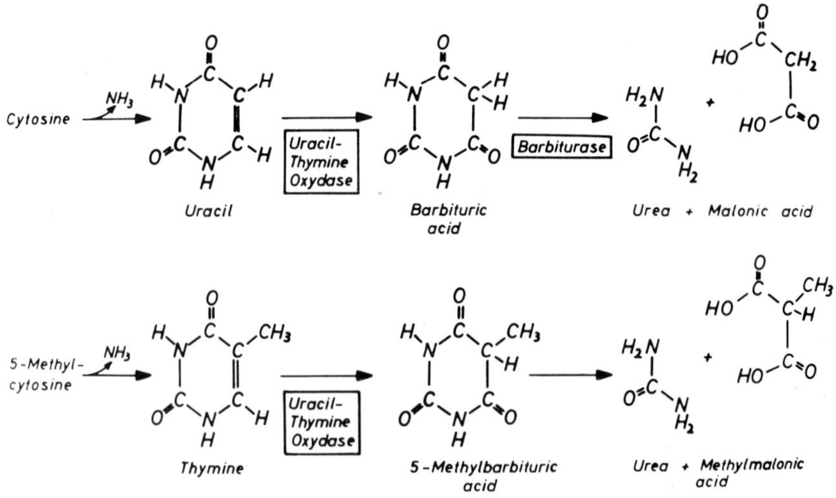

Fig. 69. Oxidative pyrimidine degradation by microorganisms.

found between the regenerating capacity and the activity of dihydrouracil dehydrogenase. Similar results with decarbamylase suggest that this enzyme also plays a role in the regulation of growth. Certain uracil analogs inhibit the enzymatic degradation of uracil and thymine (Sebesta *et al.*, 1961). In their reaction sequence the methods for a controlled chemical degradation of pyrimidine bases (Phares, 1951) resemble the reductive pyrimidine breakdown in animal tissues.

BIOSYNTHESIS OF PENTOSES

The pentoses constitute an important group of naturally occurring substances which play a decisive role in the living cell as components of nucleic acids and of numerous coenzymes. Apart from highly polymerized pentosans, found in many plants as constituents of supporting tissues, free pentoses are rarely found in biological materials. In man, the excretion of free pentoses in the urine (*pentosuria*) is brought about by overindulgence in pentose-rich fruits or as a result of a metabolic defect (cf. p. 144). Diseases which involve tissue breakdown, such as muscular dystrophy, experimental hyperthyroidism and trauma, likewise result in increased pentose levels in blood and urine. During the past ten years, numerous enzymatic reactions by which pentoses can be either synthesized or broken down have been discovered in the animal organism. The essential synthesis of coenzymes and nucleic acids is, therefore, independent of an exogenous supply of pentoses. Various microorganisms may be easily cultured on pentoses; in some cases, the production of enzymes needed for pentose degradation is induced in the course of growth (Horecker, 1962).

The carbohydrate component of nucleic acids is very uniform (cf. p. 8). D-ribose, the sugar moiety of all RNA, has been known since the beginning of this century. At first, a hexose was assumed to be the carbohydrate part of DNA, but in 1930 2-deoxyribose was identified as the only sugar of DNA. More recently, other pentoses have been found in different nucleosides and nucleotides. Thus, the antibiotic *cordycepin* contains cordycepose, a 3-deoxypentose. 3-Deoxy-3-amino-D-ribose, bound glycosidically to a purine, occurs in the antibiotic puromycin (cf. pp. 301 and 303). A uracil arabinoside and xyloside have been isolated from the sponge *Cryptotethia crypta* (Bergman and Burke, 1955; Tekman and Öner, 1963); other arabinosides have been synthesized and their metabolic behavior investigated (Schrecker and Urschel, 1966; Moore and Cohen, 1967). The 2'-O-methyl ether of adenosine (Smith and Dunn, 1959b) was isolated from liver tissue, along with the 2'-O-methyl derivatives of guanosine, uridine and cytidine, from yeast tRNA and *E. coli* tRNA (Hall, 1963b; Honjo *et al.*, 1964; Nichols and Lane, 1966).

In the following sections, pentose metabolism will be discussed only in relation to nucleic acid biosynthesis. Four different pathways are known at present for obtaining straight-chain pentoses: (1) oxidation and removal of the carbon atom in position 1 of hexoses, (2) oxidation and removal of carbon 6 of hexoses, (3) synthesis of the C_5 chain by condensation of a C_2 and a C_3 unit, and (4) dehydrogenation of a pentitol. The individual reactions in these synthetic pathways will be described first, followed by a discussion of their metabolic significance.

Oxidation and Subsequent Removal of Carbon 1 from Hexoses

In 1931, Warburg showed that glucose-6-phosphate, formed in the hexo-kinase reaction and metabolized in the glycolytic cycle leading to lactate, can also be degraded by an oxidative pathway, one reaction product of which is CO_2. NADP was found to be a cofactor in this reaction. Despite the efforts of different workers, the exact reaction mechanism remained obscure until Horecker began studying this degradative pathway in the 1950's.

The first NADP-dependent dehydrogenation begins with glucose-6-phosphate and, in a two-stage reaction, yields 6-phosphogluconate. Glucose-6-phosphate dehydrogenase,* named *Zwischenferment* by Warburg, and later crystallized (Julian *et al.*, 1961), is widely distributed in the animal, plant and microbial worlds, and catalyzes a reaction which Horecker proved to be reversible.

The further oxidation of phosphogluconate is catalyzed by an NADP-dependent phosphogluconate dehydrogenase.** The enzyme, first reported by Warburg and Lipmann as present in yeast and erythrocyte hemolysates, has been isolated from many biological materials and was recently obtained in crystalline form (Pontremoli *et al.*, 1961). CO_2 and ribulose-5-phosphate are the products of its dehydrogenating action. The reversibility of the reaction was demonstrated with an enzyme purified from yeast. An isomerase,[†] found in all animal and plant materials studied so far, maintains an equilibrium between the ribulose-5-phosphate formed in the decarboxylation and ribose-5-phosphate, with 75% of the phosphate in the aldose and 25% in the ketose form.

Long before the discovery of this isomerase reaction, studies were begun on the metabolic fate of ribose-5-phosphate. In 1938, Dische had

---

*EC 1.1.1.49 D-Glucose-6-phosphate:NADP oxidoreductase.
**EC 1.1.1.44 6-Phospho-D-gluconate:NADP oxidoreductase (decarboxylating).
[†]EC 5.3.1.6 D-Ribose-5-phosphate ketol-isomerase.

observed that ribose-5-phosphate added to erythrocyte hemolysates quickly disappeared and increased amounts of a hexose phosphate ester appeared in its place. The initial step in this transformation is an epimerization of ribulose-5-phosphate, first observed by Ashwell in spleen extracts; this establishes an equilibrium with D-xylulose-5-phosphate. This epimerase* has been obtained in highly purified form from both animal and microbial materials.

When it had been demonstrated that hexose monophosphate is formed from ribose phosphate, the existence of a cycle became probable, since a pentose phosphate can be formed from a hexose phosphate by the loss of the carbon atom in position 1; the pentose phosphate is then retrans-

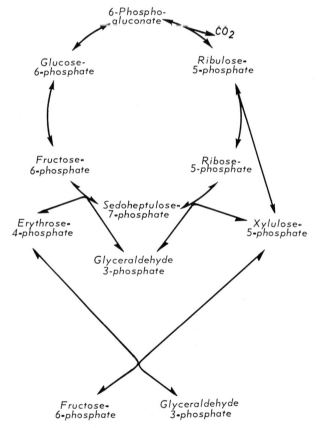

Fig. 70. Compounds involved in the pentose phosphate cycle. The enzymes are discussed in the text.

*EC 5.1.3.1 D-Ribulose-5-phosphate 3-epimerase.

formed by additional reactions to hexose monophosphate. The whole sequence of reactions, now known as the "pentose phosphate" or
"Horecker cycle", is summarized in Fig. 70.

Two new enzymes were discovered in connection with the elucidation
of the pentose phosphate cycle: a thiamine pyrophosphate-dependent
transketolase* and a transaldolase.** Phosphate esters with three, four
and seven carbon atoms are formed as intermediates in these reactions.

The function of the transketolase is to transfer "activated glycolaldehyde"; in this reaction, various ketoses and ketose phosphate esters
can function as donors, while numerous aldehydes or aldose esters serve
as acceptors. In this way, ribose-5-phosphate can be produced from
sedoheptulose-7-phosphate. The transaldolase catalyzes the transfer of
"activated dihydroxyacetone" in the reaction sedoheptulose-7-phosphate +
glyceraldehyde-3-phosphate $\rightleftarrows$ fructose-6-phosphate + erythrose-4-
phosphate.

Biosynthesis of Pentoses by Oxidation and Subsequent Removal of Carbon 6 from Hexoses

The first indication of another pentose pathway came from the work of
Salkowski and Neuberg in 1902, who incubated D-glucuronic acid with a
putrified meat mash and were able to isolate a derivative of L-xylose.
Studies of the metabolic abnormality pentosuria yielded further information. People afflicted with this disease excrete in the urine large
amounts of L-xylulose, which Touster et al. (1957) showed was formed
from D-glucuronate. When glucuronolactone labeled with C^{13} in the
1-position was administered, 5-C^{13}-L-xylulose was excreted in the
urine, but administration of glucuronolactone labeled in the 6-position
led to the excretion of unlabeled xylulose. Contrary to the reports of
earlier workers, Touster found that even normal individuals excrete
L-xylulose following high doses of glucuronate. Figure 71 shows the
series of reactions involved in this pentose pathway. Further studies by
Touster have elucidated the normal L-xylulose pathway. Liver mitochondria reduce the ketose to the five-carbon alcohol xylitol. A more detailed investigation of this reaction revealed the presence of two xylitol
dehydrogenases, an NAD-dependent, D-xylulose-forming dehydrogenase,[†]
and an NADP-dependent, L-xylulose-forming enzyme[††] (Hollmann and
Touster, 1957). Since D-xylulose is phosphorylated in the liver by a

*EC 2.2.1.1 D-Sedoheptulose-7-phosphate:D-glyceraldehyde-3-phosphate glycolaldehyde-transferase.
**EC 2.2.1.2 D-Sedoheptulose-7-phosphate:D-glyceraldehyde-3-phosphate dihydroxyacetone-transferase.
[†]EC 1.1.1.9 Xylitol:NAD oxidoreductase (D-xylulose-forming).
[††]EC 1.1.1.10 Xylitol:NADP oxidoreductase (L-xylulose-forming).

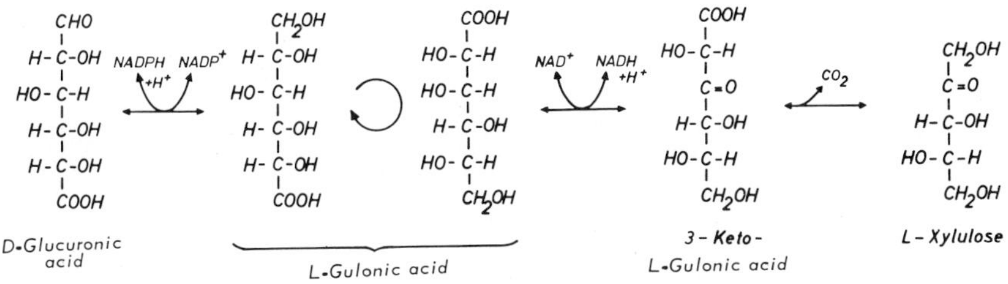

Fig. 71. L-Xylulose formation from D-glucuronate.

specific kinase and, as D-xylulose-5-phosphate, can be brought into the pentose phosphate cycle (see Fig. 70), Hollmann (1961) proposed combining these reactions with already known reactions into a "glucuronate-xylulose cycle" (Fig. 72). This cycle, like the pentose phosphate cycle, includes the hexose-pentose-hexose sequence and therefore offers the mammalian organism a second pathway for the biosynthesis of five-carbon sugars. To protect the aldehyde group of glucose during dehydrogenation at position 6, it is first coupled to UDP in a glycosidic linkage (see Fig. 72).

In addition to conversion to xylulose, there is another possibility for pentose formation from UDP-glucuronate. Hassid *et al.* (1959) showed that a direct decarboxylation of UDP-glucuronate takes place in plant tissues and gives rise to UDP-bound D-xylose. Xylose is in equilibrium with UDP-bound L-arabinose through a UDP-glucosyl-4-epimerase.* The possible pentose formation outlined is obviously limited to plants and probably plays a part only in the production of water-insoluble polysaccharides of the xylan and araban types.

Biosynthesis of Pentoses by Condensation of a C_2 Unit with a C_3 Compound

In considering the scheme of the pentose phosphate cycle on p. 143 in a counterclockwise direction, it becomes apparent that pentoses can also be formed from glucose-6-phosphate by a nonoxidative pathway. The previously mentioned transketolase catalyzes the transfer of C_2 fragments that have the structure of glycolaldehyde. The "activated glycolaldehyde" so formed has recently been isolated. It has the structure of a 2-α, β-dihydroxyethylthiaminepyrophosphate. If D-glyceraldehyde-3-phosphate serves as the acceptor aldehyde, then D-xylulose-5-phosphate is produced and is in equilibrium with ribose-5-phosphate and ribulose-5-phosphate through the reactions described above.

*EC 5.1.3.2 UDP glucose 4-epimerase.

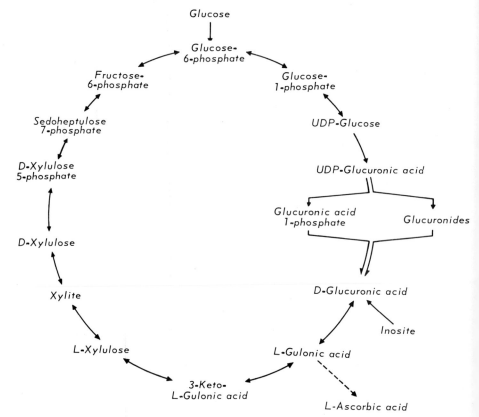

Fig. 72. The glucuronatexylulose-cycle.——=multiple step reaction (after Holl-mann, 1961).

In model experiments, pentoses were also obtained by an aldolase-catalyzed condensation of free glycolaldehyde with C_3 compounds (Hough and Jones, 1951); however, no proof has yet been found for the existence of such a reaction *in vivo*.

It has already been pointed out on pp. 118 and 134 that deoxyribonucleo-tides can be formed enzymatically from the corresponding nucleotides. There is an additional synthetic pathway for these deoxy sugars (Fig. 73).

The condensation of acetaldehyde with D-glyceraldehyde phosphate results in the formation of the 5-phosphate ester of 2-deoxyribose. The enzyme catalyzing this reaction was detected by Racker in 1951 and called deoxyriboaldolase;* it is found in both microorganisms and animal tissues.

*EC 4.1.2.4 2-Deoxy-D-ribose-5-phosphate acetaldehyde-lyase.

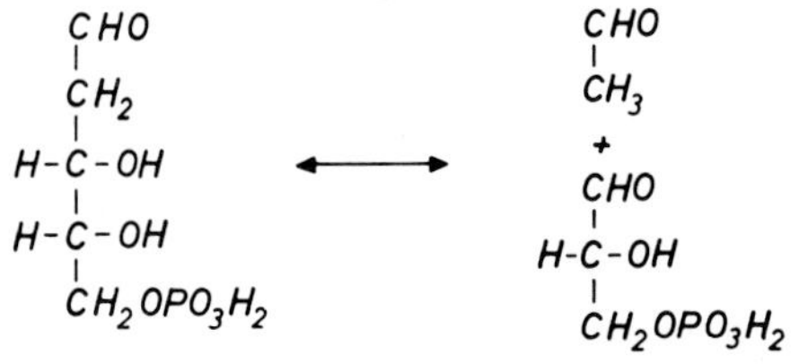

$$
\begin{array}{ccc}
CHO & & CHO \\
| & & | \\
CH_2 & & CH_3 \\
| & & \\
H\text{-}C\text{-}OH & \longleftrightarrow & + \\
| & & CHO \\
H\text{-}C\text{-}OH & & | \\
| & & H\text{-}C\text{-}OH \\
CH_2OPO_3H_2 & & | \\
& & CH_2OPO_3H_2 \\
\end{array}
$$

Fig. 73. The reaction of deoxyriboaldo-lase.

Free acetaldehyde is formed metabolically, as in the breakdown of threonine (Lin and Greenberg, 1954; Malkin and Greenberg, 1964). With purified deoxyriboaldolase (Pricer and Horecker, 1960), it has been shown that the K_s values for all reaction products have an order of magnitude of $10^{-3}M$; therefore, the normally very small acetaldehyde concentrations are sufficient for a condensation reaction. In regenerating liver and hepatomas, the activity of deoxyriboaldolase is greatly increased compared with that in normal liver (Boxer and Shonk, 1958); in vitamin B_{12} deficiency, this aldolase is present in smaller amounts (Wong and Schweigert, 1951). Polycarboxylic acids, phosphate and acetate activate the enzyme (Jiang and Groth, 1962; Roscoe and Nelson, 1964). From the results of *in vivo* and *in vitro* experiments done with normal and regenerating rat liver, Groth and Jiang (1966) postulated the deoxyriboaldolase to be essential for the biosynthesis of deoxyribonucleotides in mammals. On the other hand, Larsson and Neilands (1966) observed in their studies with double-labeled nucleosides and nucleotides that these were converted to deoxyribonucleotides without having been split; a utilization of newly formed deoxyribose-5-phosphate should have altered the ratio of the amounts of the two isotopes with which the nucleotides had been labeled.

Pentose Formation by Dehydrogenation of Five-carbon Alcohols

Enzymes that dehydrogenate open-chain polyalcohols occur widely in nature and are used industrially to carry out these reactions. Polyol dehydrogenases have also been found in mammalian tissues (McCorkindale and Edson, 1954). The reactions catalyzed by these enzymes may be divided into two groups: the formation of aldehyde sugars, catalyzed by an NADP-dependent enzyme* according to the reaction:

$$\text{polyol} + \text{NADP} \rightleftharpoons \text{aldose} + \text{NADPH}$$

and the formation of ketoses by other dehydrogenases** that use NAD or NADP:

*EC 1.1.1.21 Alditol:NADP oxidoreductase.
**EC not yet assigned.

$$\text{polyol} + \begin{matrix} \text{NAD} \\ \text{or} \\ \text{NADP} \end{matrix} \rightleftharpoons \text{ketose} + \begin{matrix} \text{NADH} \\ \text{or} \\ \text{NADPH} \end{matrix}$$

Some examples of the dehydrogenation of polyol phosphate esters, in which aldose phosphates or ketose phosphates are the reaction products, have also been described. Generally, acyclic polyalcohols with chain lengths of C_2 to C_7 can undergo these reactions. The discussion here is concerned only with those pentoses whose formation by a dehydrogenation mechanism is clearly established. These are: D-xylulose, L-xylulose, D-ribulose, L-ribulose, 7D-xylose L-arabinose and D-ribose-5-phosphate (reviewed by Touster and Shaw, 1962).

Functional Significance of Various Synthetic Pathways

Numerous attempts have been made to determine which of the four possible pentose pathways is used by a given organism. This has largely been done with radioactive precursors (specifically C^{14}-labeled glucose, $1-C^{14}$-acetate or $NaHC^{14}O_3$, which correspond in their behavior to a $3,4-C^{14}$-glucose), followed by isolation and step-wise degradation of the resulting nucleic acid pentoses (Bernstein et al., 1958; Horecker et al., 1958). In some in vivo experiments, the radioactive precursor imidazolylacetate was also administered (Tabor and Hayaishi, 1955; Hiatt, 1958); under these conditions, imidazolylacetate ribonucleoside is excreted in the urine, with the ribose showing the same isotope distribution as the sugars of the nucleic acids.

The interpretation of isotope experiments is often difficult (cf. p. 166), especially in terms of turnover rates, since recycling can lead to changes in labeling that are difficult to predict (Wood and Katz, 1958). In spite of these limitations, the work discussed below provides important information on preferred pentose pathways. Moreover, the use of sugars labeled with O^{18} in known sites appears to lead to results that can be more readily interpreted than those from C^{14}-labeled precursors (Rittenberg and Ponticorvo, 1962).

Under normal metabolic conditions, degradation via oxidation of the carbon atom in position 1 appears to play only a minor role, particularly in mammalian organisms. In the mouse, its contribution to pentose formation amounts to 5-10% (Shuster and Guldin, 1958); for E. coli, values of 25% (Bagatell et al., 1959) to 100% (Lanning and Cohen, 1954) have been reported. Variations of this kind in the same organism may be due to temporary shifts in metabolism. This appears to be the case in vitamin B_1 deficiency (Hiatt, 1958), in increased NADPH requirements (Kinoshita, 1957) and in pregnancy (Agranoff et al., 1954).

Generally, oxidative degradation with the loss of carbon 6 is unimportant, but it can easily be detected (Hiatt and Lareau, 1958). In pentose-containing plant tissues, however, it is the principal mechanism for the formation of C_5 sugars.

Having thus ruled out the two oxidative pathways, we are left with the transaldolase-transketolase catalyzed mechanism as the principal route, one that has been observed in numerous human studies (Hiatt et al., 1960), in rats (Hiatt and Lareau, 1958), mice (Shuster and Goldin, 1958), tumor cells (Elzina and Engelhardt, 1958) and also in yeast (David and Ronault, 1954). The polyol dehydrogenation observed in liver and in microorganisms seems, on the contrary, to be unimportant in the synthesis of nucleic acid pentoses.

Formation of Pentose Phosphate Esters

Ribose and deoxyribose are needed in the form of phosphate esters for the biosynthesis of nucleosides and nucleotides. In intermediary metabolism, ribose-5-phosphate can be formed directly from hexose phosphates by the reactions shown in Figs. 70 and 72; moreover, phosphorylation of free ribose is possible. A ribokinase* has been detected in various bacteria grown on ribose; a kinase, specific for ribose and 2-deoxyribose, has been purified from calf liver, and an inducible deoxyribokinase** has been isolated from lactobacilli. Deoxyribose-5-phosphate can also be produced by the aldolase reaction described on p. 146.

Ribose-1-phosphate is kept in equilibrium with ribose-5-phosphate through the action of a phosphoribomutase.** This mutase, which also acts on the phosphate esters of deoxyribose, though at a slower rate, has been found in yeast, smooth muscle and human blood (Guarino and Sable, 1955). A special phosphodeoxyribomutase has been purified from *Sarcina lutea* (Smith and Bernstein, 1961). A phosphoribokinase,[†] from pigeon liver, forms ribose-1,5-diphosphate from ribose-5-phosphate (Scarano, 1953). Another possible synthesis of ribose-1-phosphate is by a nucleoside phosphorylase[††] which catalyzes the general reaction (cf. p. 116):

$$\text{nucleoside} + H_3PO_4 \rightleftharpoons \text{base} + \text{ribose-1-phosphate}$$

Only ribosides with bases adenine, guanine, diaminopurine and uracil as bases are split in this way. 2-Deoxyribose-1-phosphate is formed in the

*EC   2.7.1.15   ATP:D-ribose 5-phosphotransferase.
**EC   not yet assigned.
[†]EC   2.7.1.18   ATP:D-ribose-5-phosphate 1-phosphotransferase.
[††]EC   2.4.2.1   Purine nucleoside:orthophosphate ribosyltransferase;
[††]EC   2.4.2.3   Uridine:orthophosphate ribosyltransferase.

mutase reaction described and, in addition, by the action of purine deoxyriboside phosphorylase[†] and thymidine phosphorylase[††] (cf. p. 128); dR-1-P can be prepared as a crystalline salt in this manner (Friedkin, (1950).

The most important phosphate ester for nucleotide formation is 5-phosphoribosyl-l-pyrophosphate (PRPP), necessary for the *de novo* synthesis of purine and pyrimidine nucleotides (see pp. 108 and 127), as well as for the utilization of free purines and pyrimidines (cf. pp. 116 and 128). The enzyme, ribose phosphate pyrophosphokinase,[*] which synthesizes PRPP from ATP and ribose-5-phosphate, has been found in many biological materials. More recently, PRPP has also been obtained synthetically (Tener and Khorana, 1958). Despite intensive efforts, the deoxyribose analog PdRPP has not yet been detected.

ROLE OF FOLIC ACID AND ITS DERIVATIVES IN NUCLEOTIDE SYNTHESIS

Folic acid, an important substance in biochemistry, was first isolated in 1943, and its structure was elucidated in 1948 by Stokstad *et al.* It functions in the transfer of single carbon moieties; such a transfer can take place at three different levels of oxidation—as formyl, hydroxymethyl or methyl groups. Like most other vitamins, folic acid also occurs in different biologically active forms. Most *in vitro* investigations have been made with the compound shown in Fig. 74. Common deriva-

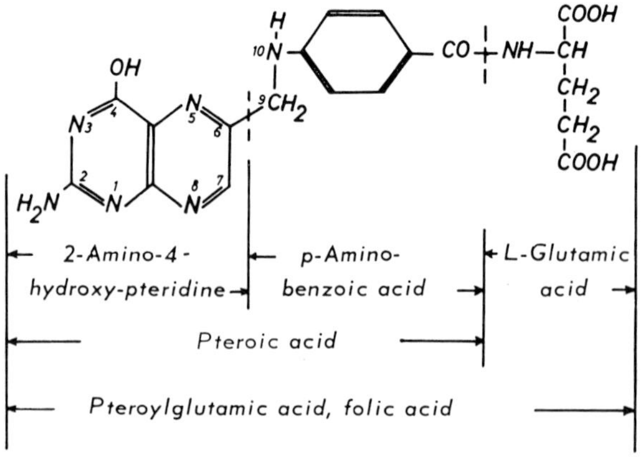

Fig. 74. Structure of folic acid.

†EC 2.4.2.6 Nucleoside:purine (pyrimidine) deoxyribosyltransferase.
††EC 2.4.2.4 Thymidine:orthophosphate deoxyribosyltransferase.
*EC 2.7.6.1 ATP:D-ribose-5-phosphate pyrophosphotransferase.

tives of folic acid carry up to six additional glutamate residues bound in a peptide-like manner to the γ-carboxyl groups of the glutamate. All derivatives of folic acid carrying single carbon moieties are derived from a hydrogenated form, 5,6,7,8-tetrahydrofolic acid (FH_4), which is formed enzymatically via the 7,8-dihydro compound (FH_2). The three compounds differ clearly in their spectral behavior. Folate reductase* and dihydrofolate reductase** have been purified from animal tissues and from microorganism. Some of these enzymes reduce free folate to FH_2, and FH_2 to FH_4 as well (e.g., Mathews and Huennekens, 1963). However, other preparations react only with FH_2, while free folate is not reduced (Blakley and McDougall, 1961; see Fig. 75). Only one enzyme has been found

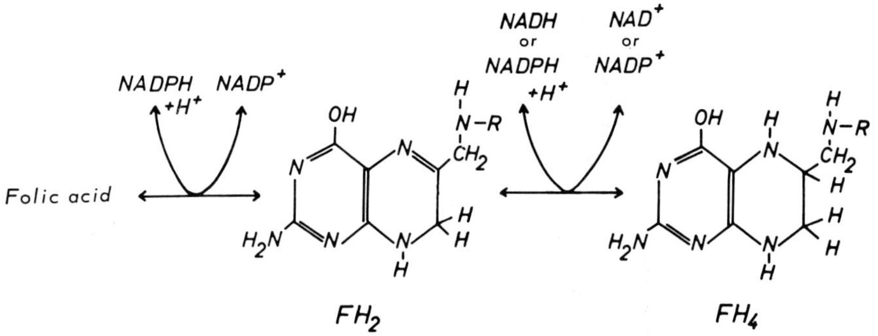

Fig. 75. Enzymatic formation of dihydrofolic acid (FH_2) and tetrahydrofolic acid (FH_4).

to reduce free folate and not FH_2 (Wright et al., 1958), and since it cannot use reduced pyridine nucleotides for hydrogen transfer, it seems to be a special case. The question of whether a specific folate reductase exists in addition to dihydrofolate reductase seems, moreover, to be unimportant since in the biosynthesis of folate, the dihydro compound is formed directly (Brown et al., 1961). Detailed studies of a highly purified FH_2 reductase from fowl liver have revealed that the substrate specificity of the enzyme is pH dependent (Mathews and Huennekens, 1963). Experiments with a partially purified dihydrofolate reductase from E. coli have shown that folate is a competitive inhibitor in the reduction of FH_2; the two substrates appear to be bound to the same site on the enzyme (Bertino et al., 1964). Small concentrations of organic mercurials lead to a considerable increase in the activity of the dihydrofolate reductase (Kaufman, 1964; Perkins and Bertino, 1964). Both reduction

*EC 1.5.1.4 7,8-Dihydrofolate : NADP oxidoreductase.
**EC 1.5.1.3 5,6,7,8-Tetrahydrofolate : NADP oxidoreductase.

steps are completely inhibited by low concentrations of folate analogs (cf. p. 308; Futterman, 1957).

It is interesting that the dihydrofolate reductase formed after phage infection in *E. coli* is NADH dependent, while the corresponding reductase of the host cell is NADPH specific (Mathews and Cohen, 1963b). A dihydrofolate reductase formed in amethopterin-resistant mutants of *Diplococcus pneumoniae* shows changed properties with regard to pH optimum, heat sensitivity and inhibitor sensitivity compared to the enzyme from normal cells (Sirotnak *et al.*, 1964). Whereas all the folate and dihydrofolate reductions mentioned so far appear to be irreversible, there seems to be a special enzyme for the reoxidation of FH_4; FH_2 is not an intermediate in this NADP-dependent reaction of the tetrahydrofolate dehydrogenase isolated from baker's yeast (Busch and Donaldson, 1965).

In 1948, Gordon *et al.* first postulated that folic acid might participate in biological transformylations. The transfer of single carbon compounds plays an important role in intermediary metabolism, as in the incorporation of the carbon atoms in positions 2 and 8 of purines and in the formation of the methyl groups of methionine, thymine and choline. As Fig. 76 shows, various enzymatic reactions lead to the production of single carbon derivatives of tetrahydrofolate. Thus, free formate can be converted to N_{10}-formyl-FH_4 by FH_4 formylase$^\diamond$ in an ATP-dependent reaction. The enzyme has been crystallized and its mode of action investigated thoroughly (Jaenicke and Brode, 1961; Lansford *et al.*, 1964; Uyeda and Rabinowitz, 1964; Himes and Wilder, 1965). Formaldehyde or the β-carbon atom of serine reacts with the formation of N_5, N_{10}-methylene-FH_4. N_5-formimino-FH_4, formed in histidine or purine degradation (Tabor and Rabinowitz, 1956), can be transformed to N_5, N_{10}-methenyl-FH_4 by a cyclodeaminase.* Finally, N_5-formyl-FH_4 is formed from N-formylglutamate (Miller and Waelsch, 1957).

The FH_4 derivatives mentioned are also in metabolic equilibrium with one another. N_5-formyl-FH_4 is transformed, with the aid of an ATP-dependent isomerase,** to N_{10}-formyl-FH_4 (Kay *et al.*, 1960), which, in turn, is in equilibrium with N_5, N_{10}-methenyl-FH_4 through the mediation of a cyclohydrolase† (Rabinowitz and Pricer, 1956); the latter can be used for the enzymatic determination of tetrahydrofolate. N_5, N_{10}-Methylene-FH_4 is dehydrogenated to N_5, N_{10}-methenyl-FH_4 by hydroxymethyl-FH_4 dehydrogenase†† (Ramasastri and Blakley, 1962, 1964;

$^\diamond$EC 6.3.4.3 Formate:tetrahydrofolate ligase (ADP).
*EC 4.3.1.4 5-Formiminotetrahydrofolate ammonia-lyase (cyclizing).
**EC not yet assigned.
†EC 3.5.4.9 5,10-Methenyltetrahydrofolate 5-hydrolase (decylizing).
††EC 1.5.1.5 5,10-Methylenetetrahydrofolate:NADP oxidoreductase.

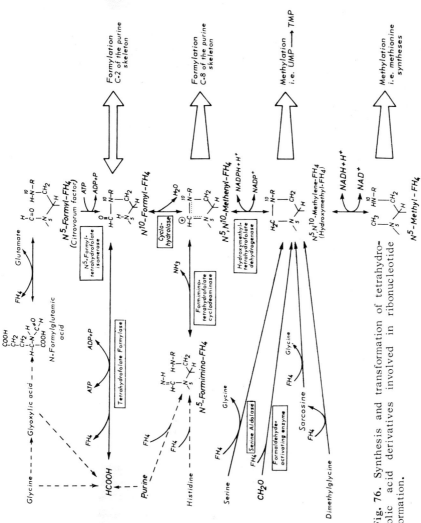

Fig. 76. Synthesis and transformation of tetrahydro-folic acid derivatives involved in ribonucleotide formation.

Donaldson *et al.*, 1965). This dehydrogenase, which has been used for the enzymatic determination of FH_4, is inhibited by purine nucleotides (Dalal and Gots, 1966). Furthermore, N_5, N_{10}-methylene-FH_4 can be converted by reduction to 5-methyltetrahydrofolate (Donaldson and Keresztesy, 1962; Larrabee *et al.*, 1963), the folate derivative found mainly in liver and serum.

In blood diseases, pathologically altered blood cells often differ in their content of various enzymes of folate metabolism (Bertino *et al.*, 1962). The single carbon derivatives of FH_4 participate in a variety of reactions which are summarized in Fig. 76. Study of the biological importance of folic acid was simplified because all of its derivatives which are of biological interest can be prepared chemically and also because a series of folic acid analogs exists which specifically and effectively inhibit the course of folic acid-dependent biological reactions (cf. p. 308). A number of reviews of the function and chemistry of folic acid have been published (e.g., Stokstad and Koch, 1967; Friedkin, 1963).

REGULATION OF NUCLEOTIDE FORMATION

The orderly course of metabolic processes is generally achieved by a self-regulating "negative feedback" system. This control mechanism can balance the interaction of different organs within a multicellular organism, for example, by means of certain hormones; it exists even in unicellular organisms and ensures that complicated biosyntheses are performed only in response to specific requirements. Anabolic reactions are controlled by their products whose concentrations determine whether or not a series of reactions is initiated; high end product concentrations inhibit further synthesis, while lowering the concentration releases this inhibition. Regulation can be visualized occurring in two ways: In one, the production of needed enzymes is repressed due to a block of the structural gene responsible for enzyme synthesis (cf. p. 248). Often this involves repression of enzymes in a given reaction sequence (Ames and Garry, 1959). An alternate possible control involves the reversible inactivation of one enzyme at the beginning of a reaction sequence by one of the end products. This mechanism, known as *end product inhibition*, or negative feedback inhibition, is not gene dependent. Metabolic regulation then results from the interaction of both mechanisms (Fig. 122). In multicellular organisms, enzyme repression can also be influenced by extracellular factors (e.g., hormones; cf. p. 252). Since the processes in enzyme induction and repression will be dealt with more thoroughly in connection with the function of nucleic acids (see p. 247), only the phenomenon of end product inhibition will now be discussed in detail.

The first indications of the existence of a feedback control came from investigations of microorganisms. When sulfonamides were added to culture media in which *E. coli* were grown, an aromatic amine was found to accumulate in the medium (AICA, cf. p. 107); this did not occur if purines were added to the medium (Gots, 1950). Similarly, in the case of an uracil-dependent mutant of *Aerobacter aerogenes*, orotate accumulated in the culture liquid; when the bacteria were grown in the presence of an excess of uracil, orotate production was repressed. Uracil—or a uracil metabolite—inhibits the synthetic pathway which terminates in orotate production (Brooke *et al.*, 1954). These mutants are incapable of total pyrimidine nucleotide synthesis, because they no longer possess the enzyme orotidine-5′-phosphate pyrophosphorylase which permits the formation of OMP (cf. Fig. 62). The resulting lack of pyrimidine nucleotides leads to a sustained triggering of the synthetic processes, which are, however, incapable of progressing beyond the orotic acid stage-they can be blocked only by the exogenous supply of uracil (utilization via "preformed pathway"; see p. 136).

The control principle just described was encountered in numerous anabolic reactions, especially those concerned with amino acid and nucleotide metabolism (reviewed by Umbarger, 1961). For technical reasons, most investigations have been carried out on microorganisms, but similar mechanisms can be shown to exist in higher organisms (e.g. Walker, 1961, 1962; Scarano *et al.*, 1963). Feedback inhibition can also be effected by numerous synthetic analogs; this is the reason why the term *"allosteric inhibition"* has been suggested in place of "end product inhibition." Many analogs have practical importance for chemotherapy, since the effect of the drug is enhanced by the cell's inability to dilute the analog with the correct natural product whose biosynthesis has been inhibited by a feedback mechanism (Smith and Sullivan, 1960; Paigen, 1962).

The first understanding of the mechanism of action of allosteric inhibition came from the studies of Pardee and co-workers on the regulation of the biosynthesis of pyrimidine nucleotides in *E. coli* (Yates and Pardee, 1956b; Gerhart and Pardee, 1962). The first specific enzyme of this synthetic pathway is aspartate transcarbamylase (see p. 125), which has been obtained from *E. coli* in crystalline form. A 70% inhibition of this highly active transcarbamylase is produced by small concentrations ($10^{-4} M$) of CTP or dCTP. Detailed investigation led to the suggestion of a second active site in the enzyme molecule. This center reversibly binds the inhibitor and thereby causes a deformation and temporary inactivation of the transcarbamylase. When the inhibitor concentration falls because CTP is used up in nucleic acid synthesis, the transcarbamylase again becomes fully active and synthesis of pyrimidine nucleotides starts again.   Active inhibitors are dCTP, CTP and, to a smaller extent, the mono- and diphosphates of cytidine (Fig. 77A). Free cytidine or uridine nucleotides, on the other hand, do not affect enzyme activity.   By a variety of chemical or physical methods, it has been possible to destroy the binding site of the inhibitor (''allosteric site'') while maintaining or even increasing the transcarbamylase activity (Gerhart and Pardee, 1962; Kleppe *et al.*, 1966). After treatment of the enzyme with *p*-mercuribenzoate, it is possible to isolate distinct subunits of the aspartate transcarbamylase: larger *catalytic subunits* (molecular weight 96,000) which are completely insensitive to the inhibitor CTP, and smaller *regulatory subunits* (molecular weight 30,000) which are catalytically inactive.   The native enzyme contains two catalytic and four regulatory subunits (Gerhart and Schachman, 1965).

A feedback control has also been found in mammalian liver (Bresnick, 1962) and yeast (Kaplan *et al.*, 1966) transcarbamylases in contrast to the bacterial enzyme; however, uridine derivatives show the most pronounced inhibition, while cytidine phosphates are scarcely active. Transcarbamylases from other sources are inhibited by still other compounds (Neumann and Jones, 1964), and a study of transcarbamylase activities in various microorganisms has shown that certain organisms

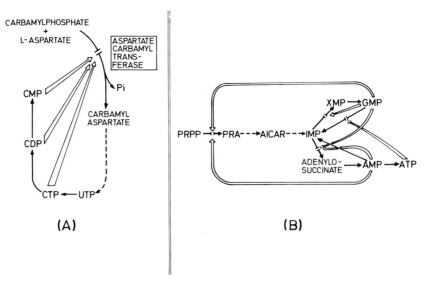

(A)   (B)

Fig. 77.   Feedback control mechanisms in the biosynthesis of ribonucleotides. ⟶ solid arrows: single enzyme-catalyzed reaction steps; ---▶ sequence of enzymatic reactions; ⟹ inhibitory effect upon ⊣ an enzyme reaction.

do not possess any feedback control mechanism for this enzyme (Bethell and Jones, 1966). In the Novikoff ascites hepatoma, the second specific enzyme in pyrimidine biosynthesis is controlled by an allosteric mechanism (Bresnick and Blatchford, 1964a).

Detailed investigations of dCMP deaminase likewise suggest the existence of two different active centers on the enzyme and a temporary reversible deformation of the feedback-inhibited protein (Scarano *et al.*, 1963; Maley and Maley, 1964, 1965, 1966a). Although dCMP deaminase does not occur in normal *E. coli*, it can be readily induced by infection with T-even phages (Fleming and Bessman, 1965; Maley and Maley, 1965, 1966b; Bessman and Fleming, 1966). dCMP deaminase is the enzyme most affected by the drug 5-iodo-2´-dexoyuridine triphosphate (cf. p. 310), which acts by an allosteric mechanism (Prusoff and Chang, 1966).

A negative feedback control for the sequence of reactions: thymidine $\longrightarrow$ dTMP $\longrightarrow$ dTDP $\longrightarrow$ dTTP has been found in rat liver and hepatoma tissue (Ives *et al.*, 1963) as well as in *E. coli* (Okazaki and Kornberg, 1964a and b).

A feedback control mechanism is involved too, in the reduction of ribonucleotides to deoxyribonucleotides (Ives *et al.*, 1963; Morris and Fischer, 1963). Figure 78 summarizes the presently known feedback effects which regulate the interconversion of various pyrimidine deoxyribonucleotides. Interesting results were obtained with the deoxyribonucleotide-forming system of *L. leichmannii* in which various allosteric effectors obviously determine the substrate specificity of the enzyme system (Beck, Goulian, Larsson and Reichard, 1966).

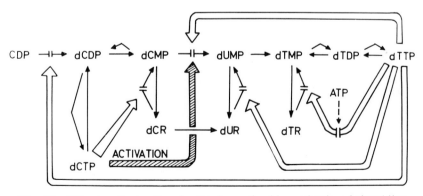

Fig. 78. Effects of negative feedback on the biosynthesis of thymidine triphosphate. Bent arrows distinguish the kinase reactions from the phosphatase reactions (straight arrows). Broad arrows designate feedback pathways.

In the *de novo* synthesis of purine ribonucleotides, there are also several allosteric controls involved (see Fig. 77B). The enzymes controlled is the phosphoribosylpyrophosphate amidotransferase (Henderson, 1962; Nierlich and Magasanik, 1965a). Moreover, some transformations of IMP, AMP, GMP and XMP are controlled by allosteric mechanisms (Magasanik, 1958).

A new mechanism for a negative feedback control has been found recently in *Salmonella typhimurium* (Dalal and Gots, 1966); here, the end product of a biosynthetic pathway regulates the available amount of a coenzyme which is essential for one of the early reaction steps. In the case mentioned earlier, the formation of methenyltetrahydrofolate is inhibited by the presence of adenosine nucleotides.

Genetic studies of *E. coli* have shown that the genes for the repression-controlled enzymes of pyrimidine biosynthesis (dihydroorotase, dihydroorotate dehydrogenase, OMP pyrophosphorylase and OMP decarboxylase) lie close together on the chromosome, while the gene for aspartate transcarbamylase—controlled by negative feedback—is located some distance away (Beckwith *et al.*, 1962).

FUNCTIONAL SIGNIFICANCE OF NUCLEOTIDES AS COENZYMES AND AS PRECURSORS FOR NUCLEIC ACID FORMATION

Free mononucleotides of all the bases of ribonucleic acid may be detected in the so-called acid-soluble fraction (cf. p. 195) in the form of nucleoside mono-, di- and triphosphates. Their proportions differ in various tissues, which, according to their metabolic functions, can be classified into "energy" and "metabolic" types.

In the course of the last two decades, many new nucleotides have been discovered (see Fig. 79), sometimes at a rate of 20–25 new compounds per year. Of these, some are subunits for nucleic acids, while others fulfill specific functions in intermediary metabolism. There are numerous excellent literature surveys on the type and functional importance of nucleotides (Henderson and LePage, 1958; Bock, 1960; Kennedy, 1960; Leloir and Cardini, 1960; Potter, 1960; Strominger, 1960; Utter, 1960; Montgomery and Thomas, 1962). At this point, the historical development will be outlined very briefly; in addition, various tables will provide a compact list of the most important nucleotides with coenzyme character.

The first naturally occurring mononucleotide, AMP, was discovered in Embden's laboratory in 1927. The intensive search for further adenine nucleotides extended over the next three decades and, as Table 6 shows,

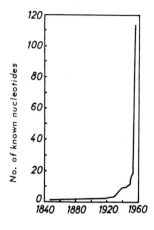

Fig. 79. Time course of the discovery of new nucleotides (from Henderson and LePage, 1958).

was richly rewarded. It was soon realized that the "muscle adenylic acid" (1) (numbers in parentheses refer to items in Table 6) isolated by Embden was not identical with "yeast adenylic acid" (6) obtained by alkaline hydrolysis of yeast RNA, even though they are composed of the same subunits (1 mole each of adenine, ribose and phosphate). ADP (2) plays an important role as a primary phosphate acceptor in oxidative phosphorylation and photophosphorylation, in addition to exercising a controlling function in cell respiration. In ATP (3), two of the three phosphate residues are attached by "energy-rich" bonds. ATP serves as the phosphate donor for a majority of phosphotransferases; in addition, the energy stored in the ATP molecule also plays an important role in the biosynthesis of non-phosphorus-containing molecules, with the "synthetases" splitting ATP into either ADP and P_i or AMP and pyrophosphate (PP). The functional significance of adenosine tetraphosphate (4) and pentaphosphate (5), and of adenosine-3'-triphosphate (7), detected in sulfur bacteria, is still unknown. Cyclic AMP (8) acts like a hormone (Orloff and Handler, 1961) or an activator of different enzymes. NAD (9) and NADP (10) are dissociable codehydrases widely distributed in nature, while FAD (11) is a structural constitutent of certain yellow respiratory enzymes. Coenzyme A (12) is a very versatile cofactor important for the activation of acids; at present, about 50 different acyl-CoA derivatives are known. Compounds (14) and (15) play a role in sulfur metabolism; PAPS (16) represents "activated" sulfate. Enzymatic synthesis of ADP-glucose (17) has been reported; the compound plays a role in the biosynthesis of starch and phytoglycogen (Frydman and Cardini, 1964; Murata and Akazawa, 1964). DPA (18) is a specific cofactor in the bioluminescence of marine plants. Compound (21) is formed by the transfer of the "active" methyl of compound (20). The

TABLE 6:   Adenosine-containing Nucleotides and Coenzymes

Compound	Abbreviation Commonly Used	Reference
(1) Adenosine-5′-phosphate	AMP	Embden and Zimmermann, 1927
(2) Adenosine-5′-diphosphate	ADP	Lohmann, 1935
(3) Adenosine-5′-triphosphate	ATP	Lohmann, 1929
(4) Adenosine tetraphosphate		Marrian, 1954
(5) Adenosine pentaphosphate		Sacks, 1955
(6) Adenosine-3′-phosphate		Doery, 1956
(7) Adenosine-3′-triphosphate		LePage and Umbreit, 1943
(8) Cyclic adenosine-3′,5′-phosphate		Sutherland and Rall, 1958
(9) Nicotinamide adenine dinucleotide	NAD	Schlenk and von Euler, 1936
(10) Nicotinamide adenine dinucleotide phosphate	NADP	Warburg et al., 1935
(11) Flavine adenine dinucleotide	FAD	Warburg and Christian, 1938
(12) Coenzyme A	CoA	Lipmann, 1954
(13) Adenine myonic acid dinucleotide		DeCaputto et al., 1961
(14) Adenine pteridine dinucleotide		Mosley and Caputto, 1958
(15) Adenosine-5′-phosphoryl sulfate	APS	Robbins and Lipmann, 1958
(16) 3′-Phosphoadenosine-5′-phosphoryl sulfate	PAPS	Robbins and Lipmann, 1958
(17) Adenosine-5′-diphosphoglucose		Recondo and Leloir, 1961
(18) Adenosine-3′, 5′-diphosphate	DPA	Cormier, 1962
(19) Adenosine-5′-diphospho-2-phospho-D-glycerate		Hashimoto and Yoshikawa, 1961
(20) S-Adenosyl-L-methionine		Cantoni, 1953
(21) S-Adenosyl-L-homocysteine		Cantoni, 1953
(22) Adenosine-5′-diphospho-L-aspartate		Hansen and Hagemann, 1956
(23) Adenosine-5′-diphospho-L-glutamate		Hansen and Hagemann, 1956
(24) Acyl-AMP		Berg, 1956a and b
(25) Adenosyl-5′-phosphorylamide		Katanuma, 1958
(26) Luciferyl-AMP		Rhodes and McElroy, 1958
(27) Adenosine-5′-aldehyde		Hogenkamp et al., 1962
(28) 8,5′-Cyclic adenosine		Hogenkamp, 1963

ADP-amino acids (22) and (23) have been found in liver and in lactating mammae; nothing is known of their significance. Different amino acids and carboxylic acids are "activated" in the form of an acyl-AMP compound (24). Adenylamidate (25) may participate in amino acid metabolism. Luciferyladenylate (26), isolated from fireflies, is used in the microdetermination of ATP. Compounds (27) and (28) were discovered as constituents of coenzyme B_{12}. A whole series of other adenine-containing compounds can be isolated from nucleic acid degradations, but most of them are artifacts.

In 1950, Leloir isolated from yeast a glucose derivative that contained 2 moles of phosphate and 1 mole of uridine (Caputto *et al.*, 1950). Within a few years, a large series of additional UDP-glycosyl compounds became known. The biosynthesis of these sugar derivatives involves two pathways. Either a pyrophosphorylase reaction occurs:

(a) UTP + sugar-1-phosphate $\rightleftharpoons$ UDP-sugar + PP

or a transferase reaction takes place:

(b) UDP-X + sugar-1-phosphate $\rightleftharpoons$ UDP-sugar + X-1-P

The sugar nucleotide, when in the "active" state, can be changed chemically with its glycosidic bond intact [e.g., epimerization (Leloir, 1951; Feingold *et al.*, 1960; Jacobson and Davidson, 1962), dehydrogenation (Strominger *et al.*, 1954), decarboxylation (Hassid *et al.*, 1959; Feingold *et al.*, 1960) or reduction to a deoxy sugar (Barber, 1962)]. Moreover, formation of other sugar compounds may involve such "activated" sugars. Examples are the formation of the disaccharides lactose (Gander *et al.*, 1956) and sucrose (Cardini *et al.*, 1955) and the polysaccharide glycogen (Leloir and Cardini, 1957). Table 7 lists 18 UDP monosaccharides and some other uracil derivatives.

The glucose and galactose derivatives [compounds (1) and (2) of Table 7] are in equilibrium by way of an epimerase (Leloir, 1951), are involved in the biosynthesis of di- and polysaccharides, and are also converted enzymatically to other hexose derivatives. Galactosemia results from a lack of galactose-1-phosphate uridylyltransferase* which catalyzes the reaction between galactose-1-phosphate and UDP-glucose to yield UDP-galactose and glucose-1-phosphate (Kalckar and Maxwell, 1958). The polymer inulin is probably synthesized in the body from UDP-fructose (3). The two UDP-pentoses [compounds (4) and (5)] are in enzymatic equilibrium with each other; both can be formed from the corresponding sugar-1-phosphates and UTP or by decarboxylation of the

*EC 2.7.7.10 UTP:α-D-galactose-1-phosphate uridylyltransferase.

TABLE 7. Uridine-containing Nucleotides and Coenzymes

Compound	*Reference*
(1) UDP-D-glucose	Caputto *et al.*, 1950
(2) UDP-D-galactose	Leloir, 1951
(3) UDP-D-fructose	Gonzales and Pontis, 1963
(4) UDP-D-xylose	Ginsburg *et al.*, 1956
(5) UDP-L-arabinose	Hassid *et al.*, 1959
(6) UDP-dihydroxyacetone	Smith *et al.*, 1961
(7) UDP-L-rhamnose	Barber, 1962
(8) UDP-D-glucuronic acid	Smith and Mills, 1954
(9) UDP-D-galacturonic acid	Smith *et al.*, 1957
(10) UDP-L-iduronic acid	Jacobson and Davidson, 1962
(11) UDP-D-glucosamine	Schmitz *et al.*, 1954
(12) UDP-D-galactosamine	Maley and Maley, 1959
(13) UDP-N-acetyl-D-glucosamine	Pontis, 1955; Cabib *et al.*, 1953
(14) UDP-N-acetyl-D-galactosamine	Pontis, 1955
(15) UDP-N-acetyl-D-mannosamine	Comb and Roseman, 1958
(16) UDP-N-acetyl-D-glucosamine-6-phosphate	Strominger, 1955
(17) UDP-N-acetyl-D-glucosamine-6-phospho-1-galactose	Gabriel and Ashwell, 1961
(18) UDP-N-acetyl-D-lactosamine	Kobata, 1962
(19) UDP-muramic acid	Park and Strominger, 1957
(20) 5-Ribosyl-UDP-D-glucose	Rabinowitz and Goldberg, 1961
(21) UMP-amino acids	Davies and Harris, 1960; Comb *et al.*, 1961

UDP-uronic acids (*8*) and (*9*). The pentose derivatives are starting materials for the formation of pentosans. UDP-dihydroxyacetone (*6*) has been found in diplococci; UDP-rhamnose (*7*) is used for transglycosidations in plants. UDP-glucuronate (*8*) can be formed in two ways: Either UTP reacts with glucuronic acid-1-phosphate, or UDP glucose is dehydrogenated enzymatically. The UDP derivative (*8*) can then be used for the formation of glucuronides, as a precursor in vitamin C synthesis, for pentose formation by decarboxylation, or for an epimerase-catalyzed conversion to UDP-galacturonate (*9*) which is a precursor of pectin. UDP-L-iduronate (*10*), a constituent of chondroitin sulfate B, is produced from UDP-D-glucuronate in an epimerization reaction. UDP-glucosamine (*11*), UDP-galactosamine (*12*) and the UDP derivatives of the N-acetylamino sugars (*13*), (*14*) and (*15*) are in metabolic equilibrium by means of epimerases. The phosphorylated derivatives (*16*) and (*17*) have been isolated from the oviduct of the hen, where they are probably synthesized in the course of eggshell formation. Lactosamine derivatives were discovered in milk; UDP-muramic acid (*19*) and UMP-amino acid derivatives are involved in the production of the bacterial cell wall.

The glucose derivative of pseudouridine diphosphate (*20*) can be formed enzymatically, but its biological significance is not yet clear.

A series of sugar-containing derivatives of guanosine diphosphate has also been discovered; these are listed in Table 8. The reduction of

TABLE 8. Guanosine-containing Nucleotides and Coenzymes

Compound	*Reference*
(1) GDP-D-mannose	Cabib and Leloir, 1954
(2) GDP-D-glucose	Pontis *et al.*, 1960
(3) GDP-D-fructose	Pontis *et al.*, 1960
(4) GDP-L-galactose	Su and Hassid, 1960
(5) GDP-L-fucose	Ginsburg and Korkman, 1958
(6) GDP-colitose (3,6-dideoxy-L-galactose)	Heath, 1960
(7) GDP-D-rhamnose	Markowitz, 1961
(8) GDP-talomethylose	Markowitz, 1961
(9) GDP-mannuronic acid	Preiss, 1962
(10) GDP-glycero-D-mannoheptose	Ginsburg *et al.*, 1962a
(11) P^1,P^4-Diguanosine-5′-tetraphosphate	Finamore and Warner, 1963
(12) P^1,P^3-Diguanosine-5′-triphosphate	Warner and Finamore, 1965
(13) Cyclic guanosine-3′,5′-phosphate	Ashman *et al.*, 1963
(14) Guanosine tetraphosphate	Gardner and Hoagland, 1965

glucose to a deoxy sugar was first observed in the case of GDP-glucose (*5*) (Foster and Ginsburg, 1961). A peculiar phosphorolysis involving a GDP-sugar has been detected in yeast extracts (Carminatti and Cabib, 1961):

$$G\text{-}P\text{-}P\text{-mannose} + P^{32} \longrightarrow G\text{-}P\text{-}P^{32} + \text{mannose-1-}P$$

Cytidine-containing coenzymes have also been found. As shown by Table 9, these include derivatives of cytidine diphosphate, deoxycytidine diphosphate and cytidine monophosphate. The CDP derivatives discovered first [compounds (*1*) through (*5*), (*9*) and (*10*)] constitute a group of activated alcohols. Compounds (*1*) to (*3*) act as coenzymes in phosphatide synthesis. Of the compounds in Table 9, CDP-glycerol (*3*) and CDP-ribitol (*5*) were first isolated from *L. arabinosus*; both participate in the biosynthesis of teichoic acids in the bacterial cell wall (Baddiley *et al.*, 1957). CDP-glucose (*6*) is the precursor for the formation of dideoxyhexoses (*7*) and (*8*) which occur in bacteria; CMP-acetylneuraminic acid (*10*) was found in various bacteria; CMP-sialic acid (*11*) was first detected in salivary glands. Compound (*12*), CMP-X, releases amino acids on hydrolysis.

TABLE 9. Cytidine-containing Nucleotides and Coenzymes

Compound	Reference
(1) CDP-colamine	Kennedy and Weiss, 1956
(2) CDP-choline	Kennedy and Weiss, 1956
(3) CDP-glycerol	Baddiley et al., 1956
(4) CDP-digyceride	Paulus and Kennedy, 1960
(5) CDP-ribitol	Baddiley et al., 1956
(6) CDP-D-glucose	Ginsburg et al., 1962b
(7) CDP-tyvelose (3,6-dideoxy-D-mannose)	Nikaido and Jokura, 1961
(8) CDP-abequose (3,6-dideoxy-D-galactose)	Nikaido and Nikaido, 1966
(9) dCDP-choline	Sugino and Sugino, 1960
(10) CMP-N-acetylneuraminic acid	Comb et al., 1959
(11) CMP-sialic acid	Roseman, 1962
(12) CMP-X	Bergkvist, 1958

The transfer of CDP-activated molecules to acceptors differs from that of other nucleoside diphosphate-activated compounds in that the reaction involves a simultaneous phosphorylation of the substrate, with CMP as the end product (CDP-Y $\longrightarrow$ YP + CMP), whereas in UDP-activated reactions, for example, the nucleoside diphosphate remains behind (UDP-X $\longrightarrow$ X +UDP). It may be helpful to recall in this connection (cf. p. 134) that in the reaction nucleotide $\longrightarrow$ deoxynucleotide, these compounds are substrates and products rather than coenzymes. Similarly, the sugar derivatives of hydroxymethylcytosine from phage DNA (see p. 261) are likely to serve structural rather than coenzyme functions.

TABLE 10. Thymidine-containing Nucleotides and Coenzymes

Compound	Reference
(1) dTDP-D-glucose	Kornfeld and Glaser, 1961
(2) dTDP-D-4-keto-6-deoxy-D-glucose	Okazaki et al., 1962
(3) dTDP-L-rhamnose	Glaser and Kornfeld, 1961
(4) dTDP-D-mannose	Baddiley et al., 1962
(5) dTDP-D-galactose	Neufeld and Iloes, 1962
(6) dTDP-amino sugars	Kornfeld and Glaser, 1962
(7) dTDP-4-acetamido-4,6-dideoxyhexoses	Gilbert et al., 1965

Table 10 lists some thymidine diphosphate-activated monosaccharides. The reduction of some sugars to deoxysugars takes place in combination with dTDP.

IDP-glucose and IDP-mannose were the first inosine derivatives to be discovered (Verachtert et al., 1964).

CHAPTER VI

Metabolism of Deoxyribonucleic Acid

STUDIES *IN VIVO* AND WITH WHOLE CELLS

The first investigations of DNA metabolism were carried out with P^{32}-labeled phosphate by von Euler and von Hevesy (1942), who showed that P^{32} incorporation in DNA was highest in tissues with a high rate of mitosis. This relationship between growth rate and the incorporation of labeled precursors into DNA has been confirmed repeatedly; an example is shown in Table 11.

TABLE 11. Ratios of the Specific Activity of DNA Phosphate to That of Inorganic Phosphate in Mouse Sarcomata 180 with Three Different Growth Rates (Each Value Average of Three Tumors)[a]

Mean Growth Rates (volume %/day)	$\dfrac{\text{Specific Activity DNA P}}{\text{Specific Activity Inorganic P}} \times 10^3$
18	3.6
26	5.1
44	8.6

[a]The animals were killed 2 hr after intraperitoneal injection of the radiophosphorus (from Harbers, 1960).

In the early studies, attempts were made to infer the rate of DNA synthesis from that of P^{32} incorporation and to relate this quantitatively to the growth rate. Results seemed to indicate that about twice as much DNA was formed as would have been expected from the increase in tissue mass (von Euler and von Hevesy, 1942). Therefore, DNA was considered to have a continuous turnover in the sense of Schoenheimer's postulate of the dynamic state of body constituents (1942). However, the identification of DNA as the cellular information storage material raised doubts about this conclusion. Indeed, studies of normal and regenerating rat liver, and of rat hepatomas by Brues and co-workers (1944), showed that DNA is a very stable cell constituent. The stability of DNA has subsequently been confirmed by numerous experiments (see also p. 178).

In following the time course of the incorporation of labeled precursors into a certain end product, one first observes an increase in the specific activity. If the experiment is continued long enough, a maximum occurs, followed by a decline in specific activity (cf. Fig. 80). This drop is usually due to catabolic processes. However, if cell multiplication

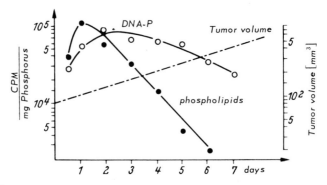

Fig. 80. Specific activity of DNA and phospholipids from mouse sarcoma 180 plotted against time after a single injection of radioactive orthophosphate. The slope of the straight line (-.-.-.-) indicates the increase in tumor volume per unit time (tumors with equal growth rates were selected). The increase in tumor mass (doubling time: 2.9 days) corresponds well to the decrease in the specific activity of the DNA (biological half-life: 2.7 days). The actual specific activity of the P-phospholipids was considerably higher than shown in the diagram (from Harbers, 1960).

takes place during the experiment, labeled cell constituents become "diluted" by newly formed cells, which also decreases the specific activity. If tissue constituents undergo no turnover, their specific activity, after having reached a maximum will be "diluted" only as a result of the growth of new cells (Fig. 80). Instead of timing the specific activity, it is also possible to follow the total activity of a tissue fraction; with no catabolism, the total activity will reach a maximum and then remain constant (cf. Fig. 94).

All "long-term experiments" carried out in this manner have shown that DNA does not undergo turnover in the sense of Schoenheimer's postulate. However, this does not exclude the possibility of exchange processes which might affect minor portions of a DNA molecule.

Support for this possibility came from studies that showed incorporation of precursors under conditions where net synthesis was impossible. For example, labeled precursors are incorporated into DNA of whole cells under certain *in vitro* conditions which do not permit growth, and 5-bromouracil replaces the natural thymine of DNA in *E. coli* cells even in stationary cultures exhibiting no DNA synthesis (Zamenhof *et al.*, 1956). DNA synthesis is completely blocked by appropriate doses of ionizing radiation, which, however, do not interfere with certain incorporation processes that are not radiosensitive (cf. p. 340). From these observations, it seemed probable that an exchange could occur in the terminal nucleotides of DNA, and perhaps also inside the chain (cf. p. 343). After it became possible to study DNA synthesis enzymatically (Fig. 81, cf. p. 173), Adler and co-workers (1958) demonstrated a terminal incorporation of deoxyribonucleoside triphosphates at preexisting DNA.

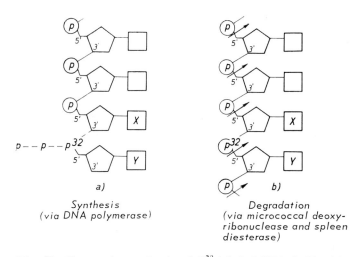

a)

Synthesis
(via DNA polymerase)

b)

Degradation
(via micrococcal deoxy-
ribonuclease and spleen
diesterase)

Fig. 81. Enzymatic synthesis of P^{32}-labeled DNA (left) with subsequent enzymatic degradation to 3′-deoxyribonucleotides (right). The arrows indicate the ester bonds which are split by micrococcus deoxyribonuclease and spleen diesterase (from Josse, Kaiser and Kornberg, 1961).

The problems discussed here have an important practical corollary: *The incorporation of a labeled precursor in a high molecular weight end product does not necessarily mean that net synthesis has occurred.* Unfortunately, the possibility of exchange with labeled precursors was frequently overlooked (cf. also p. 137), especially in autoradiographic studies, and thus eventually led to false conclusions.

The rate of incorporation of labeled precursors into the DNA of mitochondria is considerably higher than that of the nuclear DNA in the same cell (maximum ratio of specific activities 30 : 1; Neubert *et al.*, 1965;

Schneider and Cuff, 1965). The radioactivity of labeled mitochondrial DNA is lost very rapidly; thus mitochondria appear to have only a short life span. In week-old seedlings of tobacco, chloroplast DNA was also found to be replicated several times faster than nuclear DNA (Green and Gordon, 1966).

ENZYMATIC SYNTHESIS OF DNA

Even though a large number of compounds had been found which served as DNA precursors *in vivo*, the enzymatic mechanisms involved in DNA synthesis remained unknown until the experiments of Arthur Kornberg and his collaborators. Using extracts of *E. coli*, these investigators were able to show how labeled thymidine was converted to thymidine monophosphate (dTMP), thymidine diphosphate (dTDP), thymidine triphosphate (dTTP), and finally to an acid-insoluble form, similar to or identical with DNA. By fractionation of the original extract, it was possible to isolate several of the enzymes involved and to study the conditions needed for the uptake of thymidine phosphates by DNA. It immediately became apparent that a small amount of high molecular weight DNA (a "primer") and ATP were needed for the formation of new DNA. At first it was not clear whether the direct precursors of DNA were the nucleoside tri- or diphosphates (analogous to the conditions of the polynucleotide phosphorylase; cf. p. 198). However, when the other deoxyribonucleosides were also made available as triphosphates, these were shown to be the immediate precursors of DNA, since the simultaneous supply of all four deoxyribonucleoside triphosphates considerably increased the incorporation rate. This system no longer required ATP but still required primer DNA. Hence, the formation of DNA was shown to be catalyzed by a single enzyme, designated as *DNA polymerase,** whose action may be described as follows (Lehman *et al.*, 1958):

$$
\begin{array}{c}
\alpha n d\mathrm{TTP} \\
+ \\
n d\mathrm{GTP} \\
+ \\
n d\mathrm{CTP} \\
+ \\
\alpha n d\mathrm{ATP}
\end{array}
+ \mathrm{DNA} \;\underset{}{\overset{\mathrm{Mg}^{++}}{\rightleftharpoons}}\; \mathrm{DNA} +
\begin{bmatrix}
\alpha \mathrm{dTMP} \\
\mathrm{dGMP} \\
\mathrm{dCMP} \\
\alpha \mathrm{dAMP}
\end{bmatrix}_n
+ 2n(1 + \alpha)\,\mathrm{PP}
$$

This DNA polymerase isolated from *E. coli* is now often referred to as the Kornberg enzyme.

*EC 2.7.7.7 Deoxynucleosidetriphosphate:DNA deoxynucleotidyltransferase.

For DNA synthesis to occur, all four deoxyribonucleosides must be present as triphosphates. In addition, some Mg^{++} ions and a small amount of primer DNA are required. "Primer" originally meant a starter molecule ("initiator") which initiated polymerization by acting as a nucleus. Generally, in enzymatic polynucleotide synthesis, primer also refers to a template which directs the synthesis of a complementary polynucleotide chain (see the following text).

E. coli proved to be particularly useful for studies of DNA polymerase. Because of their extremely high growth rate (generation time about 20 minutes under optimum conditions), the cells must contain large amounts of DNA polymerase. Yet the enzyme yield was small, 1 kg of E. coli yielding only 10 mg of purified enzyme. Further purification raised the specific activity to 2000 times that of the starting material (up to 4000). Unfortunately, the enzyme activity was always affected by the presence of deoxyribonucleases, which partially degraded the primer as well as the newly formed DNA. Thus far, three different deoxyribonucleases have been detected in E. coli. In 1964, Richardson et al. succeeded in obtaining more highly purified DNA polymerase from E. coli. In addition, the DNA polymerase of B. subtilis was isolated almost free of exonucleases (Okazaki and Kornberg, 1964). Large concentrations of pyrophosphate inhibit enzymatic DNA synthesis. P^{32}-labeled pyrophosphate exchanges with the terminal phosphate of deoxyribonucleoside triphosphate, but only in the presence of DNA. While formation of new DNA does not occur in the absence of one or more of the deoxyribonucleoside triphosphates, there is only slight inhibition of the splitting of pyrophosphate groups under these conditions.

At first, DNA polymerase activity could be detected only as a result of incorporation of labeled precursors into DNA; later it was possible to achieve net synthesis of DNA (leading to an increase by a factor of 10-20 in the amount of DNA compared with the primer). Enzymatically produced DNA exhibited the same chemical properties as that formed in intact cells. Since the presence of a primer was required and DNA polymerase itself had no effect on the base sequence in the newly formed DNA, it seemed likely that the primer DNA was identical with the template, i.e., that structure which made possible the exact reproduction of the DNA. This would be in agreement with the mechanism of replication suggested by Watson and Crick (1953b). In experiments with primer DNA of various origins (M. phlei, A. aerogenes. E. coli, T2 bacteriophage and calf thymus), it was observed that the molar base ratio adenine : thymine and guanine : cytosine in the enzymatically produced DNA always amounted to 1; this gave some evidence for the presence of a double helical structure. On the other hand, the molar base ratio $(A + T)/(G + C)$ (α in the previous reaction) was like that of the primer (Table 12). It was not possible to influence the composition of newly formed DNA by varying the relative quantitaties of the four precursors; instead it was found that the deoxyribonucleoside triphosphate present in the lowest concentration was rate limiting.

In Table 12, the ratio $(A + T)/(G + C)$ of enzymatically synthesized DNA is either equal to, or greater than (but never smaller than), that of

TABLE 12. Molar Ratios of Purines and Pyrimidines in Primer and Newly Formed DNA. The Newly Formed DNA Always Contains a Small Amount of Primer (after Lehman, 1959)

DNA Species	Number of Analyses	A	T	G	C	$\dfrac{A+T}{G+C}(=\alpha)$	$\dfrac{A+G}{T+C}$
M. phlei							
Primer DNA	3	0.65	0.66	1.35	1.34	0.49 (0.48–0.49)	1.01 (0.98–1.04)
Newly formed DNA	3	0.66	0.80	1.17	1.34	0.59 (0.57–0.63)	0.85 (0.78–0.88)
A. aerogenes							
Primer DNA	1	0.90	0.90	1.10	1.10	0.82	1.00
Newly formed DNA	3	1.02	1.00	0.97	1.01	1.03 (0.96–1.13)	0.99 (0.95–1.10)
E. coli							
Primer DNA	2	1.00	0.97	0.98	1.05	0.97 (0.96–0.99)	0.98 (0.97–0.99)
Newly formed DNA	2	1.04	1.00	0.97	0.98	1.02 (0.96–1.07)	1.01 (0.96–1.06)
Calf Thymus							
Primer DNA	2	1.14	1.05	0.90	0.85	1.25 (1.24–1.26)	1.05 (1.03–1.08)
Newly formed DNA	6	1.19	1.19	0.81	0.83	1.46 (1.22–1.67)	0.99 (0.82–1.04)
T2 Phage							
Primer DNA	2	1.31	1.32	0.67	0.70	1.92 (1.86–1.97)	0.98 (0.95–1.01)
Newly formed DNA	2	1.33	1.29	0.69	0.70	1.90 (1.82–1.98)	1.02 (1.01–1.03)

primer DNA. This increase of adenine-thymine is probably due to the formation of a small quantity of polyd(AT) (see p. 176).

The molar base ratio is only a fairly crude characteristic of a DNA. A correct comparison would require analyses of the base sequences in both the newly formed and the primer DNA. At present, there are no suitable methods available for this type of work, so investigations have been restricted to studying the statistical distribution of adjoining bases. Comparative analyses of this kind have confirmed that the frequencies of nearest-neighbor base sequences in the primer and in enzymatically formed DNA are indeed the same; furthermore, they have demonstrated the complementarity and the antiparallel nature of the newly synthesized strands (Josse *et al.*, 1961; Swartz *et al.*, 1962). With the help of nucleoside-5′-triphosphates labeled with P^{32} in the α-phosphate, it is possible to determine which bases are adjacent to one another in an enzymatically produced DNA. The incorporation of a single P^{32}-deoxyribonucleotide together with nonlabeled triphosphates, followed by degradation of this DNA to 3′-mononucleotides (Fig. 81), provides information about nearest-neighbor frequencies; the necessary degradation is accomplished by the action of deoxyribonuclease from *Micrococcus pyogenes* or by a calf spleen phosphodiesterase. Studies of this kind also indicate that DNAs of similar base composition may have different dinucleotide frequencies.

The DNA of T2 phages contains 5-hydroxymethylcytosine in the place of cytosine (cf. p. 260). In systems lacking 5-hydroxymethylcytosine deoxyriboside triphosphate as precursor, DNA from T2 phages proved to be an effective primer. It thus appears that the guanine in the template DNA allowed base pairing with cytosine instead of 5-hydroxymethylcytosine which normally occurs in this phage DNA. Similarly, it has been possible to achieve enzymatic production of DNA that contains unnatural bases, provided the bases are present as deoxyribonucleoside triphosphates and are added to the reaction mixture along with the other three deoxyribonucleotides. Thus, it was found that thymine could be replaced by uracil or 5-bromouracil, cytosine by 5-methylcytosine or 5-bromocytosine, and guanine by hypoxanthine. However, xanthine was not incorporated into DNA. From these observations it follows that *DNA polymerase catalyzes the biosynthesis of DNA*, but the enzyme *cannot* recognize the DNA molecule over any appreciable region and it *cannot act on the base sequence which is determined by the primer template.* In this way, exact copies of the primer DNA are formed. *The template,* however, *cannot distinguish between some bases;* e.g., guanine can form hydrogen bonds with cytosine, 5-methylcytosine or 5-bromocytosine. *Specific kinases bring only the correct bases up to the deoxyriboside triphosphate level* and thus essentially determine that the newly formed DNAs always have identical base sequences. The specificity of the kinases is an important means for securing precise replication of the DNA molecules (cf. p. 130). Certain synthetic base analogs, e.g., some pyrimidines halogenated in the 5-position, can get to the deoxyriboside triphosphate level and will then be incorporated into the DNA molecule (see p. 310).

The results of comparative studies with normal and a mutant DNA polymerase indicate that the normal enzyme rejects noncomplementary base pairs much more effectively than the mutant (Freese and Freese, 1967). These observations support a model according to which DNA polymerase can recognize position and structure of the pentose phosphate moieties of the incoming deoxyribonucleoside triphosphates, but not the various bases. Correct base pairing should be achieved by hydrogen bonding, incorrectly paired DNA precursors being rejected by the enzyme (Freese and Freese, 1967).

Uracil occurs exclusively in RNA (the only known exception is a DNA-containing virus; cf. p. 261), apparently because it is not converted to deoxyuridine di- and triphosphates. Recently, however, enzymes have been obtained which can catalyze the *in vitro* formation of deoxyuridine triphosphate.

Since sufficient characterization of DNA by chemical means is not yet possible, attempts have been made to prepare biologically active DNA by means of the DNA polymerase system, as a direct demonstration of the template effect. In experiments with transforming DNA (cf. p. 273) from *Haemophilus influenzae* and *Diplococcus pneumoniae* as primer, the amount of enzymatically produced DNA increased by a factor of 6–8, but the original transforming activity did not increase. Since it was not possible at that time to prepare DNA polymerase free from deoxyribonucleases, the newly formed DNA was probably inactivated by these enzymes. Later, Litman and Szybalski (1963) succeeded in demonstrating enzymatic synthesis of DNA that had some transforming activity. In these experiments, transforming DNA, in which part of the thymine in both strands had been replaced by 5-bromouracil, was introduced as primer; moreover this primer DNA had undergone thermal denaturation and was, therefore, less sensitive to inactivation by deoxyribonuclease. The enzymatically formed DNA contained 5-bromouracil in only one strand or—after further synthesis—none at all (cf. p. 178). The resulting density difference permitted the separation of the newly formed DNA from the bromine-containing primer DNA in CsCl (see p. 84) and also allowed its transforming activity to be tested.

DNA multiplication is not possible unless all four deoxyribonucleoside triphosphates are supplied in the *E. coli* DNA polymerase system. However, small amounts of a single deoxyribonucleotide can be incorporated into primer DNA in a terminal position (Fig. 81). This kind of incorporation implies the extension of an already present polynucleotide chain, but not net synthesis, i.e., the formation of new DNA strands. An enzyme was recently found in calf thymus nuclei which catalyzes the terminal incorporation of deoxyribonucleotides into preexisting DNA. This so-called nuclear enzyme can be separated from DNA polymerase and therefore is not identical with it (Krakow *et al.*, 1962).

Terminal incorporation can be demonstrated by means of enzymatic hydrolysis of DNA. If DNA is degraded by enzymes that give rise to 3′-deoxyribonucleoside phosphates, then the terminal unit of the DNA is split off in the form of a nucleoside (Fig. 81). When *E. coli* DNA polymerase was used to incorporate C^{14}-labeled dCTP into DNA which was then hydrolyzed with spleen diesterase, 67% of the C^{14} activity was in the form of a nucleoside (which, as the 5′-deoxyribonucleoside phosphate, comprised the end of the polynucleotide chain) and 33% was in the form of a nucleotide. With the aid of P^{32}-labeled deoxyribonucleoside triphosphate, it can also be established which bases (Y in Fig. 81) make up the terminal segment of the primer DNA where addition has taken place (cf. p. 167). In this way, deoxyadenosine can be shown to be preferentially added to deoxyadenosine (57%), whereas thymidine is added preferentially to thymidine (52%); on the other hand, after incorporation of deoxycytidine triphosphate, the P^{32} activity was distributed relatively uniformly over the four 3'-deoxyribonucleotides (Adler et al., 1958).

After Kornberg and his group had succeeded in preparing DNA polymerase from E. coli and thus obtained the first insight into the mechanism of enzymatic DNA formation, various attempts were made to detect and isolate DNA polymerase from other species, especially mammalian tissues. These investigations, unlike those with E. coli, were very difficult since mammalian DNA polymerase showed comparatively little activity. Thus it was not possible to demonstrate net synthesis of DNA or to exclude the possibility of mainly terminal incorporation of precursors. However, there was some evidence for real DNA synthesis, because the rate of incorporation was highest when all four deoxyribonucleoside triphosphates were supplied simultaneously. The mechanism of action of mammalian DNA polymerase seems to be similar to that of the enzyme from E. coli. Studies of regenerating rat liver and of leukemic L-5178-Y lymphoblasts showed that mammalian DNA polymerase can differ in its properties depending on its origin. The enzyme obtained from leukemic cells was inactivated by dialysis, unlike the DNA polymerase from rat liver, whereas the Ca^{++}- or Mn^{++}-induced activation of the enzyme from liver was not observed with the DNA polymerase from the L-5178-Y lymphoblasts.

The highest DNA polymerase activity and the most intensive synthesis of DNA occur in growing tissues. Table 13 compares the relative DNA polymerase content of various tissues of the rat. When mammalian cells were fractionated, DNA polymerase was first found only in the soluble cytoplasm and not in the nucleus. Apparently the enzyme is very easily lost from the nucleus, even though it is largely located there in vivo as can be seen when the nuclei are isolated in nonaqueous media (Siebert, 1963).

DNA polymerases from various sources differ with respect to the physical state of the necessary template DNA. One group is represented by the E. coli DNA polymerase which uses double- as well as single-stranded DNA as primer; it includes the enzymes of B. subtilis, sea urchin and regenerating rat liver. The other type of DNA polymerase requires preferentially single-stranded DNA as primer (ascites tumor cells,

TABLE 13. Activity of DNA Polymerase in
Various Organs of the Rat, with
Thymus Tissue Taken as 100
(after Bollum, 1959)

Tissue	Relative Activity
Thymus	100
Small intestine	30
Regenerating liver	28
Spleen	22
Testes	9
Kidneys	8
Normal liver	7
Lung	4
Brain	3
Heart	3
Pancreas	1
Skeletal muscle	1
Flexner-Jobling carcinoma	130
Walker 256 carcinoma	300

calf thymus, T2-infected *E. coli*). Bollum (1959a) observed that native calf thymus DNA is completely inactive as primer and must be heat denatured before DNA synthesis can be initiated by the DNA polymerase from the same tissue. Denaturation of DNA by treatment with dilute acid or with deoxyribonuclease, on the other hand, leads to a loss of priming activity. In contrast, pretreatment of a primer DNA with a very small amount of crystalline pancreatic deoxyribonuclease results in increased DNA synthesis. Later investigations by Bollum (1962) showed that oligodeoxyribonucleotides with more than two monomers and a free 3'-hydroxyl group can act as primers. This role of the free 3'-hydroxyl group is confirmed by the observation that a DNA phosphatase from *E. coli*, which removes the terminal 3'-phosphate, can increase the priming activity of DNA two- to tenfold. The priming efficiency of native rat liver DNA for DNA polymerase of regenerating rat liver is fairly low; it increases with thermal denaturation. However, when native DNA is preincubated in biological media, the priming efficiency becomes higher than that observed after thermal denaturation (de Recondo, 1966). This effect is due to the action of an unknown enzyme which causes an activation of the primer.

In certain instances, DNA polymerase from *E. coli* is capable of repairing damage to DNA, as well as catalyzing DNA synthesis (cf. p. 343). If the free 3'-hydroxyl groups at the ends of native DNA are partially degraded by exonuclease III, the DNA polymerase, using the single strands obtained as template, effects a reformation of the complete double-helical structure (Richardson *et al.*, 1964). From *E. coli* cells infected with T4 phages, a polymerase could be isolated that catalyzes the covalent joining of two segments of an interrupted strand in a native DNA (Weiss and Richardson, 1967). The enzyme, which requires the presence of ATP, leads to the formation of phosphodiester bonds, AMP and pyrophosphate. In some cases, it is apparently possible for RNA to serve as a template for DNA synthesis. Thus, poly(A + U) as primer leads to the formation of poly(dA + dT) (Lee-Huang and Cavalieri, 1963).

The problems of enzymatic repair of DNA will be discussed in more detail in a later chapter (see p. 343).

The autoradiographic investigations of Cairns (cf. p. 180) demonstrated that the replication of DNA in *E. coli* begins at a fixed point and then proceeds through a fork mechanism involving the simultaneous replication of both antiparallel strands. However, the Kornberg enzyme synthesizes DNA exclusively by chain elongation at the terminal 3'-hydroxyl group [Fig. 81(a)]; therefore this enzyme cannot synthesize two new polynucleotide chains in the same direction starting from one end of a DNA template whose two strands have opposite polarity (Schildkraut *et al.*, 1964). Thus, the possibility was considered that more than one polymerase might be involved in the process of DNA replication. Recently, Hori *et al.* (1966) did indeed find two kinds of DNA polymerase in extracts of *Alcaligenes faecalis*. According to the initial observations, one enzyme prefers single-stranded, the other double-stranded, DNA as primer for the polymerization reaction. At present, it is not known whether or how these two enzymes are involved in DNA replication.

Studies by Schildkraut *et al.* (1964) revealed that the DNA synthesized by the Kornberg enzyme has some unusual properties. It is easily renatured after denaturation by heat or by alkali, and—in contrast to DNA isolated from cells—a multi-branched structure is seen in electron micrographs. These observations agree with the conclusion that the DNA polymerase isolated from *E. coli* is indeed able to synthesize new DNA with a DNA primer as a template. However, this formation of DNA is not identical with DNA replication as it takes place in the living cell. Thus the possibility has been considered that the DNA polymerase from *E. coli* is only a "repair enzyme."

For normal DNA polymerase action the presence of primer DNA is necessary. However, if incubation continues for a sufficient time, a DNA that contains only adenine and thymine can be formed without primer (Lehman, 1959; Schachman *et al.*, 1960); if this artificially produced

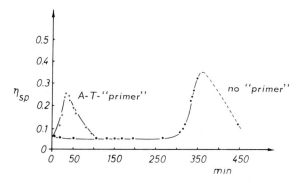

Fig. 82. Enzymatic synthesis of polydAT with and without dAT primer (from Lehman, 1959).

polydAT is then used as primer, DNA formation begins at once (Fig. 82), utilizing only the deoxyribonucleoside triphosphates from adenine and thymine, but not those from guanine and cytosine.   5-Bromouracil, a thymine analog, along with adenine, can be used instead of thymine to form an artificial DNA (dABU polymer), provided the two bases are in the form of deoxyribonucleoside triphosphates (Trautner *et al.*, 1962).   In the enzymatic synthesis of this type of DNA, but not in the case of polyd(AT,) a small amount of deoxyguanosine triphosphate can also be utilized at a rate of about 1 in 10^4 nucleotides. Model experiments of this kind are helpful in elucidating the mechanism of mutagenic base analogs (cf. p. 242).

Polyd(AT) is extremely sensitive to the action of deoxyribonucleases and is therefore degraded rapidly when incubated over a long period of time (see Fig. 82). The preparation of polyd(GC) is even more difficult, because of its still higher sensitivity to nucleases. Contrary to the results given in Fig. 82, the polyd(AT) produced by a DNA polymerase isolated from *B. subtilis* (which is nearly free of exonuclease activity; see p. 169) does not undergo rapid, spontaneous degradation.

A detailed analysis of polyd(AT) has shown that it has a double helical structure (molecular weight up to several million), with the molar ratio of adenine to thymine equal to one. With the help of P^{32}-labeled deoxyadenosine triphosphate and thymidine triphosphate (cf. p. 170), it has been demonstrated that within each polynucleotide strand of polyd(AT), adenine and thymine alternate with one another (Schachman *et al.*, 1960). In only GC-containing DNA does this alternating arrangement of purines and pyrimidines not occur (Josse *et al.*, 1961). While DNA polymerase can normally react only with deoxyribonucleoside triphosphates, on replacement of Mg^{++} ions by Mn^{++}, it is also possible to incorporate ribonucleoside triphosphates in enzymatically synthesized DNA (Berg *et al.*, 1963). If three bases are supplied as deoxyribonucleoside triphosphates and the fourth as ribonucleoside triphosphate, the resulting DNA is mixed with respect to its pentose content, but its base sequence apparently mimics the primer. Because of its ribose content, this DNA is alkali labile (cf. p. 33). Since alkaline hydrolysis occurs at the 5′-phosphoester bond of ribose, this DNA can be split selectively. In the future, this type of degradation may be helpful for base sequence analysis of DNA.

The action of DNA polymerase is greatly inhibited by the presence of RNA polymerase (Berg *et al.*, 1965). Apparently helical DNA to which RNA polymerase is bound is not available as a template for DNA replication. In contrast to RNA polymerase, DNA polymerase forms a readily dissociable complex with the template (cf. p. 201).

The DNA template alone is not sufficient to insure exact replication; in addition, specific kinases are needed to bring only the correct bases up to the deoxyribonucleoside triphosphate level. The specificity of kinases is especially important for the metabolic processes after bacteriophage infection. T2 phage DNA contains 5-hydroxymethylcytosine in the place of cytosine. After penetration of the host cells by the phage DNA, all of the deoxycytidine triphosphate is degraded by a specific phosphatase and is thus unavailable for use by the DNA polymerase in the formation of new phage DNA. At the same time, 5-hydroxymethylcytosine is formed by appropriate enzymes and is brought to the deoxyriboside triphosphate level (Fig. 129; cf. p. 266). These observations raised the question of whether the rare bases detected in the DNA of different species are located in some specific DNA strands in which they completely replace one of the normal bases, or whether they occur sporadically along with the usual purines and pyrimidines. As far as the second alternative is concerned, it was difficult to imagine how consistent reproduction could occur via the Watson-Crick mechanism. The observation that with T-even phages (T2, T4 and T6), glucose incorporation at hydroxymethylcytosine takes place only in the completed DNA, and not at the deoxyribonucleotide level, suggested that methylation might also take place at the completed polynucleotide. Gold, Hurwitz and Anders (1963) have demonstrated that methylation of high molecular weight DNA can indeed be catalyzed by means of species-specific enzymes.* Such *in vitro* methylations are usually possible only with heterologous DNA, i.e., with enzymes and DNA from different organisms; similar conditions apply to the methylation of RNA (cf. p. 210). It is interesting in this connection that guanine preferentially follows 5-methylcytosine in the base sequence of various kinds of DNA.

Gold and Hurwitz (1964a and b) later achieved a 400-fold purification

*EC_1 not yet assigned.

of an enzyme from *E. coli* W which methylates heterologous DNA using S-adenosylmethionine as C_1 donor. 5-Methylcytosine and 6-methylamino-purine are end products of the reaction, which is apparently irreversible. S-Adenosylhomocysteine is an active competitive inhibitor. Methylation of the DNA is impaired by heat denaturation, but not by sonication or the presence of small amounts of nucleases. The methylation of DNA is in-hibited by actinomycin (cf. p. 294) and proflavin. Previous enzymatic methylation does not appear to have any effect on the priming efficiency of the DNA for the DNA polymerase.

REPLICATION OF DNA

On the basis of the Watson-Crick model of DNA, each new DNA mole-cule is thought to be composed of an "old" and a "new" polynucleotide strand, with the "old" serving as the template for the new ("semi-conservative" replication). Labeled DNA precursors have been used to test this hypothesis experimentally.

If a labeled precursor is supplied just before DNA synthesis begins, it is incorporated into the newly formed DNA. According to the Watson-Crick hypothesis, the label should be in only one DNA strand (Fig. 83). Following mitosis, the DNA is uniformly divided among the daughter nuclei; the Watson-Crick hypothesis predicts that equal parts of the labeled DNA would occur in the daughter cells, with labeled and un-labeled nuclei in later generations (Fig. 83). In these studies, radio-active DNA is usually detected by means of autoradiographic and histo-chemical procedures.

Plaut and Mazia (1956) were the first to apply this principle to studies of the doubling mechanism of DNA, but they were unable to confirm the Watson-Crick hypothesis. When C^{14}-labeled thymidine was supplied to the meristem cells in the root tips of *Crepis capillaris*, the labeled DNA was found to be unequally distributed in the daughter nuclei. The dif-ference in C^{14} content of the DNA amounted on the average to 1.5:1, with a maximum of 2.5:1. The sum of labeled DNA in the daughter cells agreed quite well with the total amount of radioactive DNA in cells still in prophase. Shortly thereafter, however, Taylor, Woods and Hughes (1957) found that when H^3-thymidine was provided to *Vicia faba* seedlings, the radioactive DNA was uniformly distributed in the daughter cells. By use of the very weak β-emitter H^3, it was possible to achieve high reso-lution and therefore to assign the labeled DNA to single chromosomes. Moreover, when the first cell division was blocked by colchicine, all of the chromosomes were found to be uniformly labeled, while in the next generation only one of two daughter chromatids was radioactive (see Fig.

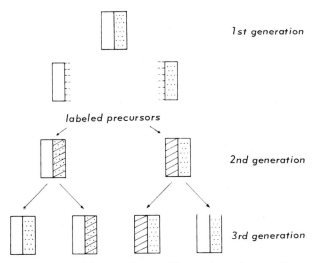

Fig. 83. Simplified scheme illustrating the application of isotopes in the study of the mechanism of DNA replication. The top line shows the two complementary strands of a DNA molecule in the first generation. If a labeled precursor is supplied before replication, it is incorporated into the newly formed strands of the two daughter molecules (second generation, cross-hatched part). If during the subsequent DNA synthesis, labeled precursors are no longer available, then two labeled and two unlabeled DNA molecules are formed in the third generation. This pattern may be modified by crossing-over.

83). The findings of Plaut and Mazia are not easy to explain. Possibly the unequal distribution of the radioactive DNA was due to mitosis having occurred more than once between the time labeled thymidine was taken up and the experiment was begun. Moreover, the distribution of label can be affected by crossing-over.

The behavior of *E. coli* DNA, as observed by Meselson and Stahl (1958), is also consistent with the Watson-Crick hypothesis. The cells were grown in N^{15} medium until completely labeled and were then cultured further in a normal N^{14} medium. By centrifugation in a CsCl density gradient, it was possible to separate N^{14}- and N^{15}-containing DNA as well as N^{14}-N^{15} hybrids. After the first doubling of the uniformly labeled cells, the label was found to be evenly distributed in N^{14}-N^{15} hybrids; in the next generation, N^{14}-N^{15} hybrids and unlabeled N^{14} molecules were found to occur in a ratio of $1:1$. These results show that the DNA molecules in *E. coli* are composed of two subunits; on replication, each daughter molecule receives one subunit.

These results (Meselson and Stahl, 1958) did not provide definite proof that the label was in fact associated with only one strand of the DNA of the first filial generation in which the ratio of $N^{14}:N^{15}$ was 1. In repeating these experiments with C^{13} and N^{15}, and partially degrading DNA by sonication, Rolfe (1962) was unable to detect in the CsCl density gradient any fragments completely labeled with C^{13} or N^{15}. Thus the possibility of hybrid molecules formed by chain elongation could be excluded.

For the replication process, the double helix of DNA must be unwound. No details are presently known about the course of unwinding and the subsequent double helix re-formation. Levinthal and Crane (1956) suggested that the biosynthesis of the new complementary DNA halves begins at the same time as the unwinding of the original helix; the formation of daughter molecules should be accomplished simultaneously with the complete separation of the two original polynucleotide strands. These views received experimental support from studies by Cairns (1963a), who found that the circular DNA of *E. coli* duplicates by forming a fork. The limbs of the fork each contain one strand of new and one of old material. During the period of replication, the distal ends of the two daughter molecules are joined by the two starting points (Fig. 84) which always appear to be at the same locus in the DNA molecule. In view of the number of turns in helical DNA of *E. coli* and the short time of synthesis, replication must result in rapid rotation ($\sim$ 10,000 rpm) of the DNA molecules. By countercurrent distribution, Kidson (1966) was able to isolate newly synthesized DNA of *E. coli* and *B. subtilis*. This DNA in the region of

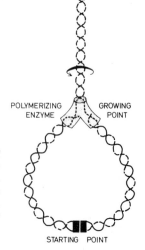

POLYMERIZING ENZYME   GROWING POINT

STARTING POINT

Fig. 84. Scheme of DNA replication. Breakage of hydrogen bonds and unwinding precedes the synthesis of the two new strands, each of which forms a double helix with one of the "old" DNA strands. The two new DNA molecules are joined at the common starting point during the process of replication (from Cairns, 1963a; modified from Delbrück and Stent, 1957). (Reproduced with permission of Academic Press)

the replication point was found to be more denatured than the bulk DNA; apparently the newly replicated DNA is associated with partially denatured regions and gradually becomes more completely hydrogen bonded.

The entire genome of *E. coli* exists in the form of a single circular molecule which, when extended, is almost 1 mm long (Fig. 85). During

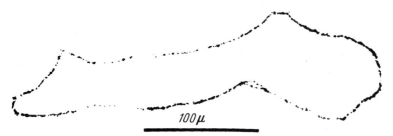

Fig. 85. Autoradiograph of a DNA molecule from *E. coli* labeled with H[3]-thymidine which illustrates the circular structure of DNA (from Cairns, 1963b).

rapid cell division accompanying exponential growth, the resulting rate of new DNA formation is 33 μ (chain length)/min.   The replication mechanism which Cairns (1963a) deduced from the circular form of *E. coli* DNA is shown diagrammatically in Fig. 86. Under normal growth conditions, a new round of replication cannot begin until completion of the previous one.   The velocity of DNA production seems to have the same order of magnitude in all biological materials (Table 14).

The biosynthesis of single-stranded DNA by ϕX174 phage (cf. p. 260) has been investigated by Sinsheimer *et al.* (1962). Infection of the host cell (*E. coli*) was shown to be followed by formation of a two-stranded

Fig. 86. Two phases in the duplication of a circular chromosome (from Cairns, 1963a).

TABLE 14. Duration and Replication Rate of DNA from T2 Phage, *E. coli* and Human Chromosomes. Numbers in Parentheses Represent Calculated Values (after Cairns, 1963b)

	Total DNA Content (chain length in μ)	Duration of DNA Replication at 37°C (min)	Rate of DNA Replication (μ/min)
T2 phage	50	1–2	(25–50)
E. coli	1,000	30	33
Human chromosomes	(20,000)	~400	(~50)

"replicative" form of DNA (RF), which then made possible the synthesis of normal phage DNA (see p. 267); this RF DNA of DNA was also found to be infectious.

DNA SYNTHESIS AND CELL DIVISION

The time course of DNA replication within the cell cycle has been studied mainly with histochemical and autoradiographic methods. Early results obtained with plant material showed that DNA synthesis occurs only within a relatively short period during interphase. There is a period following mitosis without DNA synthesis (G_1); after DNA replication (S), there is another period during which no DNA is synthesized (G_2; see Fig. 87). The duration of these periods is fairly constant for a certain type of cell. In cell cultures, G_1 and G_2 may be extended by chemical

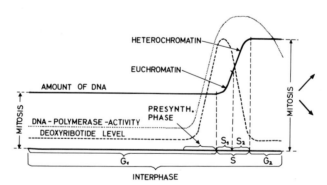

Fig. 87. Illustration of the time course of DNA synthesis (S phase) preceded by synthesis of DNA polymerase and DNA precursors within the interphase. During the S period, first the DNA of euchromatin is replicated (S_1), followed by the DNA of heterochromatin (S_2).

or physical treatment which then causes a synchronization of the cell population. In mammalian tissues, DNA doubling is observed during the last part of interphase, shortly before mitosis; in plant cells, on the other hand, DNA synthesis has sometimes been detected during the first half of interphase (e.g., Harbers, 1958). In the synchronously dividing plasmodia of the slime mold *Physarum polycephalum*, DNA replication begins immediately after telophase (without a G_1 period; Nygaard *et al.*, 1960). The prerequisites for DNA synthesis are an adequate supply of the necessary precursors and a sufficient amount of DNA polymerase. DNA replication is preceded by a substantial rise in the concentration of deoxyribonucleotides (Foster and Stern, 1958) and a severalfold increase in DNA polymerase activity. The increase in deoxyribonucleotide concentration is due to a rapid production of the necessary enzymes (especially thymidine kinase; cf. p. 131), this, in turn, precedes the biosynthesis of a specific RNA which apparently contains the information for these enzymes. The initial stages of liver regeneration in partially hepatectomized rats are accompanied by a brief period of almost synchronous cell division during which the increased DNA polymerase level is maintained beyond completion of DNA replication. The doubling of histones apparently takes place simultaneously with DNA replication. In mammalian cells (Bloch, 1966), as well as in *Physarum polycephalum* (Mohberg and Rusch, 1965), the ratio of histones to DNA remains constant during the S period. When DNA synthesis is inhibited by 5-fluorodeoxyuridine (Flamm and Birnstiel, 1964) or by ionizing radiation (Lehnert and Okada, 1964), histone synthesis may continue and thus cause an increase of the histone:DNA ratio.

Chromosomal DNA is not replicated sequentially along the entire chromosome, starting at a specific point and then proceeding uniformly along the two arms; instead, replication occurs at numerous points on each chromosome at a given time. Furthermore, the rate of DNA synthesis varies at different times in specific parts of some chromosomes (Stubblefield and Mueller, 1962; Taylor, 1963). The results of various autoradiographic experiments indicated that certain DNA molecules may by synthesized at a specific time during the S period. Biochemical studies on *Physarum polycephalum*, an organism with naturally synchronous mitosis, showed this indeed to be the case (Braun *et al.*, 1965). More recent observations have supplied some evidence that during the S period, the DNA of euchromatin is replicated first (S_1), then, DNA synthesis of heterochromatin occurs (S_2) (Baer, 1965; Marshall, 1966; Harbers and Spaar, 1968; Fig. 87). Thus, incorporation of 5-bromodeoxyuridine into the early-replicating DNA ("active" euchromatin) is much more harmful for the cell than incorporation into late-replicating DNA ("inactive" heterochromatin) (Müller and Kajiwara, 1966). The two X chromosomes in the somatic cells of females are not doubled simultaneously (Morishima *et al.*, 1962); the last-formed X chromosome remains detectable in interphase as a heterochromatin called sex chromatin or Barr's body (Barr and Bertram, 1949). This now appears to be a special example of late-replicating heterochromatic DNA.

Thus far nothing is really known about the control of differentiation, i.e., the mechanism which insures that from a certain kind of somatic cell, identical daughter cells always arise with exactly the same pattern of genetic information in euchromatin and heterochromatin. Apparently this control occurs via the time sequence of DNA replication. The duration of the S period was found to be identical in three different strains of *Tetrahymena pyriformis*, although they varied with respect to the amount of DNA per nucleus (Cameron and Stone, 1964). Thus it seems that DNA synthesis is regulated in a precise temporal manner independent of the amount of DNA in the nucleus and perhaps also of the ploidy condition of the nucleus. In the cytoplasm of *Tetrahymena pyriformis*, DNA synthesis (of mitochondria) appears to occur uniformly during the whole interphase, with a temporary increase during the S period (Cameron, 1966).

In the G_2 period (see Fig. 87), an exchange of genetic material may occur. DNA polymerase and certain deoxyribonucleoside triphosphates seem to be required for restoration of normal chromosome structure. The ability of chromosome fragments of the root cells of *Vicia faba* to recombine has been found to be greatly reduced as a result of treatment with 5-fluorodeoxyuridine just before mitosis (Taylor *et al.*, 1962); this effect can be reversed by adding thymidine (cf. p. 314). There are indications that exchange between DNA subunits and genetic recombination take place only in the diffuse euchromatin and not in heterochromatin (Roberts, 1965).

The scheme shown in Fig. 87 is not universally valid. In synchronous cultures of *E. coli* B, Abbo and Pardee (1960) observed a continuous DNA synthesis over the entire interphase, as evidenced by the uniform increase in the amount of DNA (Fig. 88) and the incorporation of labeled thymidine. This behavior explains the variation of DNA content per cell in these organisms compared with the relatively constant ratio in multicellular systems.

DNA synthesis and cell division have frequently been considered to be directly related events. The often observed higher radiation sensitivity of the mitotic process in comparison with DNA synthesis suggests two independent processes which are normally coordinated. DNA synthesis probably controls the time of mitosis either directly or indirectly (Sachsenmaier, 1962). Experimental results by Bucher and Mazia (1960) provided

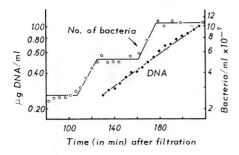

Fig. 88. Continuous DNA synthesis in growing cultures of *E. coli* synchronized by filtration (from Abbo and Pardee, 1960).

further support of this hypothesis. A similar interpretation may apply to the observation that DNA replication maintains synchrony longer than mitosis in artificially synchronized cells (Lark, 1962).

For cell division to take place, an intact nucleolus appears to be required; damage to the RNA-rich nucleolus blocks mitosis (Gaulden and Perry, 1958). In the slime mold *Physarum polycephalum*, it was observed that initiation of cell division depends on a specific RNA fraction which is synthesized 6 hours before mitosis. Apparently the function of this RNA is to control the synthesis of specific enzymes which act as "starter" (Sachsenmaier, 1962). When fresh cells from an organism are placed in a tissue culture, protein synthesis and RNA formation begin immediately, while DNA synthesis and cell division occur with delay. Finally, blocking of RNA or protein metabolism also inhibits DNA synthesis (cf. p. 213).

CONTROL OF DNA SYNTHESIS AND GROWTH

Early attempts to understand the control of growth in multicellular organisms focused on the substances which were believed to stimulate growth and increase the rate of cell division. Later observations indicated that growth and differentiation are controlled by the level of normally present inhibitors, which result in a form of negative feedback (Rose, 1958; Stich and Florian, 1958). Thus, the normal inhibition of DNA synthesis and mitosis is absent in regenerating rat liver following partial hepatectomy. The inhibition is restored as soon as normal liver size is regained. The organ-specific inhibitor, whose chemical nature is as yet unknown, is found in liver and blood serum, and can delay liver regeneration (Stich and Florian, 1958). In carcinogenesis, normal growth control appears to be lost (cf. p. 281).

The regeneration of rat liver after partial hepatectomy can be inhibited by the injection of liver homogenates. Fractionation of the homogenates revealed that the histone components were mainly responsible for the inhibitory effects (Sluyser *et al.*, 1965); the various histones showed the same activity.

Until now, no details about the mechanism of growth control have been known. There seems to be a periodic induction and repression of specific enzymes during the cell cycle. Thus, the enzymes necessary for preparing mitosis are induced in the G_2 period when DNA replication is finished. The total cell cycle appears to be controlled by a sequential activation and inactivation of specific groups of genes; generally, only when one step in the cell cycle has been completed can the signal for the next be given (Sachsenmaier, 1966). In analogy to the control of protein synthesis, Jacob (1966) has postulated that in microorganisms, each genetic element constitutes a unit of replication, or *replicon*, which determines a circuit controlling its own replication in coordination with cell division. The control of DNA replication appears to be located in the bacterial membrane which ostensibly is the site of DNA synthesis (Ganesan and Lederberg, 1965). It is assumed that all cellular replicons are attached to the bacterial membrane. Thus the necessary segregation of DNA after replication and its distribution in the two daughter cells should be due to synthesis of the membrane between the points of attachment of the newly formed DNA molecules.

Analysis of the different events during the cell cycle is facilitated by the use of inhibitors with specific effects on certain metabolic processes (cf. p. 294); in addition, synchronized systems are necessary. Synchrony can be obtained by various kinds of treatment in cell cultures; in the multinuclear plasmodia of the slime mold *Physarum polycephalum*, DNA replication and mitosis are spontaneously synchronized (Howard, 1932; Güttes *et al.*, 1961).

Thus far, it has not been possible to study directly the primary factors in growth control. Instead, only secondary processes can be examined, such as changes in the activity of certain enzymes in relation to cell cycle and growth rate, as well as the effects of substrates and tissue extracts.

The formation of DNA polymerase at the end of the G_1 period is accompanied by a rise in the activity of the specific kinases which convert the deoxyribonucleosides to the triphosphate level prior to DNA synthesis (Weisman *et al.*, 1960). Kinases which catalyze the phosphorylation of deoxyribonucleoside monophosphates occur in high concentration in proliferating tissues such as Ehrlich ascites tumors, regenerating rat liver and the bone marrow of rabbits. In contrast, these enzymes are present only in small amounts in "resting" tissues like liver and kidney. Addition of liver or kidney extracts to enzyme systems from proliferating tissues (e.g., regenerating liver or tumor cells) leads to a decrease in the rate of phosphorylation of thymidine monophosphate and of DNA synthesis (Gray *et al.*, 1960). The phosphorylation of the remaining deoxyribonucleotides, dAMP, dCMP and dGMP, seems to be unaffected under these conditions. The activity of thymidylate kinase appears to be determined by the tissue concentration of dTMP, for, after *in vivo* administration of thymidine, which is rapidly phosphorylated intracellularly, there is a rise in enzyme activity in a variety of organs of the rat which can be inhibited by the addition of 5-fluoropyrimidines (Hiatt and Bojarski, 1960; cf. p. 312).

In some cells (e.g., Ehrlich ascites carcinoma), deoxyadenosine inhibits DNA formation, an effect which can be reversed by deoxyguanosine (G. F. Maley and Maley, 1961; Overgaard-Hansen and Klenow, 1961). Similarly, thymidine, in high concentration, inhibits DNA synthesis; this inhibition is overcome by the addition of low concentrations of deoxycytidine (Morris and Fischer, 1963). Apparently, the phosphorylated thymidine inhibits the formation of deoxycytidine triphosphate, whose deficiency leads to a blocking of DNA replication (Morris *et al.*, 1963). The inhibitory effect of deoxyadenosine seems to be based on a similar interaction of the two purine deoxyribonucleotides. In cultures of *A. faecalis*, the addition of all four deoxyribonucleosides leads to the loss of a previously induced synchronization of the cell cycle (Lark, 1962). Brewer and Rusch (1965) succeeded in isolating nuclei from *Physarum polycephalum* which retained their physiological control of DNA replication. The isolated nuclei were unable to initiate DNA synthesis, but they could carry on active synthesis after initiation. In *Stentor* and in *Amoeba proteus*, DNA replication is apparently controlled by a cytoplasmic initiator (de Terra, 1967; Prescott and Goldstein, 1967).

FUNCTIONAL ACTIVITY OF DNA

In the giant chromosomes of various diptera, local swellings termed "puffs" are found at specific loci (Beermann, 1952, 1956). This puffing of the chromosome structure occasionally goes so far that loop-like single strands protrude from the chromosomal arrangement to form a ring (*Balbiani's rings*). The distribution of puffs is equal in all cells of a given tissue; on the other hand, chromosomes of cells from different tissues show different patterns. From this it has been concluded that puffs represent regions of high gene activity. By using labeled precursors, attempts have been made to obtain insight into the metabolic behavior of puffs. Although H^3-thymidine was not incorporated into the chromosomes uniformly, there was no marked increase in the labeling in the puffs compared to other parts of the chromosome (Ficq, Pavan and

Brachet, 1958; Rudkin and Woods, 1959); therefore, puff formation does not appear to have resulted from a local increase in DNA synthesis. On the other hand, the puffs showed considerably increased incorporation of H^3-cytidine and H^3-uridine, as well as of C^{14}-adenine, in chromosomal RNA (Ficq et al., 1958; Pelling, 1959; Rudkin and Woods, 1959). These observations support the current view that DNA functions by transcribing its information via RNA (cf. p. 216). Inhibition of the biosynthesis of DNA-dependent RNA by drugs such as actinomycin (cf. p. 294) causes the puffs in the giant chromosomes to undergo involution, which suggests that puff formation at an active chromosomal locus is not only related to its capacity to synthesize RNA but also dependent on it (Izawa et al., 1963). The role of puffs as functional phenomena of gene activity at given chromosomal loci seems well established now. The puffing patterns are specific for a certain tissue or stage of development, making available only those parts of the genetic information needed for the functions of the corresponding somatic cell. The mechanism of puff formation is not yet known.

Puffing can be induced experimentally by the molting hormone "ecdysone" (Fig. 89). In *Chironomus* larvae, a minimal dose of this hormone leads within 30 minutes to the formation of a puff at a specific site of chromosome 1, and 30 minutes later to the formation of another puff in chromo-

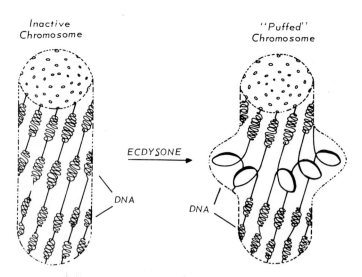

Fig. 89. Ecdysone-induced puffing in giant chromosomes from the salivary glands of *Chironomus tentans*. The normal compact structure (left) is expanded, making the DNA available as template for RNA synthesis (from Karlson, 1963).

some 4 (Clever, 1961). This is followed—in normal development as well as experimentally—by a chain of secondary puffing reactions. Apparently the hormone acts as a direct inducer of activity at the first two loci, and the chain of secondary events is then triggered.

While polytene giant chromosomes can be seen in the interphase, chromosomes can usually be observed only during mitosis. In the interphase, the chromosomal material appears in the cell nuclei as chromatin. Analogous to the puffs in giant chromosomes, RNA synthesis seems to be limited to (or takes place mainly at) the DNA of extended euchromatin (cf. p. 88), while the more densely packed heterochromatin contains functionally inactive DNA (Frenster *et al.*, 1963). There is nothing known as yet about the way in which the formation of euchromatin and heterochromatin is controlled (cf. p. 183) and the forces which hold heterochromatin together in clumps.

Ord and Stocken (1966) found a greater proportion of disulfide in dense chromatin than in euchromatin. Littau *et al.* (1965) observed that removal of the lysine-rich histones from isolated calf thymus nuclei loosened the structure of heterochromatin; it was concluded that the dense structure of heterochromatin might be due to cross-linking of this histone fraction. On the other hand, the ratio of (total) histones to DNA is apparently identical in euchromatin and heterochromatin (cf. p. 90). The two kinds of chromatin differ in their melting behavior, heterochromatin having a higher T_m value (Frenster, 1965).

The nonavailibility of genetic information in heterochromatin (which appears to be permanent) must be distinguished from repression and derepression as postulated by Jacob and Monod (see p. 248), the action of which is supposed to be limited to the DNA of euchromatin. Therefore, to avoid confusion, the term "repressed chromatin" should not be used in referring to heterochromatin.

Cytologically and functionally, different kinds of euchromatin and heterochromatin can be distinguished (cf. Brown, 1966). The chromatin in the region of the nucleolus is apparently involved in the synthesis of rRNA (Ritossa and Spiegelman, 1965), or perhaps even more generally in some of the fundamental metabolic processes which take place in all cells. Carcinogenesis may be accompanied by an increase in heterochromatin, i.e., a decrease in functional, available DNA (see p. 284).

Like ecdysone, various hormones can cause increased RNA synthesis in mammalian tissues. It is not known whether hormone action results in the conversion of heterochromatin to euchromatin (in analogy to the puffing phenomenon). It now seems that certain hormones produce an enzyme induction via DNA of euchromatin and mRNA formation (cf. Sekeris, 1966); however, hormone action does not always involve gene activation.

The administration of hydrocortisone to adrenalectomized rats caused an increase in the template activity of liver chromatin for enzymatic RNA synthesis (Dahmus and Bonner, 1965). This effect was interpreted as an increased availability of the genetic material for transcription. Because removal of histones from isolated nucleohistones leads to a much higher priming efficiency for the DNA-dependent RNA polymerase (cf. p. 204), histones are often considered as regulators of RNA synthesis.

According to one theory, only DNA which is not associated with histones should be active. Observations by Bonner and Huang (1963), indicated that pea embryo chromatin contained DNA some of which was complexed with histones. However, in mammalian tissues, histones were found in the (active) euchromatin as well as in the heterochromatin. If a regulatory role is postulated for histones, a certain turnover might be expected. In long-term experiments (cf. p. 166) on relative turnover rates, it was found that most of the histones and DNA turn over at approximately the same rates in various tissues (Byvoet, 1966); this was regarded as evidence for the metabolic integrity of the DNA-protein complex. Pogo *et al.* (1966), who came to similar conclusions, observed that acetylation of preexisting histones (mainly the arginine-rich fraction), but not histone synthesis, increased the capacity for nuclear RNA formation. From these experimental results it appears that histone acetylation is a process preceding RNA synthesis, perhaps causing a change in the conformation of the DNA primer and thus increasing its template activity.

ENZYMES DEGRADING DEOXYRIBONUCLEIC ACID

Enzymes able to liquefy gels made of DNA were found in various bacteria around the beginning of this century, but a detailed study of this special group of phosphodiesterases was not possible until pure enzyme preparations became available. Numerous deoxyribonucleases (DNases) have now been obtained in purified form. Most of these DNA-specific enzymes act as endonucleases, starting their attack on the substrate in the middle of the DNA chain rather than at one of the ends. According to the products formed as a result of their action, DNases have been divided into two categories: those which produce 5'-monoesters and those which produce 3'-monoesters. Another classification of DNases is concerned primarily with the optimum conditions for their reaction and is related to historical developments in this field: In 1948, a pancreatic DNase was isolated, which was activated by Mg^{++} ions and showed a pH optimum around 7. This enzyme was termed "DNase I" in order to distinguish it from "DNase II" found a few years later in spleen and thymus, which had an acid pH optimum and was inhibited by magnesium ions. In general, it can be said that most DNase I type enzymes form 5′-mono-esters, whereas DNase II preparations lead to the formation of 3′-monoesters. These specificities, however, are not always clear-cut, and there are even some highly purified DNases which will also act on RNA.

DNase I, first crystallized from pancreatic tissue (Kunitz, 1948), has been found in various animal tissues. This protein has a molecular weight of 60,000, shows a broad pH optimum around pH 7, and is activated by various divalent cations whose optimum concentration is dependent on the concentration of substrate (Kunitz, 1950). Mononucleotides and oligonucleotides of variable size are formed in the reaction; the enzymatic attack shows no preference for certain base combinations. Under normal conditions, the cleavage of DNA is slowed down by the formation of low molecular weight products, a phenomenon designated as "autoretardation" (Laskowski, 1966). However, with a large amount of enzyme, and manganese ions substituted for magnesium, the degradation of DNA yields a mixture of 2% mononucleotides, 63% dinucleotides and 25% trinucleotides (Vanecko and Laskowski, 1961). Changes in the structure of DNA make the derivative a poorer substrate for the enzyme (Tamm *et al.*, 1952), and free DNA is more susceptible to the action of DNase I than is the nucleohistone from which it was prepared (Klamerth, 1957). Various animal organs contain a protein which acts as a specific inhibitor for DNase I (Cooper *et al.*, 1950) and which must be destroyed before the DNase activity of the tissue is revealed. Within the cell, the DNase I activity seems to be bound to subcellular structures.

The same holds true for *DNase II*, first isolated from spleen and thymus, but later found in many other tissues. This enzyme, which has a pH optimum between 4.5 and 5.5, has now been obtained free of contaminating RNase (Maver *et al.*, 1959). DNase II is inhibited by magnesium ions and various polyvalent anions such as sulfate; monovalent cations in a concentration of 0.2–0.3M are effective activators. The action of this enzyme leads to the formation of 3′-oligonucleotides. The enzymatic reaction seems to proceed in a biphasic sequence: The first fast phase which results in the cleavage of about 10% of the internucleotide linkages is followed by a slower phase which cleaves 25–30% of the remaining bonds (Koerner and Sinsheimer, 1957). The existence of a specific DNase II inhibitor of unknown structure has been postulated (Kowlessar *et al.*, 1957).

Microbial DNases have been found in many microorganisms, and their occurrence in all organisms is quite probable; any negative results obtained are most likely due to the use of non-optimal assay conditions. In *Staphylococcus aureus*, it was found that the most pathogenic

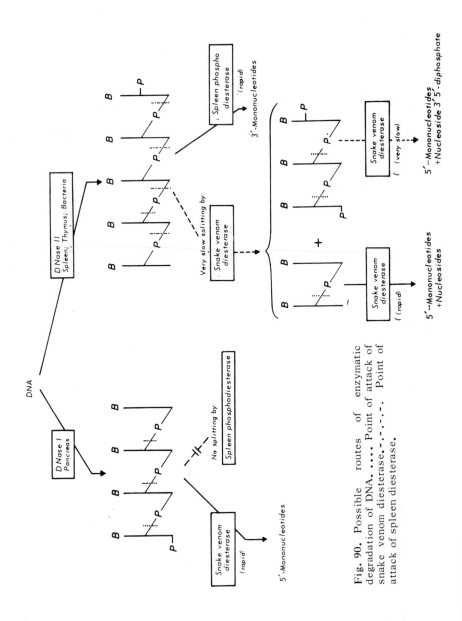

Fig. 90. Possible routes of enzymatic degradation of DNA. ..... Point of attack of snake venom diesterase. –·–·–·–. Point of attack of spleen diesterase.

strains were the best producers of DNase (Weckman and Catlin, 1957). A DNase I type enzyme has been purified from streptococci and named streptodornase; in contrast to the animal DNase I, its natural inhibitor is a specific RNA rather than a protein (Bernheimer, 1953). In other streptococci, three different kinds of DNase have been found—partly extracellular, partly intracellular, and with a different substrate specificity (Wannamaker, 1958). The DNases of *E. coli* have been studied in great detail by Lehman and co-workers (Shortman and Lehman, 1964, and previous publications). An interesting observation has been made with DNase from *A. faecalis*, which rapidly attacks thymus DNA but acts very slowly on DNA of the same organism (Catlin and Cunningham, 1958). Other DNases have been purified from yeast (Zamenhof and Chargaff, 1949) and from different plant tissues (e.g., Shuster, 1957). For additional information, excellent review articles are available (e.g., Laskowski, 1961, 1966; Siebert, 1966b).

Recently, various *DNA exonucleases* have been obtained from microorganisms and snake venom, some of which are free of endonuclease activity (Lehman and Richardson, 1964; Okazaki *et al.*, 1966). Other exonucleases degrade DNA as well as RNA, yielding either 3′- or 5′-mononucleotides (Williams *et al.*, 1961; Korn and Weissbach, 1964; Kerr *et al.*, 1965).

Figure 90 shows how the reaction products obtained after the action of animal DNases can be degraded further. The specific modes of action of these enzymes have been very helpful in elucidating DNA structures (cf. p. 35).

Metabolism of Ribonucleic Acids

CHAPTER VII

INVESTIGATIONS IN VIVO AND WITH WHOLE CELLS; NUCLEUS-CYTOPLASM RELATIONSHIPS

Early tracer studies showed that the incorporation of radioactive phosphate is far greater in RNA than in DNA. This suggested that RNA has a higher turnover rate than DNA. However, these early experiments dealt only with the incorporation into total RNA. When it became possible to separate cell nuclei from the cytoplasm, it was observed that the most intensive uptake of labeled phosphate occurs in nuclear RNA. The specific activity of nuclear RNA soon reaches a high maximum, while that of the cytoplasmic RNA is still slowly increasing (Fig. 91). This supported the hypothesis that some or all of the cytoplasmic RNA is formed in the cell nucleus and then migrates into the cytoplasm.

In 1955, Goldstein and Plaut demonstrated directly the migration of RNA from the nucleus to the cytoplasm of *Amoeba proteus*. With the aid of a micromanipulator, nuclei which had incorporated P^{32} were implanted in nonradioactive enucleated amoeba. Initially, all of the label remained in the nuclei, but after 12 hours much of the radioactive phosphorus had moved to the cytoplasm where it was identified as an RNA constituent. However, when nonradioactive nuclei were implanted in radioactive enucleated amoeba, no incorporation of radioactive material in the nuclei was

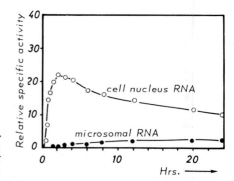

Fig. 91. Incorporation of P^{32}-orthophosphate into RNA from cell nuclei and microsomes as a function of time (from Barnum and Huseby, 1950).

observed. Similarly no incorporation of labeled precursors into the RNA of the cytoplasm was detected in enucleated cells (Prescott, 1959). Later, a migration of labeled RNA from the nuclei into the cytoplasm was demonstrated in mammalian cells (Goldstein and Micou, 1959). Goldstein (1958) observed the release of a nuclear protein into the cytoplasm in *Amoeba proteus*; in contrast to the behavior of RNA, this protein returned to the cell nucleus after a short time. At present a clear interpretation of this behavior (feedback control?) is not possible.

Recently, more detailed kinetic studies on the incorporation of P^{32} into different kinds of rat liver RNA have been made by Noll (1965). Figure 92 shows the relative specific activities of mRNA, rRNA, and acid-soluble P^{32} phosphate as a function of time. The activity of nuclear mRNA rises to a maximum 4 hours after the injection of P^{32}-labeled

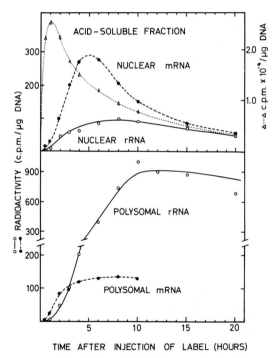

Fig. 92. Kinetics of mRNA and rRNA labeling in nucleus and cytoplasm from rat liver. The specific activities of the nuclear and polysomal fractions were normalized with respect to DNA, on the assumption that the mass ratio of total rRNA to total DNA is equal to 3 (from Noll, 1965). (Reproduced with permission of Williams & Wilkins Co.)

orthophosphate, then falls exponentially due to the decline of the acid-soluble precursor pool. In polysomes, labeled mRNA appears with an extremely short delay. Most of the cytoplasmic P^{32} activity is found in the rRNA fraction as a constituent of polysomal ribosomes. The kinetic findings displayed in Fig. 92 are consistent with the idea that rRNA must pass through several precursor stages within the nucleus before it is assembled into complete ribosome particles (see Fig. 52 and p. 98). The synthesis of rRNA apparently takes place in the chromatin associated with the nucleolus (Perry *et al.*, 1964), and the assembling of the ribosome particles occurs within the nucleolus (Allfrey and Mirsky, 1962). Labeled pseudouridine exhibits a high turnover rate in the nucleolus (Sirlin *et al.*, 1961). This is considered as an indication that tRNA is formed there. Chipchase and Birnstiel (1963) demonstrated a DNA-dependent biosynthesis of tRNA in isolated cell nuclei from pea seadlings. On the basis of kinetic studies, it may be assumed that the nucleoli form a pool for the chromosome-produced RNA, part of which can evidently function in the protein synthesis that occurs near the nucleolus (Schultze *et al.*, 1958); another part migrates into the cytoplasm, bypassing the nucleolus. This interpretation derives support from the observation that when the DNA template is blocked by actinomycin, the nucleoli fade and eventually disappear (Bierling, 1960). Figure 93 summarizes the processes of nuclear formation of mRNA and rRNA, and their subsequent transfer to the cytoplasm. It seems quite possible that

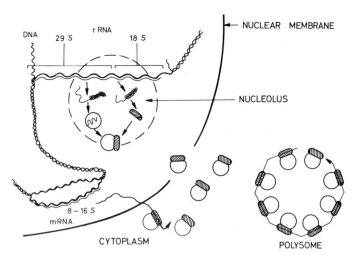

Fig. 93. Nuclear synthesis and transfer into the cytoplasm of mRNA and ribosomes (from Noll, 1965). (Reproduced with permission of Williams & Wilkins Co.)

polysomes are already formed in the nucleus or at the nuclear membrane and then migrate into the cytoplasm as a unit (Bach and Johnson, 1966).

For some time, there was considerable doubt as to whether the cytoplasm could synthesize RNA independently of the cell nucleus. In *in vitro* experiments with isolated cytoplasm from rat liver, incorporation of labeled precursors had been observed (Heidelberger *et al.*, 1956; Potter *et al.*, 1956), but this was later proved to be a terminal addition to RNA already present (cf. p. 208), without net synthesis. The finding that DNA is a constituent of mitochondria and plastids (cf. p. 91) and that mitochondria contain an RNA polymerase (Neubert *et al.*, 1965) later indicated that RNA synthesis does indeed take place in the cytoplasm, apparently at a lower rate than in the nucleus. In the green alga *Acetabularia*, a small increase (i.e., net synthesis) of RNA was observed after removal of the cell nucleus (Brachet, 1959a; Schweiger and Bremer, 1961).

The contradictory results obtained by various authors working with *Acetabularia* can partially be attributed to differences in techniques and in the metabolic state of the cells at the time of the experiment. Under normal conditions, the capacity for RNA synthesis in *Acetabularia* is evidently lost with the removal of the nucleus, but if the plants are kept in the dark for 10 days prior to enucleation, RNA formation may occur (Schweiger and Bremer, 1961). Apparently, the cytoplasmic RNA synthesis in *Acetabularia*, which takes place at the DNA-containing chloroplasts, is controlled by the nucleus. Under certain conditions, amoeba can also form RNA after enucleation (Plaut, 1958).

Since the rate of incorporation of labeled precursors into RNA is generally higher than that of DNA, it seemed reasonable to assume that RNA is not a permanent cell component but undergoes continuous syn-

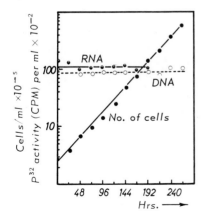

Fig. 94. Constant total activity of P^{32}-labeled RNA and DNA during growth of L-mouse cells in tissue culture (from Siminowitch and Graham, 1956).

thesis and breakdown; this view was supported by the observation that the total RNA content per cell may vary due to exogenous factors (Thomson *et al.*, 1953). By means of long-term experiments, it was shown that part of the RNA, like DNA (see p. 166), is very stable and under certain conditions (high growth rate) does not turn over once it has been synthesized (Fig. 94). This evidence was first obtained from microorganisms and later also from mammalian cells (Graham and Siminovitch, 1957; Scott and Taft, 1958). In *E. coli*, rRNA is preserved during growth; RNA labeled with C^{13} and N^{15} does not release any of the label to newly synthesized RNA (Davern and Meselson, 1960).

ENZYMATIC RNA SYNTHESIS

Study of the enzymatic synthesis of RNA has yet to yield a uniform picture of this process. RNA synthesis may be classified according to whether DNA or RNA is required as primer. Because it is often difficult to determine net synthesis of RNA, the requirement of all four bases as precursors (in the form of ribonucleoside triphosphates) has served as a criterion for biosynthesis. Otherwise, incorporation may be attributed to terminal addition to already existing RNA molecules. Futhermore, enzymatic synthesis has been observed of an RNA whose composition is a function only of the kind of precursor supplied. Whether such nucleotide-dependent processes do not in fact need a primer can be determined only when the enzymes involved have been purified. Finally, structural alterations in the existing RNA molecule can occur through enzymatic reactions at its bases.

The first enzymatic production of RNA-like polynucleotides was reported in 1955 by Grunberg-Manago, Ortiz and Ochoa. An enzyme called polynucleotide phosphorylase* which requires ribonucleoside diphosphates as substrates, was isolated from *Azotobacter vinelandii*. The reaction may be described by the following general equation:

$$nX—R—PP \rightleftharpoons (X—R—P)_n + nP$$

where X represents organic bases, R is ribose, P is orthophosphate and PP is pyrophosphate. The bases can be adenine, guanine, hypoxanthine, uracil or cytosine; later it was shown that thymine riboside diphosphate can also be used as a substrate. The amounts of orthophosphate formed are equimolar to the amounts of ribonucleoside diphosphate utilized.

---

*EC   2.7.7.8   Nucleosidediphosphate:polynucleotide nucleotidyltransferase.

The primary structure of enzymatically synthesized polynucleotide was found to be like that of normal RNA (Heppel *et al.*, 1957a and b). Contrary to the behavior of DNA polymerase, synthesis of polyA, polyC, polyU, polyG and also poly(rib)T was possible even if only some and not all of the naturally occurring bases were supplied in the form of riboside diphosphates. The base composition of this synthetic RNA is therefore a function of the type of precursors supplied. At first it seemed as if the polynucleotide phosphorylase required no primer at all. After improved purification of the enzyme, however, it was found that biosynthesis of polynucleotides began slowly and only after a fairly long lag period. Addition of oligoribonucleotides immediately initiated synthetic reaction (Mil and Ochoa, 1957) with the required primer concentration decreasing as the chain length increased. In relatively high concentration, even dinucleotides were effective as primers. Polynucleotide phosphorylase has been detected in only a few microorganisms outside of *Azotobacter vinelandii* (Ochoa and Heppel, 1957). In *Pseudomonas aeruginosa* it is associated with the ribosome fraction. 6-Mercaptopurine, in the form of its riboside diphosphate, causes competitive inhibition of the polynucleotide phosphorylase and cannot be used as a substrate (cf. p. 319). Heppel (1963) observed that some homopolymers specifically inhibit the polynucleotide phosphorylase from utilizing certain ribonucleoside diphosphates. Thus polyA prevents the polymerization of UDP and polyC inhibits the utilization of IDP. The polynucleotide phosphorylase from *Agrobacterium tumefaciens* cannot utilize GDP for RNA production because this ribonucleotide competitively inhibits the incorporation of ADP. A similar inhibition has been observed with the enzyme from *Micrococcus lysodeikticus*.

Based on the observations described above, it was first thought that RNA biosynthesis might proceed via the ribonucleoside diphosphate stage. However, the presence of the polynucleotide phosphorylase could not be demonstrated in multicellular organisms. Instead it was found that normal RNA synthesis requires ribonucleoside triphosphates as precursors. The requirement of all four bases as riboside triphosphates was considered an indication of biosynthesis (in contrast to merely terminal incorporation). Finally, cell-free extracts which catalyzed a rapid incorporation into RNA were successfully obtained from mammalian cells, microorganisms and plants. In addition to the presence of all four bases as riboside triphosphates, divalent cations (Mg^{++} or Mn^{++}; Mg is usually more effective than Mn) and a certain amount of DNA are required. The key role of DNA in RNA formation had long been conjectured. The first support for such a proposal came from virus research.

After infection with DNA-containing phages (cf. p. 257), a new and specific RNA appeared in the host cells and apparently formed the template for the production of enzymes and viral proteins induced by the viruses (e.g., Volkin, 1960). The observations of Furth, Hurwitz and Goldmann (1961a and b) yielded the first insight into the relationship between DNA and DNA-dependent RNA synthesis. Furth and co-workers isolated a DNA-dependent RNA polymerase* from *E. coli* which catalyzes the synthesis of RNA from the four ribonucleoside triphosphates in the presence of DNA. RNA formation was inhibited by deoxyribonuclease, but was not much affected by ribonuclease. The molar base ratios of the newly formed RNA coincided to a large degree with the analogous ratios in the primer DNA (with uracil in RNA corresponding to thymine in DNA). Similar results were obtained by Hartmann and Coy (1961) and (in experiments with a RNA polymerase from *M. lysodeikticus*) by Weiss and Nakamoto (1961).

These initial observations suggested that primer DNA acts as a template for RNA formation and thus determines the base sequence of RNA. Direct proof came from experiments in which synthetic polydeoxyribonucleotides with only one or two bases were added as primers instead of complete DNA with all four bases. With polydT as primer, polyA resulted, with only ATP being utilized. The use of polydAT as primer (cf. p. 57) led to the utilization of approximately equimolar amounts of ATP and UTP, whereas CTP and GTP were not polymerized. P^{32} labeling showed that adenine and uracil occur in alternating sequence in the newly formed RNA, which corresponds to the sequence of adenine and thymine in the primer (cf. p. 176). PolydGC primed only cytidine and guanosine triphosphates for RNA synthesis. These findings demonstrate that DNA serves as a template for RNA production in a way similar to DNA replication (Fig. 95). In DNA doubling, the template and its product are identical and combine to a new double helix, while the RNA synthesized at the DNA is finally

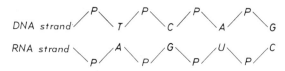

Fig. 95. Template effect of DNA in the biosynthesis of DNA-dependent RNA. Abbreviations: A = adenine, C = cytosine, U = uracil, G = guanine, P = orthophosphate, T = thymine.

*EC 2.7.7.6 Nucleosidetriphosphate:RNA nucleotidyltransferase.

released from its template. The enzymatic synthesis of RNA may be described as follows:

$$
\begin{matrix} n\text{ATP} \\ n\text{UTP} \\ n\text{CTP} \\ n\text{GTP} \end{matrix} + \text{DNA} \xrightarrow{\text{Mg}^{++}\text{ or Mn}^{++}} \text{DNA} + \begin{bmatrix} \text{AMP} \\ \text{UMP} \\ \text{CMP} \\ \text{GMP} \end{bmatrix}_n + 4n\text{PP}
$$

This simplified equation ignores the fact that the individual RNA bases generally are not present in equimolar amounts. If double-stranded DNA is supplied as primer, the molar base ratios A:U and G:C of the enzymatically synthesized RNA are often equal to 1—analogous to the ratios in the primer DNA. The two complementary RNA strands which have been synthesized can then form a double helix *in vitro*.

In the living cell, only one polynucleotide chain of the DNA double helix serves as the template for RNA synthesis. Only this "sense" (codogenic) strand of the DNA contains the correct genetic information, while the complementary strand merely stabilizes the DNA and permits its replication. Under *in vitro* conditions, the RNA polymerase isolated from *E. coli* generally copies both strands of the DNA. However, if the primer DNA has a sufficiently high molecular weight, only one strand is used as a template (Zillig, personal communication). It is not known as yet what structural features in the DNA cause RNA synthesis to proceed exclusively in one direction and thus copy the sequence of only one of the two strands.

DNAs of different origins may vary with respect to their priming efficiency for the RNA polymerase from *E. coli*. Heat denaturation of DNA results in a considerable loss of the priming activity. The single-stranded DNA of the ϕX174 phage is also effective as a template; at low concentrations, it is even more effective than double-stranded DNA.

RNA polymerase from *E. coli* forms a poorly dissociable complex with native and denatured DNA. Enzyme-bound DNA is not available as a template for DNA replication or as a substrate for exonuclease action. This is interpreted as an indication of strong binding of the RNA polymerase at or near the 3′-hydroxyl terminus (Berg *et al.*, 1965). In contrast, DNA polymerase forms a readily dissociable complex (cf. p. 177). Under usual *in vitro* conditions, RNA synthesis proceeds at maximal rate for only a short time, finally resulting in the accumulation of a ternary complex of DNA, RNA polymerase and RNA. Throughout the reaction, the number of nascent RNA molecules is constant; apparently the synthesis of all RNA molecules starts during the first minutes of incubation, and during the rest of the reaction the molecules merely increase in size (Bremer and Konrad, 1964). Recently, Fuchs, Millette *et al.* (1967) observed that with higher concentrations of Mg^{++} and NH_4Cl, a release of newly formed RNA from the primer DNA and thus a repeated use of the template are possible. Fuchs *et al.* (1964) found a molecular weight of 6×10^5 for the purified RNA polymerase from *E. coli*. Enzyme-DNA complexes could be demonstrated in electron micrographs (Fig. 96). Electron microscopic examination of the isolated enzyme

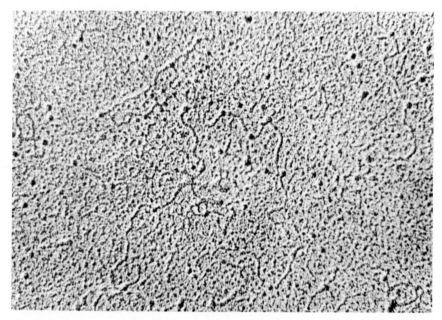

Fig. 96. RNA polymerase-DNA complexes prepared for electron microscopy by the technique of Kleinschmidt *et al.* (1962; from Fuchs *et al.*, 1964). (Reproduced with permission of Academic Press, Inc.)

particles showed a hexagonal arrangement of six subunits surrounding an empty core (diameter ~125 Å; Fig. 97). The remainder of the particle appears quadrangular with the core seen from the side. A tentative model of the enzyme molecule in which six subunits form a short hollow cylinder is shown in Fig. 98. From pulse-labeling experiments, it appears that synthesis of the RNA strands proceeds from the 5′- toward the 3′-end (Bremer, Konrad *et al.*, 1965), i.e., in the same direction as the process of translation (cf. p. 221).

Short-chain oligodeoxyribonucleotides and also oligoribonucleotides can act as templates for RNA polymerase and stimulate the incorporation of complementary nucleotides. The presence of two complementary oligodeoxyribonucleotides together results in greater nucleotide incorpora-

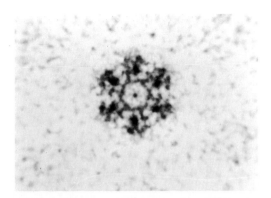

Fig. 97. Electron micrograph of an RNA polymerase particle showing the hexagonal arrangement of the six subunits. Magnification
800,000X. In order to demonstrate the hexagonal symmetry more distinctly, the original
picture was turned rotated through ‿60° six
times and exposed after each rotation. The
sample was prepared by negative staining with
phosphotungstic acid as described by Horne
and Brenner (1958; from Zillig *et al.*, 1966).
(Reproduced with permission of Harper & Row)

tion than the sum of the incorporations when the single-stranded primers
are present individually (Clark and Jaouni, 1965). Synthetic and natural
copolymers of the DNA type were found to cause the formation of ribonucleotide copolymers. However, the chain lengths and composition of
the synthesized RNA did not exactly correspond to the template. For
example, polydT with a chain length between 7 and 14 deoxyribonucleotides primed the synthesis of polyA, consisting of 50–100 nucleotides
(Falaschi *et al.*, 1963), apparently due to a kind of slipping mechanism.
Enzymatically synthesized oligoribonucleotides of known base sequence
are a helpful tool in investigating the early steps of protein synthesis
and in analyzing the genetic code (cf. p. 231).

Uncharged tRNA is a potent inhibitor of *E. coli* RNA polymerase
in vitro (Tissières *et al.*, 1963); aminoacyl-tRNA seemed to provide a
chemical mechanism for the regulation of RNA synthesis (Stent and
Brenner, 1961). However, the difference in polymerase inhibition between amino acid-free and aminoacyl-tRNA was found to be too small to
support this model of regulation of bacterial RNA synthesis (Bremer,
Yegian and Konrad, 1965).

In contrast to *in vitro* conditions with DNA primer, the template activity in living cells is restricted by certain control mechanisms (cf. p. 247).
Nucleohistones, isolated from somatic cells, generally have a much

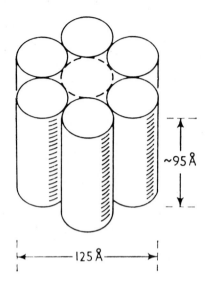

Fig. 98. Tentative model of the RNA polymerase molecule (from Fuchs *et al.*, 1964). (Reproduced with permission of Academic Press, Inc.)

lower priming efficiency than pure DNA. Removal of the histones causes a severalfold increase in the DNA-dependent RNA formation at nucleohistones (Huang and Bonner, 1962) or in isolated nuclei (Allfrey *et al.*, 1963), while saturation with histones results in inhibition or blocking of RNA synthesis (e.g., Huang and Bonner, 1962; Hindley, 1963). Observations of this kind are often considered as evidence that histones can modify the template function of DNA. Another interpretation is that firm aggregation of isolated nucleohistones is primarily responsible for their low priming activity (Sonnenberg and Zubay, 1965). Nucleohistones from Ehrlich ascites tumor cells were found to have equal or even better priming efficiency for the *E. coli* RNA polymerase than pure DNA of the same origin. However, cautious removal of most of the histones led to a material with a still higher priming activity (Harbers and Vogt, 1966). Thus it seems that histones may indeed suppress DNA-dependent RNA synthesis. On the other hand, pure DNA is apparently not the most effective primer; the best primer seems to be a DNA-protein complex from which most of histones have been removed. Evidently during the process of RNA formation, the helical structure of the DNA template is maintained. However, the hydrogen bonds between the DNA bases are certainly broken temporarily; thus, during transcription, a structure with a small zone of strand separation akin to melting must pass along the two-stranded molecule, somewhat like the fastener of a zipper. The short-time "melting," which makes the "sense" strand of the DNA available as a template, is also possible for the DNA-protein complex remaining after removal of the histones. Nucleohistones prepared from euchromatin showed a priming efficiency for the *E. coli* RNA polymerase several times higher than that of heterochromatin (Vogt and Harbers, unpublished; cf. p. 189).

DNA-dependent RNA polymerase also catalyzes the incorporation of the triphosphates of pseudouridine (Kahan and Hurwitz, 1962) and of some synthetic bases. Pseudo-UMP, thymine riboside phosphate, 5-fluoro-UMP, and 5-bromo-UMP can replace the normal UMP in RNA; likewise IMP, 6-azaguanosine phosphate and 5-bromo-CMP can replace GMP or CMP, respectively. 4-Azauridine phosphates and xanthosine phosphates, on the other hand, are not incorporated into RNA (see p. 171 for similar findings in enzymatic DNA synthesis). The molar base ratios in RNA-containing analogs correspond to those of RNA synthesized under similar conditions with normal substrates.

In addition to biosynthesis of RNA, terminal incorporation of ribonucleotides into DNA was observed (Hurwitz, 1959; Krakow *et al.*, 1961). In an extract obtained from calf thymus nuclei, base utilization was found not to be uniform (GTP > CTP > ATP > UTP), the nucleotides forming the end of a polydeoxyribonucleotide chain. The functional importance of these processes is not yet known.

An RNA polymerase from *Azotobacter vinelandii* is reportedly able to utilize both DNA and RNA as primer (Krakow and Ochoa, 1963), with the possibility of a two-enzyme system excluded by the competitive action of the two types of nucleic acids. Poly-5-fluorouridylic acid (cf. p. 312) inhibits both DNA- and RNA-dependent synthesis, while actinomycin (see p. 294) blocks only the DNA-dependent RNA formation. An RNA polymerase obtained from *Micrococcus lysodeikticus* can also use both types of nucleic acid as primer (Fox *et al.*, 1964a and b). DNA competitively inhibits the RNA-dependent RNA synthesis in this system; conversely, the DNA-dependent reaction is inhibited competitively by RNA.

While a DNA-dependent RNA synthesis normally requires the presence of all four ribonucleoside triphosphates, it was found that the RNA polymerase from *E. coli* can also catalyze the formation of polyA if only ATP is supplied as a substrate (Chamberlin and Berg, 1963). Such a DNA-dependent reaction was not possible with the other ribonucleoside triphosphates.

Recent investigations have indicated that virtually all normal RNA synthesis in the cell is carried out at a DNA template. Nevertheless, several systems of cytoplasmic origin are known which do not appear to be DNA dependent and which incorporate ribonucleotides into polyribonucleotides. Even more recent studies by Wykes and Smellie (1966) led

to the conclusion that only homopolyribonucleotide synthesis is catalyzed by microsomal fractions and that there is little or no synthesis of
RNA-like heteropolymers.

Thus Burdon and Smellie (1961) have obtained extracts from Ehrlich
ascites tumor cells which catalyze an RNA-dependent incorporation of
labeled UTP into RNA. Attempts at the further purification of the enzyme yielded two fractions: fraction A, in which UTP utilization was
stimulated by RNA and also by the presence of the remaining bases as
riboside triphosphates, and fraction B, which catalyzed only end group
reactions whose turnover was increased by the addition of amino acids
(cf. p. 218). Both systems were found in the cytoplasm, but only fraction A was found in the cell nuclei. Straus and Goldwasser (1961) have
also reported an RNA-dependent uptake of labeled UTP into the microsomal RNA in a pigeon liver system. This incorporation was stimulated
by the addition of the other three bases as riboside triphosphates and
therefore probably represents true RNA synthesis; the incorporation was
inhibited by pretreatment of the microsomes with ribonuclease or trypsin,
or by heating of the microsomes. Carr and Grisolia (1964) have reported
the utilization of dihydrouridine monophosphate for RNA synthesis in rat
liver and brain homogenates. Dihydrouracil appeared to be incorporated
as such in microsomal RNA and tRNA with the aid of an RNA-dependent
RNA polymerase. Reddi (1961) has described an enzyme, polyribonucleotide synthetase, prepared from the soluble cytoplasm fraction of spinach
leaves, which requires all four bases as riboside triphosphates along
with an RNA-protein complex. The amount of original primer RNA was
nearly doubled as a result of the action of this enzyme. Nakamoto and
Weiss (1962), using an RNA polymerase from *Micrococcus lysodeikticus*,
observed that the base sequence in the newly formed polynucleotides is
probably complementary to that in the primer RNA. Rapid RNA-dependent RNA synthesis has only been observed after infection with RNA-
containing viruses (cf. p. 268). An enzyme which requires all four ribonucleoside triphosphates as substrates has been isolated and purified
from *E. coli* infected with RNA phages (Weissmann *et al.*, 1963). At
present it is not clear whether this kind of RNA synthesis occurs exclusively during the process of multiplication of RNA-containing viruses,
or whether RNA-dependent RNA synthesis also takes place in normal
cell metabolism (Cline *et al.*, 1963).

After infection of *E. coli* with RNA-containing phages, a RNA polymerase is formed whose properties have been studied by various groups;
the enzyme is termed RNA synthetase (when isolated together with the
required template) or replicase (without template). In general, viral RNA
is single stranded (cf. p. 269). The first step in replication is the syn-

thesis of complementary minus strands which then provide the template for the formation of new parental-type plus viral RNA (Feix *et al.*, 1967; see p. 269). This mechanism of synthesis is unique because a single-stranded nucleic acid is formed at a single-stranded template.

Haruna and Spiegelman (1965a) isolated replicases from *E. coli* cells that had been infected with the RNA bacteriophages MS-2 or Q_β. It was found that each enzyme can use as a template only the viral RNA of its origin. Even under optimal conditions, these RNA-synthesizing enzymes are virtually inactive with heterologous RNA, including rRNA and tRNA of the host cells. Thus the enzymes seem to recognize the RNA genome of their origin.

This discriminating selectivity of replicases is of obvious advantage to a virus attempting to direct its own replication in a cellular environment replete with other RNA molecules. The purified enzymes can generate identical copies of viral RNA added as a template. In dilution experiments, it was shown that the newly synthesized RNA was again fully competent to program the synthesis of viral particles (Spiegelman, Haruna *et al.*, 1965). The normal functioning of replicases requires intact homologous RNA and is inhibited by Mn^{++}. The enzyme cannot employ fragments of its own genome as templates (Haruna and Spiegelman, 1965b).

Edmonds and Abrams (1960, 1962) have found an enzyme in calf thymus cell nuclei which can use ATP and, to a lesser extent, CTP but cannot utilize GTP or UTP for the biosynthesis of polyribonucleotides. It was first thought that the reaction required no primer and the addition of other ribonucleoside triphosphates had no effect. Later it was found that two enzymes were involved. One catalyzed the formation of polyA and was able to use CTP to a limited extent (1 mole per 500 moles ATP). The other catalyzed the synthesis of polyC and utilized small amounts of ATP (1 mole per 50 moles CTP). Both enzymes require specific polynucleotide primers. Chambon *et al.* (1963) detected in nuclei from chicken liver cells a DNA-dependent RNA polymerase which utilized only ATP as substrate and required the presence of nicotinamide mononucleotide. It is not clear what role this and other enzymes, which synthesize homopolymers, play in cell metabolism. One possibility seems to be that these polynucleotides serve as emergency supplies for the rapid formation of free nucleotides.

The enzymatic processes described so far generally catalyze the *de novo* synthesis of polyribonucleotide chains. In addition, there are some reactions in which nucleotides are added to existing RNA molecules. A terminal incorporation of this kind was first observed in cytoplasmic RNA (Heidelberger *et al.*, 1956; Canellakis, 1957a). The best-known example is the formation of the end group sequence of tRNA which accepts activated amino acids for the process of translation in polypeptide synthesis (cf. p. 218). The following equation is applicable to this terminal nucleotide incorporation.

$$XTP + tRNA \rightleftharpoons XMP\text{-}tRNA + PP$$

where XTP is a ribonucleoside triphosphate and PP a pyrophosphate. The free pyrophosphate can be used for a pyrophosphorolysis of tRNA; thus the nucleotides degraded from the polynucleotide chain appear as ribonucleoside triphosphates (Canellakis and Herbert, 1961).

The terminal incorporation of nucleotides into tRNA produces the appropriate functional base sequence at the end of the polynucleotide chain [Fig. 99(a) and 100]. The extent of incorporation, as observed ex-

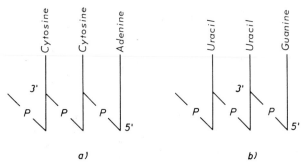

a)   b)

Fig. 99. End group structure of two kinds of RNA: (a) terminal nucleotide sequence of tRNA (cf. Fig. 55); (b) terminal sequence of an RNA from rat liver cytoplasm after incorporation of 5'-UMP and 5'-GMP (from Harbers and Heidelberger, 1959b).

perimentally, is therefore not merely an indication of enzyme activity, but also of the extent to which a given RNA lacks the terminal bases of the polynucleotide chain. If, in the course of the isolation of tRNA from tissue (Hecht et al., 1959), the extract is either stored for a longer period of time (Schneider and Potter, 1958) or briefly treated with snake venom phosphodiesterase, a partial decomposition of the tRNA occurs, which, under suitable in vitro conditions, results in increased nucleotide incorporation, leading to the restoration of the complete structure. Terminal incorporation of CTP and ATP appears to be effected by a single enzyme with two active centers (Anthony et al., 1963).

Hurwitz and co-workers have isolated an enzyme from thymus cell nuclei which utilizes only CTP as substrate and incorporates it terminally in tRNA. This incorporation requires a specific RNA, which evidently

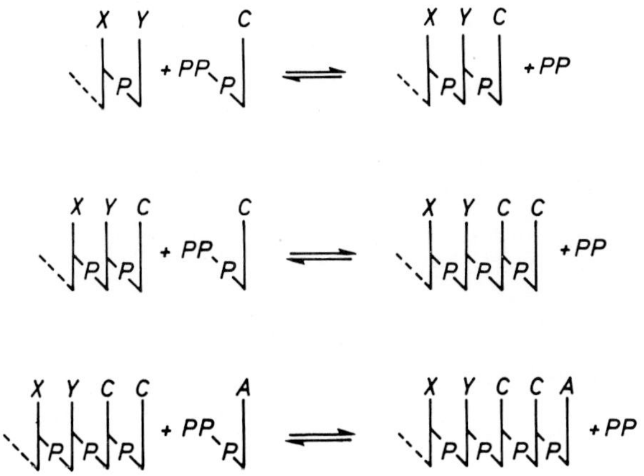

Fig. 100. Sequential addition of terminal nucleotides to tRNA (from Hoagland, 1960).

is not active as an amino acid acceptor. Daniel and Littauer (1963) purified an enzyme from rat liver which catalyzed the terminal incorporation of the nucleotides of cytosine, adenine and uracil into sRNA. The nucleotides had to be supplied as triphosphates; GTP did not react. It was possible to prepare different end group sequences with this enzyme (e.g., —XCA, —XCU, —XUC). The exchange of uracil for cytosine blocked the addition of terminal AMP to this RNA.

The end group sequence of tRNA, as shown in Fig. 99(a), is essential for its participation in protein synthesis (cf. p. 219). The transport of amino acids to the ribosomes involves the addition of the complete tRNA molecule and not merely a fragment (Hoagland and Comly, 1960). This process requires the normal terminal sequence of tRNA. From this, it is conjectured that the binding of tRNA to the ribosomes takes place via the —CAA end group [Fig. 99(a); Bloemendahl and Bosch, 1962]. In addition to the terminal incorporation of CTP and ATP in tRNA, an incorporation of uridine and guanosine phosphates has been observed in the cytoplasm of rat liver; this evidently results in the end group sequence represented in Fig.99(b); (Harbers and Heidelberger, 1959b).

Such a terminal incorporation of guanosine phosphate could also be demonstrated in cytoplasmic RNA of Ehrlich ascites tumor cells which contain an tRNA with the —CAA end group sequence; however, its functional significance remains unknown. From the pH 5 supernatant of the soluble cytoplasm of rat liver, Klemperer (1963a and b) isolated an enzyme which catalyzed the specific terminal incorporation of 5'-UMP (supplied as UTP) into ribosomal polyU in the presence of Mg^{++}. This incorporation also occurred with synthetic polynucleotides and with tRNA, and was inhibited by polyA and heat-denatured DNA. Another rat liver enzyme catalyzed the terminal incorporation of adenosine phosphate with the formation of polyA chains. At present, nothing is known about the functional significance of these reactions.

The methylation of RNA bases, of both the purine and the pyrimidine types, occurs on preexisting polynucleotide chains and is mediated by an RNA methylase* using S-adenosylmethionine as C_1 donor (Fig. 101; e.g., Biswas *et al.*, 1961; Mandel and Borek, 1961; Fleissner and Borek, 1963; Srinivasan and Borek, 1963). This kind of methylation is limited to rRNA and tRNA (cf. p. 101); other RNA is not effective as a C_1 acceptor. The formation of pseudouridine in tRNA appears to take place intramolecularly, i.e., at the completed RNA chain. From the observations of Gold, Hurwitz and Anders (1963) and of Svensson and co-workers (1963), it seems that the methylating enzymes which insure the correct arrangement of the rare bases in tRNA are strongly species specific.

Thus the pathway of the synthesis of thymine in RNA differs from that of thymine in DNA in two respects: (a) methylation does not occur at the nucleotide level, and (b) tetrahydrofolate is not involved in the reaction. Hurwitz and co-workers (1964a and b) have prepared six enzymes from *E. coli* which methylate tRNA via S-adenosylmethionine. The following products were found to be formed on tRNA: (1) 1-methylguanine (two enzymes); (2) 7-methylguanine; (3) 3-methyladenine, 6-methylaminopurine and 6-dimethylaminopurine, whose formation takes place via an adenine-methylating fraction; (4) 5-methylcytosine; (5) thymine. The sites on the tRNA molecules at which methylation occurs are unknown. The reactions are very specific for tRNA, but not completely species specific, because tRNA from other bacteria can also be methylated. Partial degradation of tRNA with snake venom phosphodiesterase decreased the acceptor ability, but did not destroy it completely. Comb (1964) suggests that the pattern of the methyl groups in tRNA may be determined partly by association with the ribosomes during the process of methylation. A more detailed study of the methylation reactions will be possible when the different kinds of tRNA are available in a methyl-free form. So far, no appreciable differences in the transfer ability of methyl-deficient and complete tRNA have been observed (Peterkofsky *et al.*, 1964).

The methylation of the bases of rRNA also occurs as a sequel to primary RNA synthesis; the interactions between rRNA and the methylating enzymes were found to be species specific (Srinivasan *et al.*, 1964; Hurwitz *et al.*, 1965). The 16S particles have more methyl groups than the 23S particles. In intact cells, methylation probably takes place at

*EC not yet assigned.

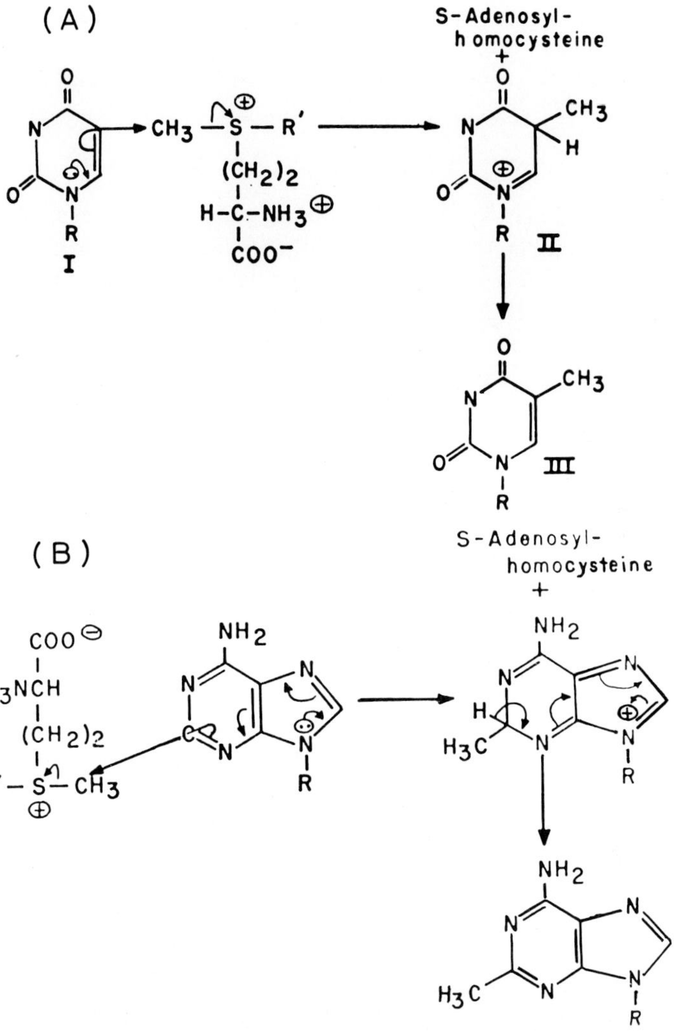

Fig. 101. Formation of methylated pyrimidine ribosides (A) and purine ribosides (B) by methyl transfer from S-adenosyl-methionine (from Borek and Srinivasan, 1966). (Reproduced with permission of Annual Reviews, Inc.)

the level of ribonucleoproteins. A transfer of methyl groups from methionine to methyl-deficient ribonucleoprotein particles which had been produced by methionine starvation of E. coli W6 cells was demonstrated by Gordon and Boman (1964). The observations of these authors suggest that rRNA methylating enzymes are bound to particles and that the methylation of rRNA is a step in the biosynthesis of ribosomes (see p. 98).

In addition to methylated purines and pyrimidines, tRNA can contain 2'-O-methyl ribosides (cf. p. 101). How the O-methyl ribosides are formed is not yet known; it is tempting to speculate that O-methylation might also occur at preexisting RNA chains.

RNA SYNTHESIS AND GROWTH

Thus far, little is known about the relationship between RNA metabolism and growth. Increased DNA synthesis during growth is accompanied by an increased incorporation of labeled precursors into RNA. In synchronized Hela cells, two peaks and two minima of RNA synthesis were observed during the mitotic cycle (Kim and Perez, 1965). The first peak occurred about 1 hour before DNA replication (S phase); the second one preceded mitosis. One minimum occurred after DNA synthesis; the other one, during the process of mitosis. These observations agree with the findings of Mittermayer et al. (1964) who also observed two peaks of RNA synthesis during the mitotic cycle of Physarum polycephalum. More recently, Mittermayer et al. (1966) were able to confirm this biphasic pattern of incorporation into RNA with isolated nuclei. It seems that almost no RNA synthesis occurs during cell division (Prescott and Bender, 1962); furthermore, no RNA synthesis appears to be possible on replicating DNA (Taylor, 1960; Prescott and Kimball, 1961). In general, DNA replication is evidently preceded by the synthesis of a specific RNA (Lieberman et al., 1963). Several observations suggest that mitosis must also be preceded by the production of a specific RNA (cf. p. 185).

In rat liver cell nucleoli, an immediate increase in RNA turnover was observed following partial hepatectomy (Tsukada and Lieberman, 1964a). Maximum incorporation of labeled precursors into the nucleoli occurred 6 hours after the operation, while the amount of nucleolar RNA reached its maximum 20 hours after hepatectomy. The extent of these changes was a function of the amount of liver tissue removed. Shortly after partial hepatectomy, the activity of DNA-dependent RNA polymerase in the liver increased (Tsukada and Lieberman, 1964b), reaching maximum activity 12 to 18 hours after the operation (Busch et al., 1962), i.e., prior to the optimum activity of DNA polymerase. The increase in RNA polymerase activity seems to be due to de novo synthesis and does not appear to be the result of activation of a preexisting enzyme (Tsukada and Lieberman, 1964b).

ENZYMES DEGRADING RIBONUCLEIC ACIDS

Enzymes which degrade RNA are called ribonucleases*,** (RNases); they are widely distributed in nature, and thus far, in no cell or tissue

*EC 2.7.7.16 Ribonucleate pyrimidine nucleotido-2'-transferase.
**EC 2.7.7.17 Ribonucleate nucleotido-2'-transferase.

studied carefully has this enzyme been absent.[†] The richest source of RNase is the digestive juice of the pancreas from which it was first isolated in 1920. Since then, many RNases have been purified from various sources, and their mechanism of action is rather well understood. All RNases act as phosphodiesterases with varying substrate specificity. Like proteases, the RNases may be classified as endonucleases and exonucleases. In 1951, it was determined that pancreatic RNase preferentially cleaves the pyrimidine ribonucleoside 3′-phosphate ester bonds (Schmidt *et al.*, 1952). Since cleavage of a polynucleotide chain includes a transfer mechanism with the formation of cyclic intermediates (Markham and Smith, 1952), the RNases should be considered phosphotransferases rather than hydrolases.

As mentioned above, *pancreatic RNase** was the first enzyme of this class to be studied in detail. It was crystallized by Kunitz in 1940, and the complete primary structure of this enzyme, consisting of 124 amino acids and including four essential disulfide bonds, has been elucidated (Anfinsen and White, 1961). Pancreatic RNase, like some other RNases, is unusually stable to various denaturing agents. Thus its standard preparation begins with extraction of the pancreatic tissue with $0.25N$ H_2SO_4. In the pH range of 2–4.5, the enzyme may even be boiled without loss of activity. Its pH optimum is 7.7; in experimental use, however, pH values between 5 and 6 are preferred in order to prevent spontaneous hydrolysis. No metal ions are required for activation. By chromatographic methods, it has been possible to divide what appeared to be pure crystalline pancreatic RNase into various fractions (Hirs *et al.*, 1953; Taborsky, 1959). Since pancreatic RNase attacks only the phosphodiester bonds of pyrimidine ribonucleosides, an undigested core rich in purine bases is formed during its action. The cleavage of the susceptible bonds appears to occur biphasically: first 2′,3′-cyclic phosphate esters of the pyrimidine nucleosides are formed as terminal groups on intermediate oligonucleotide fragments. Then, free mononucleotide cyclic phosphates are split off and hydrolyzed to the 3′-monoesters if the preceding unit is a pyrimidine derivative; 2′-esters do not appear.

The RNase activities found in other organs are at least one order of magnitude lower than in the pancreas. Spleen RNase has been studied in some detail (Kaplan and Heppel, 1956); this enzyme has a pH optimum of 6.0–6.5, but its catalytic action seems to be identical to the pancreatic enzyme. RNase from rat liver can be separated into fractions with an acid and an alkaline pH optimum, respectively (Roth, 1957); a third type

---

[†] So far no individual EC numbers have been attributed to RNases purified from various sources.

of liver RNase has recently been detected by Rakman (1966). One RNase isolated from liver nuclei leads to the formation of oligonucleotides with 5′-phosphate end groups (Heppel, 1966). The activities of some of the animal RNases seem to be controlled by specific RNase inhibitors— proteins which have been isolated from animal and plant tissues (Bernheimer and Steele, 1957; Higashi and Hasegawa, 1960; Shortman, 1961). In addition, it was observed that RNases may be inhibited by rather low concentrations of ordinary nucleotides (Gamble and Wright, 1965; Yamasaki and Arima, 1967).

Various plant RNases have been purified, and their mechanism of action has been studied (e.g., Tuve and Anfinsen, 1960; Wilson, 1967). One of these enzymes splits purine and pyrimidine esters and thus does not lead to the formation of a core. The purines appear as the free 3′-mononucleotides, whereas the pyrimidine bases yield the 2′-,3′-cyclic phosphates.

Among the microbial RNases, two enzymes isolated from *Aspergillus oryzae* are particularly interesting. Takadiastase powder, which is obtained from that organism, has been used as starting material for the preparation of takadiastase T_1 and takadiastase T_2 (Rushizky and Sober, 1962). The former enzyme is also called "ribonuclease T_1" and shows an unusual substrate specificity: Only internucleotide bonds between 3′-guanylic acid and the 5′-hydroxyl groups of adjacent nucleotides are split (McCully and Cantoni, 1961). This base preference was an important factor in the elucidation of the nucleotide sequences of tRNAs (cf. p. 103). The same enzyme has now been isolated from other microorganisms. Some of the microbial RNases attack DNA as well as RNA.[*],[**]

In addition to the RNases mentioned thus far, there are other *nonspecific phosphodiesterases*[†] whose action is not limited to phosphate ester bonds within polynucleotides. These enzymes, when incubated with oligonucleotides, will always act as exonucleases. Their specificity toward the sugar moieties of nucleotides is weak, so that ribonucleotides as well as deoxyribonucleotides may be split. Moreover, there is no preference for either purine or pyrimidine bases.

The activity of snake venom phosphodiesterase was determined in 1932 by Japanese workers. Some snake species contain only diesterase, but no monoesterase, activity in their venom; these, therefore, constitute a particularly good starting material for the enzyme preparation. Accord-

---

[*]EC 3.1.4.7   Ribonucleate (deoxyribonucleate) 3′-nucleotidohydrolase.
[**]EC 3.1.4.9   Ribonucleate (deoxyribonucleate) 5′-nucleotidohydrolase.
[†]EC 3.1.4.1   Orthophosphoric diester phosphohydrolase.

ing to the products formed from the action of the enzyme, the phosphodiesterases have been classified as follows. *Phosphodiesterase I** leads
to the formation of 5′-mononucleotides. This enzyme is found in snake
venom, in various animal tissues (Razzell, 1961a and b) and in microorganisms, of which the *E. coli* "exonucleases I, II and III" (Lehman,
1963) are well-known examples. *Phosphodiesterase II* leads to the formation of 3′-mononucleotides. This type of enzyme has been found in
various animal tissues (Razzell, 1961a) and in bacteria. Good sources
for its preparation are calf spleen (Hilmoe, 1960) and *Lactobacillus
acidophilus* (Fiers and Khorana, 1963a and b). Special *phosphodiesterases for cyclic mononucleotides* have been found, which will attack
either 2′,3′-(Drummond *et al.*, 1962) or 3′,5′-cyclic esters (Butcher and
Sutherland, 1962).

For more detailed information on RNases, the following reviews may
be consulted: Anfinsen and White (1961), Khorana (1961), Elson (1965),
Siebert (1966c).

CHAPTER VIII

Protein Synthesis

INTRODUCTION

More than twenty years ago Brachet (1942) and Caspersson (1941) inde-pendently concluded that *protein synthesis* in the cell *is associated with RNA.* Subsequently, much experimental evidence for this relationship was obtained (e.g., Brachet, 1959b). As shown in the preceding chapter, cellular RNA synthesis is a DNA-dependent process. The relationship between the genetic information of DNA and the protein can be summar-ized in the so-called *central dogma* (Crick):

$$\text{DNA} \xrightarrow{\text{(transcription)}} \text{RNA} \xrightarrow{\text{(translation)}} \text{protein}$$

Although DNA primarily determines the amino acid sequence of proteins, protein synthesis can take place in cell-free systems in the absence of DNA. In fact, the process of protein formation is generally studied in systems without DNA which itself is usually not effective as a direct template for ordering amino acid sequences (but cf. p. 236). Instead, the nucleotide sequence of DNA is first transcribed to form a complementary RNA chain (a messenger) which then carries the information to the sites of protein synthesis. Here, translation of specific nucleotide sequences into the amino acids of particular proteins takes place. This translation or coding will be discussed in the next chapter.

Detailed investigations of the mechanisms of protein formation could not be undertaken until radioactively labeled amino acids became avail-able, making it possible to trace their incorporation into proteins. Re-sults of some of the first tracer experiments with such labeled amino acids indicated that protein formation occurs in several steps. The first step is the activation of the amino acids; this is followed by their at-tachment to specific tRNAs (Hoagland, 1955; Hoagland *et al.*, 1956). In the process of translation, i.e., the ordered formation of peptide bonds, the amino acids cannot interact directly with the mRNA tem-plate, because no specific affinities exist for the purine and pyrimidine bases of RNA. Thus, for coding, an adaptor is required as a link be-tween the amino acids and the template (Crick, 1958; Hoagland, 1959). It has been determined that tRNA is the adaptor because of its ability to attach specifically to the free amino and keto groups on the bases of

the mRNA template. The final formation of polypeptide chains occurs at
the ribosomes, after the individual amino acids have been transported to
these particles by the tRNAs, and is accompanied by a temporary binding
of the aminoacyl-tRNA to the ribosomes which contain the machinery for
the formation of peptide bonds. The amino acid sequence of the peptides
is determined by the mRNA template which is moved along the ribosome
so that coding takes place like a tape mechanism. Because each ribosome has its one site for peptide bond formation, synthesis of several
identical polypeptides can occur simultaneously on an aggregate of various ribosomes (i.e., a polysome), as shown schematically in Fig. 104.
The details of these processes will be discussed in the following sections.

ACTIVATION OF AMINO ACIDS AND ATTACH-
MENT TO ₁RNA

The activation of amino acids for protein synthesis takes place according to the reaction:

amino acid + ATP + activating enzyme $\rightleftharpoons$ aminoacyl-AMP-enzyme + PP

The activating enzymes (aminoacyl-tRNA synthetases*) catalyze an
amino acid-dependent exchange between ATP and pyrophosphate. The
amino acids, together with AMP and the appropriate enzyme, form an intermediate (Figs. 102 and 103) necessary for the next step, the attachment of the amino acids to their specific type of tRNA. There was early
evidence for the presence of at least one specific tRNA for each amino
acid (Hecht et al., 1958). More recently several of these have been purified and their complete nucleotide sequence has been established (see
p. 103 and Fig. 55). Each amino acid requires a specific aminoacyl-
tRNA synthetase, some of which have been purified (e.g., Berg, 1956a and b;
Davie et al., 1956). Activation is generally analyzed by the liberation of
pyrophosphate.

Activation is followed by the attachment of amino acids to tRNA molecules which then transport the amino acids to the mRNA template at the
ribosomes. This may be described by the following reaction:

aminoacyl-AMP-enzyme + tRNA $\rightleftharpoons$ aminoacyl-tRNA + AMP + enzyme

This requires only the activating enzyme and a specific tRNA for each
amino acid. Aminoacylation of tRNA occurs with the stoichiometric
participation of the activating enzyme (Zillig et al., 1960). Apparently
the loss of some methyl groups at the rare bases of tRNA does not seri-
ously disturb the process of aminoacylation (Peterkofsky et al., 1964).

*EC not yet assigned.

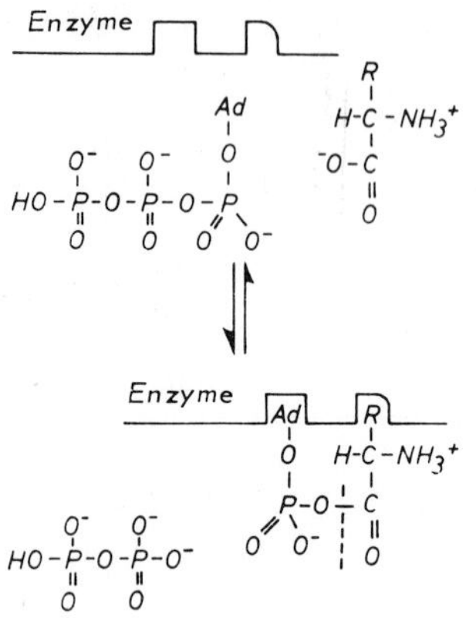

Fig. 102. Mechanism of amino acid activation. A complex is formed from the enzyme, ATP and the amino acid, with the liberation of pyrophosphate. Ad = adenine (from Hoagland *et al.*, 1956).

For the binding of amino acids to tRNA, a specific terminal sequence is required (Figs. 99 and 100). This sequence –CCA appears to be the same for all amino acids. Zachau *et. al.* (1958) established that the amino acids combine in an ester-type linkage with the hydroxyl on carbon 2′ or 3′ of the ribose in the terminal adenosine. According to later investigations by Feldmann and Zachau (1964), the amino acid attachment seems to occur mainly at the hydroxyl group on carbon 3′. These results imply that only one amino acid can be attached to one molecule of tRNA.

Since the attachment of an amino acid to tRNA requires the terminal sequence –CCA, if one or more of the terminal nucleotides are removed the tRNA can no longer accept amino acids. Under appropriate conditions, the complete—and therefore functional—end group sequence can be restored by the addition of ATP and CTP to an *in vitro* system. The reversible reaction of the terminal nucleotide incorporation is inhibited by the presence of amino acids (Hecht *et al.*, 1959). The binding of amino acids to tRNA is very stable and is maintained even at low pH and high temperatures.

Because of their selective function, the activating enzymes have key positions in the process of protein synthesis. Each enzyme must be "recognized" (a) by one specific amino acid and (b) by a specific tRNA molecule which codes for this particular amino acid (Fig. 103). It is conceivable that the single-stranded looped regions of tRNA (cf. Fig. 55) play an important role in the specificity of the interaction with the activating enzymes. Nothing is known as yet about the kind of binding (which portion of the enzyme and which bases of the tRNA are involved). Some observations seem to indicate that the association with tRNA causes a conformational change of the enzyme (Loftfield and Eigner, 1965) which might also take place due to the association with amino

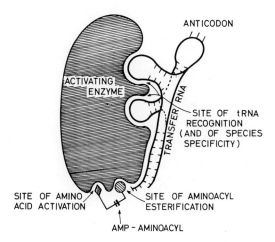

ANTICODON

ACTIVATING ENZYME

TRANSFER RNA

SITE OF tRNA
RECOGNITION
(AND OF SPECIES
SPECIFICITY)

SITE OF AMINO
ACID ACTIVATION

SITE OF AMINOACYL
ESTERIFICATION

AMP – AMINOACYL

Fig. 103. Model representing the different sites of specific interaction between activating enzyme and tRNA (modified from Zamecnik, 1966). (Reproduced with permission of American Association for Cancer Research, Inc.)

acid and ATP; furthermore, conformational differences between free tRNA and aminoacyl-tRNA have been observed (Sarin and Zamecnik, 1965). Charged tRNA seems to be less helical than the uncharged molecule. This difference in the conformation may be important for preferential binding and subsequent release of tRNA to and from the ribosome during the translation process (cf. p. 224).

For peptide formation, aminoacyl-tRNA must be bound by the ribosomes. Tracer studies have shown that this binding can also occur with uncharged tRNA if it possesses the complete end group sequence –CCA. The transfer process is not affected by puromycin (an antibiotic which blocks protein synthesis; cf. p. 301), as demonstrated by Bloemendal and Bosch (1962). The role of tRNA as an adaptor is similar to that of a cofactor (Lipmann *et al.*, 1959): There is first a very specific, species-dependent interaction with the aminoacyl-AMP complex (Fig. 103), followed by interaction with the ribosomes and—for the process of translation—with the mRNA template.

Direct evidence for the adaptor function of tRNA was obtained in the following manner: cysteine, bound to its tRNA, was chemically changed to alanine while maintaining its RNA bond. In a cell-free system capable of protein synthesis, this alanine, bound to a cysteine-tRNA, behaved like cysteine (Chapeville *et al.*, 1962). This observation proved that the reading of the message is performed by the different tRNAs and not by the amino acids.

Species relationships between the different building blocks for protein synthesis can be studied in crossing-over experiments, in which tRNA reacts with activating enzymes or ribosomes of different origin. So far only a few such experiments have been reported.

Studies by Zillig and co-workers (1960) have shown that activating enzymes from rat liver cannot transfer amino acids to *E. coli* tRNA. On the other hand, transfer can be accomplished between charged tRNA and ribosomes from various species. Thus charged tRNA from *E. coli* was able to provide the amino acids for hemoglobin synthesis when appropriate ribosomes were supplied. However, the different amino acids and their activating enzymes did not exhibit consistent behavior in such experiments. Thus, Benzer and Weisblum (1961) found that the arginine-activating enzyme from rabbit liver could interact with tRNA from *E. coli*, but the corresponding enzyme and tRNA from yeast were inactive. Tyrosine-activating enzymes from yeast and rabbit liver could substitute for one another in the acceptance of amino acids by yeast tRNA, while *E. coli* enzymes were not active under these conditions (Berg *et al.*, 1961; Rendi and Ochoa, 1961).

PEPTIDE FORMATION AT THE RIBOSOME

The last steps in protein synthesis occur at the ribosome where the reading out of the message proceeds through a series of interactions involving charged tRNA, GTP, two ribosomal transfer enzymes and an mRNA template (together, these are termed the "active complex"; Watson, 1963).

For a while it was supposed that rRNA constitutes the template which determines the amino acid sequence of the newly synthesized proteins. However, it became difficult to reconcile this hypothesis with certain experimental observations. For example, the number of ribosomes is much too small to account for the multiplicity of proteins formed in the cell. Moreover, the possibility of very rapid initiation of the biosynthesis of specific proteins (enzyme induction; cf. p. 248) is hardly compatible with the fact that ribosomes are relatively stable particles that do not undergo rapid turnover. These difficulties were resolved with the discovery of mRNA, whose existence had already been postulated on theoretical grounds.

The work of Watson (1963) suggested that each ribosome contains one specific site for the formation of peptide bonds (Fig. 104). These conclusions were based on various experimental observations, some of which will be mentioned here.

Some of the newly synthesized proteins in a cell-free system remain attached to ribosomes (Risebrough *et al.*, 1962). mRNA which is very

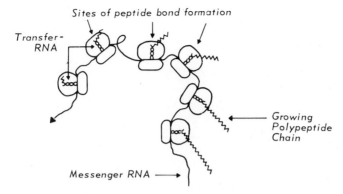

Fig. 104. Attachment of several ribosomes by mRNA to form a polysome. Each ribosome contains one site for the formation of peptide bonds along which mRNA is guided for translation of the template (cf. Fig. 107; from Watson 1963).

sensitive to ribonuclease does not appear to be involved in this attachment, because treatment with the enzyme does not release the peptide. Evidently the peptides are bound to the ribosome via tRNA molecules which are linked to the carboxyl end of the last amino acid. Prolonged dialysis causes a degradation of ribosomes into 30S and 50S subunits. Under these conditions, the peptides with tRNA are still bound to the 50S particles. These 50S particles can be further degraded by treatment with a detergent, which releases the peptide together with the terminally attached tRNA (Gilbert, 1963; Fig. 105). If the Mg^{++} concentration is raised to 0.01M, some of the free polypeptides will again be bound by the ribosomes (Watson, 1963).

Observations by Spirides and Lipmann (1962) indicated that mRNA can combine with several ribosomes. Somewhat later, various experimental results led to the conclusion that aggregates of ribosomes (i.e., polysomes; cf. Fig. 53) are formed, held together by a single mRNA strand; each ribosome moves along the mRNA, thereby synthesizing a polypeptide chain (Gierer, 1963; Gilbert, 1963; Warner et al., 1963; Watson, 1963; Wettstein et al., 1963). Polysomes can be produced in cell-free systems in the presence of appropriate mRNA. A peptide chain is initiated when a ribosome is attached to the 5'-hydroxyl end of an mRNA; the direction of reading the message thus is from 5' to 3'. The formation of peptides starts at the N-terminal amino acid. A ribosome can probably be attached to mRNA only at the point corresponding to the normal beginning of the message (cf. p. 234). GTP is required for the transport of mRNA through the ribosomal site. When the peptide chain is com-

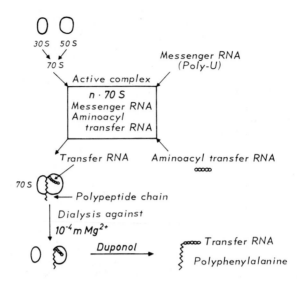

Fig. 105. Summary of experimental results demonstrating the participation of ribosomes in protein synthesis. In the actual experiment, dissociation of 70S ribosomes by dialysis against a dilute Mg++ solution had been observed in a polyphenlalanine-synthesizing system, using polyU as a messenger (from Watson, 1963).

pleted and released, the ribosome becomes detached from the 3′-hydroxyl end of the mRNA and is available for attachment of another mRNA strand. Successive attachment of various ribosomes to a single messenger leads to the formation of polysomes. Studies on the progressive breakdown of polysomes by gradual removal of magnesium ions indicate that the site of mRNA attachment is at the small subunit of the ribosome (Noll, 1965), while all nascent polypeptide chains, linked via tRNA, are bound to the 50S subunits, one chain per ribosome.

The observed requirement for two different ribosomal enzymes (Keller and Zamecnik, 1956) suggested that peptide synthesis takes place in at least two steps, one of which requires energy. Peptide bond formation is catalyzed by peptide synthetase* (transferase I). This reaction seemingly requires no GTP, since chain elongation occurs at the C-terminal which is in an activated state because of its ester linkage to a tRNA (Gilbert, 1963).

Brenner *et al.* (1950) prepared oligopeptides from synthetic acylesters of amino acids and observed that the reaction was energetically favor-

*EC not yet assigned.

able. They concluded that the active precursor in protein biosynthesis might well be an ester of the amino acid formed in an ATP-dependent reaction. Later, these predictions, which meanwhile had been widely forgotten, were confirmed by biochemists (cf. Noll, 1965).

The other transfer enzyme, transferase II or translocase* (Traut and Monro, 1964), is GTP dependent and responsible for the transport of the messenger. A model for the cooperation of both enzymes in the process of peptide formation and mRNA transport is given in Fig. 106. From an

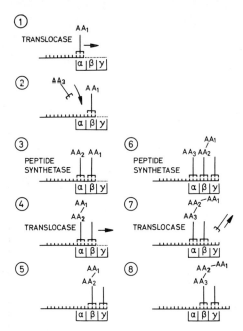

Fig. 106. The postulated sequence of events during initiation and extension of polypeptide chain at the ribosome according to Noll (1965). Explanations in the text.

analysis of various experiments, the following sequence of events at the ribosomal surface is postulated (Noll, 1965):

(1) The messenger enters the ribosomal groove with its 5′-hydroxyl end, until the first codon (cf. p. 229) arrives at the decoding site (α in Fig. 106). When a charged tRNA matching this codon is inserted into the corresponding cavity of the large ribosomal subunit, the messenger is locked in by combining with the decoding site.

*EC not yet assigned.

(2) The translocase then interacts with the mRNA-tRNA complex and moves it just the length of three nucleotides (1 codon). By this process, the -CCA terminal of the tRNA with the bound amino acid comes into the condensing site (β). The most plausible explanation for the transport mechanism seems that conformational changes at the ribosome take place, triggered by a GTPase reaction. It is postulated that the condensing site is accessible only via the decoding site.

(3) The free decoding site is now available for another charged tRNA matching the next codon. Thus two charged tRNA molecules have come into the correct position for the peptide synthetase reaction. The ester bond between the carboxyl group of the first amino acid and its tRNA is cleaved, and the carboxyl group is linked to the amino group of the second amino acid which is still bound to its tRNA in the decoding site.

(4) By the action of the translocase, the mRNA is again advanced for the length of one codon. In this way, the second tRNA is moved into the condensing site, while the first (now discharged) tRNA goes to the exit site. It seems that the peptide synthetase remains attached during this process.

The formation of a polypeptide chain is accomplished by repeating this process as many times as there are codons to be read. In the scheme of Fig. 106, a second cycle of the process (6, 7, 8) is illustrated. According to this model by Noll (1965), the ribosome oscillates between two states I and II, which differ by the presence of a peptidyl-tRNA in either the condensing or the decoding site. In state I, the decoding site may be vacant or just occupied by a charged tRNA, the latter probably being very short-lived. At present, little is known about the initiation of peptide synthesis at the ribosome or the mechanism of releasing the completed chain. From *in vitro* studies with *E. coli* systems, it was concluded that nascent proteins start with N-formylmethionyl residues and that the source of these is N-formylmethionyl-tRNA (Clark and Marcker, 1966) which is considered to be the major peptide chain initiator (e.g., Viñuela *et al.*, 1967; cf. p. 234). These ideas are supported by observations with RNA phage-infected bacterial systems (Adams and Capecchi, 1966; Webster *et al.*, 1966). Because the N-terminal amino acid of proteins generally is not formylmethionine, an as yet unknown mechanism must be postulated for removing the formyl or the formylmethionyl group from the nascent protein. So-called nonsense codons seem to be able to stop the reading of a message (see p. 234). Consequently, a specific tRNA which cannot accept any amino acid at all should be required for the release of the completed polypeptides (Gilbert, 1963). However, there is still no real proof that nonsense codons are involved in the normal release process. Another open question is how the one-dimensional polypeptide chains are converted to three-dimensional proteins with specific tertiary structures. It seems—but has not been proved—that the tertiary structure is determined by the amino acid sequence and thus develops spontaneously with polypeptide synthesis.

Investigations of the mechanism of protein formation and the detailed analysis of experimental observations were facilitated by the use of drugs which effect certain steps in the sequence in a specific manner. Thus puromycin which inhibits protein synthesis appears to block the peptide synthetase reaction (cf. p. 301), while actidione (cycloheximide) is an inhibitor of translocase (see p. 301). Streptomycin, on the other hand, causes the message to be misread, probably by changing the conformation of the ribosomes (cf. p. 302). This then permits single-stranded DNA to act directly as a template for protein synthesis without transcription of the genetic information (McCarthy and Holland, 1965).

The average distance between the adjacent polysomal ribosomes along an mRNA molecule is 60–90 nucleotides (200–300 Å); (Noll et al., 1963). The number of ribosomes contained in one polysome increases with increasing molecular weight of the mRNA. While mRNA in microorganisms has a short lifetime (Levinthal et al., 1962), mammalian mRNA is more stable. More recent observations by Stewart and Farber (1967) indicate that in the cytoplasm of rat liver cells, even free mRNA without attachment to ribosomes may remain functionally active for many hours. The absence of smaller polysomes in reticulocytes (Warner et al., 1962) suggests that its mRNA can be reutilized for a longer time; thus a certain steady state seems to exist between entry and exit of mRNA in the ribosomes of these cells. Observations by Ohtaka and Spiegelman (1963) have indicated that a viral mRNA can contain the information for several different proteins. This should necessitate another control mechanism at the level of peptide formation (cf. p. 252) to permit the synthesis of various proteins at different rates coded by one (polycistronic) messenger. It is an open question whether polycistronic messengers are limited to RNA viruses or whether they also occur in normal cell metabolism. According to Noll (1965), extremely large polysomes consisting of 30–40 ribosomes, which therefore could be polycistronic, constitute a very small fraction in preparations from rat liver.

Although protein synthesis can now be achieved in cell-free systems, with a variety of RNA preparations as messengers, thus far no specific proteins such as active enzymes have been produced in vitro, at least not in amounts comparable to in vivo conditions. The reasons for the imperfection of the present cell-free systems are not yet known. One is tempted to consider the possibility that an interaction of the polysomes with the surface of some membranous components of the cell (e.g., endo-plasmic reticulum) might play a role in the final steps of protein forma-tion. In mitochondria, protein synthesis apparently takes place by the same mechanism. However, the enzymatic equipment of mitochondria seems to be synthesized outside of the particles via nuclear DNA (cf. p. 92); it is then transferred into the mitochondria (Kadenbach, 1966).

In Fig. 107, the whole sequence of events from transcription to poly-peptide synthesis is summarized in a simplified scheme. The mechan-isms for the control of protein synthesis will be discussed in Chapter 10.

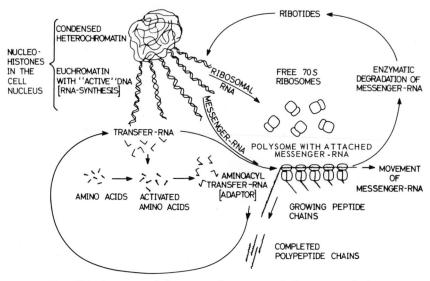

Fig. 107. Summary of the events from transcription to translation.

AGING AND MEMORY

Aging of an organism apparently affects the processes shown in Fig. 107. The template activity of native DNA for RNA synthesis seems to decrease with increasing age (e.g., Devi *et al.*, 1966). In general, it is more difficult to extract DNA in high yields from organs of older mam-mals than from those of younger animals. Wulf *et al.* (1962), who dis-cussed the few experimental observations made in this area, suggest that metabolic changes due to aging may be attributed in part to a "wearing out" of the DNA template. There are various hypotheses about the causes of aging. One of the more recent considers aging to be a genetically programmed process succeeding the development of an adult organism. It seems—and this would agree with the hypothesis—that in certain tissues of old rats, the ratio of heterochromatin to euchromatin (cf. p. 189) is somewhat higher than in those of younger animals, indicating a decrease in the availability of DNA templates for metabolic functions. Hydén (1962) observed that the RNA content at the anterior horn cells of the human spinal chord increases continuously until the age of 50–60 years and then falls again (Fig. 108).

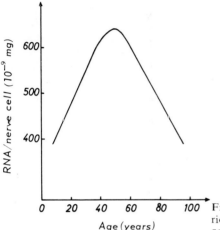

Fig. 108. The amount of RNA per anterior horn cell of man (in $\mu\mu$g) plotted against age (from Hydén, 1962).

More recently, investigations on the possible relationships between RNA-protein metabolism and memory were undertaken by several groups. Hydén and Egyházi (1962, 1963) found that changes in the base ratios of nuclear RNA in the Deiters' nerve cells of rats after intensive learning experiments persisted for at least 48 hours after the completion of the experiments and also involved the surrounding glia cells.* In the planaria *Dugesia dorotocephale*, a response to light and electric shock— which corresponds to a conditioned reflex—appeared to be retained after the organism had been cut in half and regeneration occurred. It made no difference whether regeneration took place from the tail or from the head (Corning and John, 1961). However, if ribonuclease was added to the medium in which regeneration occurred, only the heads retained their conditioning while that of the tails was eliminated. Although the authors' comment that their observations did not identify the substance responsible for the conditioned behavior of the regenerated planaria, it is tempting to assume that it might be RNA. In earlier ex-

---

*The results of newer experiments indicate that the observed base ratio changes are not specific to learning but do also occur in animals subjected to a pseudo-conditioning procedure [Hydén, H., Egyházi, E., Gohn, E. R., and Bartlett, F., *J. Neurochem.* (1967), in press].

periments, McConnell *et al.* (1959) claimed that feeding of trained planaria to untrained animals caused a certain transfer of the acquired abilities.   Later Jacobson *et al.* (1966) observed that such a transfer was also possible by injecting RNA isolated from trained planaria. Finally, it seemed that a memory transfer might even be possible with mammals: The injection of RNA-containing fractions from the brains of trained rats (donors) was reported to transfer a learned behavior to untrained animals (recipients) (Babich *et al.*, 1965).   However, much doubt has recently arisen about the validity of these experiments which could not be reproduced by other groups (cf. Byrne *et al.*, 1966); in addition, it was observed by Luttges *et al.* (1966) that P^{32}-labeled RNA isolated from the brain of rats could not be detected in the brain of recipient animals after intraperitoneal injection.

The importance of RNA-protein metabolism for the memory processes was also studied by the use of inhibitory and stimulating agents. Dingman and Sporn (1961) observed that intracisternal injection of 8-azaguanine, a guanine analog (cf. p. 318), into rats severely interfered with the maze learning. Intracerebral injection of puromycin (see p. 301) into the hippocampal and caudate cortices of mice caused loss of memory (Flexner *et al.*, 1963), apparently due to inhibition of protein synthesis. On the other hand, the drug magnesium pemoline, which is supposed to cause a selective stimulation of RNA polymerase in the brain (Glasky and Simon, 1966), was found to enhance the acquisition and retention of a conditioned avoidance response in rats (Plotnikoff, 1966).   Improvement of the learning capacity of rats was also reported to occur after the administration of yeast RNA (Cook *et al.*, 1963).   In a similar vein, attempts have been made to influence cerebral sclerosis clinically by the use of yeast RNA or a mixture of the four RNA ribonucleotides (cf. Cook *et al.*, 1963; for further discussion see Field and Abbott, 1963; Gaito, 1966).

Although there is now certain evidence that RNA and a group of proteins are indeed involved in the molecular processes of memory, nothing is known as yet about their actual roles.   Various models have been proposed in which RNA or specific proteins are considered as candidates for the organic substrate of memory.   So far, the relationship between the electrical response of a neuron and these macromolecules which are supposed to be altered (via neosynthesis?) is a complete mystery (cf. Brazier, 1964; Gaito, 1966; Hydén, 1967).

CHAPTER IX   *The Genetic Code*

INTRODUCTION

The biosynthesis of proteins with specific structures requires that the language of the genetic material (DNA) be translated into amino acid sequences. The translation process takes place at the ribosomes, using mRNA (not DNA directly) as a template. Thus the genetic code (which corresponds somewhat to a dictionary) can be defined as the relationship between the base sequences of an mRNA chain and the amino acid sequence of the particular protein which is synthesized at this RNA template. Detailed investigation of the genetic code became possible with the use of cell-free systems capable of protein synthesis. Earlier studies on mutants of tobacco mosaic virus had been begun by Wittmann; these investigations resulted in certain correlations between base changes in the viral RNA and changes in the amino acid composition of the viral coat protein. Before experimental approaches were available to determine the genetic code, certain theoretical considerations were possible; these will be discussed first.

The language of the nucleic acids has four symbols, the two purine bases adenine and guanine (now abbreviated A and G), and the pyrimidines uracil and cytosine (U and C). In order to determine one amino acid, the bases at the RNA template must be present in a specific combination (specific with respect to base composition and sequence). Such groups of bases are termed coding units or *codons*. The coding ratio is the ratio of the number of nucleotides in the RNA template to the number of coded amino acids. Since a protein may be made up of a maximum of 20 different amino acids, a two-base codon would not be sufficient, because only 16 (4^2) different combinations of the four bases are possible. This then suggested the triplet hypothesis, according to which each amino acid is determined by a combination of three bases (a triplet) in the mRNA. The triplet hypothesis received considerable support from the results of genetic studies, especially by Crick *et al.* (1961). An analysis of the observations on nitrite mutants of TMV (cf. p. 236) indicated that the coding ratio cannot be much higher than three, nor can it be a multiple of three (Wittmann, 1962). Later, Staehelin *et al.* (1964)

determined the coding ratio directly in polysomes; the results obtained agreed well with the triplet hypothesis.

Another question concerns the arrangement of the triplets in the template. Triplets may simply be arranged sequentially [Fig. 109(*a*)], or two triplets may also share one base [overlapping code, Fig. 109(*b*)] or

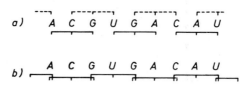

Fig. 109. Diagram of a code with (a) non-overlapping and (b) overlapping triplets. The dotted lines in (a) represent a frame shift of the original code.

even two. In a simple overlapping code, half of the bases would share in coding for two amino acids. If such a joint base sequence were altered, changes should result in two adjacent amino acids in the protein synthesized at this template. However, experiments with chemical mutants of TMV (Tsugita *et al.*, 1960; Wittmann, 1960, 1962) never led to such a coupled replacement of adjacent amino acids. Moreover, Brenner (1957) showed that an overlapping code would involve marked limitations in possible amino acid sequences, limitations that apparently do not occur in proteins whose amino acid sequences are known. Thus it was concluded that the code is not overlapping.

The triplet arrangement of bases permits $4^3 = 64$ different combinations. Because only 20 amino acids have to be coded, the question arose as to whether certain amino acids can be determined by more than one triplet (degenerate code) or whether there are only 20 functional triplets, the remaining 44 being nonsense codons (non-degenerate code). It has been determined that the code is highly degenerate (cf. Table 15).

A correct translation of the code from the template requires that the anticodon of the tRNA does not shift for one or two bases along the mRNA. Since the triplets on the RNA template apparently are not delimited (comma-less code), a phase shift must result in a completely wrong reading. Thus, it was tempting to assume (as a safety device) the existence of a non-degenerate code which cannot be read incorrectly because a phase shift, as shown in Fig. 109(a), would create only inactive (nonsense) codons. Such a hypothesis had been discussed by Crick *et al.* (1957), but it was contradicted later by experimental observations and withdrawn by the authors. Thus the correct reading of the information in an mRNA molecule depends on the step of chain initiation (cf. p. 234).

EXPERIMENTAL INVESTIGATIONS OF THE CODE

The discovery which initiated the modern developments in the study of the genetic code was made by Nirenberg and Matthaei (1961). In a cell-free system from *E. coli* which contained washed ribosomes, tRNA and a "supernatant" fraction, it was observed that the addition of polyU caused an enormous increase in the utilization of phenylalanine for protein synthesis and the formation of polyphenylalanine. Thus, only one type of nucleotide in the template, namely uridylic acid, appeared to be necessary for the coding of phenylalanine, the responsible triplet being UUU. Subsequently polyribonucleotides with different base compositions were introduced as messengers under similar conditions in the laboratories of Nirenberg and of Ochoa (cf. Jones and Nirenberg, 1962; Wahba *et al.*, 1963). Specific stimulation of the utilization of one or several amino acids by various artificial messengers enabled the investigators to correlate the base content of the triplets with the coded amino acids. However, the base sequences in these synthetic messengers which had been prepared mainly with polynucleotide phosphorylase (cf. p. 198) were unknown and assumed to be random. This approach, therefore, gave information only about the composition of the codons, while the sequence of bases within the triplets could not be determined.

An important step toward the definitive elucidation of the genetic code was the observation by Nirenberg and Leder (1964) that a trinucleotide, i.e., a single triplet, can interact with a ribosome and then specifically bind the tRNA with the appropriate anticodon. In this way it was possible to determine most of the 64 triplets (Brimacombe *et al.*, 1965; Nirenberg *et al.*, 1965). However, in certain instances the binding test did not give clear-cut results (very weak or multiple binding). Thus, the confirmation of the list of known codons by using messengers with strictly defined base sequences was of great value. By a combination of chemical methods and enzymatic synthesis, Khorana's group succeeded in preparing long RNA chains with various repeating sequences of bases. So far, all triplets determined this way (Nishimura *et al.*, 1965; Jones *et al.*, 1966; Jones and Nirenberg, 1966; agree with those found with the binding test. All 64 known triplets are compiled in Table 15. With the exception of methionine (which may play a role in chain initiation; cf. p. 224 and 234) and tryptophan, all amino acids are coded by several triplets (two, three, four or six). In general,

TABLE 15.   The 64 Possible Triplet Combinations (RNA Codons) and
Their Corresponding Amino Acids[a]

5′   3′

AAA	Lysine	CAA Glutamine	GAA Glutamic acid	UAA Nonsense (Ochre)
AAG	Lysine	CAG Glutamine	GAG Glutamic acid	UAG Nonsense (Amber)
AAC	Asparagine	CAC Histidine	GAC Aspartic acid	UAC Tyrosine
AAU	Asparagine	CAU Histidine	GAU Aspartic acid	UAU Tyrosine
ACA	Threonine	CCA Proline	GCA Alanine	UCA Serine
ACG	Threonine	CCG Proline	GCG Alanine	UCG Serine
ACC	Threonine	CCC Proline	GCC Alanine	UCC Serine
ACU	Threonine	CCU Proline	GCU Alanine	UCU Serine
AGA	Arginine	CGA Arginine	GGA Glycine	UGA Nonsense
AGG	Arginine	CGG Arginine	GGG Glycine	UGG Tryptophan
AGC	Serine	CGC Arginine	GGC Glycine	UGC Cysteine
AGU	Serine	CGU Arginine	GGU Glycine	UGU Cysteine
AUA	Isoleucine	CUA Leucine	GUA Valine	UUA Leucine
AUG	Methionine	CUG Leucine	GUG Valine	UUG Leucine
AUC	Isoleucine	CUC Leucine	GUC Valine	UUC Phenylalanine
AUU	Isoleucine	CUU Leucine	GUU Valine	UUU Phenylalanine

[a]According to observations reported by the groups of Khorana and Nirenberg. Explanations are given in the text.

the various triplets for one amino acid differ only in the last base. Only arginine and serine have two groups of triplets which also differ in the first or in the first and second bases respectively. Thus an amino acid is largely determined by the first two bases of its triplets.

The observed similarity of the various triplets which code the same amino acid suggested that the number of different tRNAs might be limited due to a multiple codon recognition. Crick (1966) has developed a hypothesis for codon-anticodon base pairing in which the bases in the first two positions of a triplet are read very strictly, while there may be some wobble in the pairing of the third base ("wobble hypothesis"). If the wobble hypothesis is correct—and in its present form it agrees with all the available data—it enables one to predict the probable nature of the anticodon of a particular tRNA according to certain rules for pairing the third base of a codon with the corresponding base on the anticodon. The rules are summarized in Table 16 which demonstrates the following: (1) It is not possible to code for either C or A alone. (2) If a tRNA contains inosine in a position to pair with the third base of the codon, it can pair with U, C and A. This excludes the possibility that tRNAs for amino acids such as cysteine or histidine (having C_1 or U as the third

TABLE 16. Base Pairing at the Third Position
of the Codon (from Crick, 1966)

Base on the Anticodon	Bases Recognized on the Codon
U	A G
C	G
A	U
G	U C
I	U C A

base of their codon) can have inosine in the corresponding place in the anticodon. (3) Although the wobble hypothesis cannot predict the number of different kinds of tRNA in a cell for a particular amino acid, it indicates that if an amino acid is coded by all four bases in the third position, at least two types of tRNA are necessary with the following recognition pattern

$$\left.\begin{matrix} U \\ C \end{matrix}\right\} \text{ plus } \left\{\begin{matrix} A \\ G \end{matrix}\right. \text{ or } \left.\begin{matrix} U \\ C \\ A \end{matrix}\right\} \text{ plus } G$$

The experimental evidence for the wobble hypothesis is based on the analysis of the nucleotide sequences of several tRNAs (cf. p. 103). If it is assumed that the codon-anticodon base pairing takes place in an anti-parallel manner, the available data are in excellent agreement with the wobble hypothesis (Crick, 1966):

Amino acid	Alanine	Serine	Tyrosine	Valine	Phenylalanine
Probable anticodon	$3'$CGI$^{5'}$	$3'$AGI$^{5'}$	$3'$AψG$^{5'}$	$3'$CAI$^{5'}$	$3'$AA 2'OMeG$^{5'}$
Codon	GCC $_{5'}$ A$3'$	UCC $_{5'}$ A$3'$	UA$\begin{smallmatrix}U\\C\end{smallmatrix}$ $_{5'}$ $_{3'}$	GUC $_{5'}$ A$3'$	UU$\begin{smallmatrix}U\\C\end{smallmatrix}$ $_{5'}$ $_{3'}$

For the valine tRNA from yeast, Ingram and Sjöquist (1963) found that its one inosine occurs in the position given above. The Ψ in the anticodon of tyrosine can form the same base pairs as U. Because all anti-codons must be similarly oriented on the tRNAs, Crick (1966) has predicted that the anticodon triplet is always at (or very near) positions 36-37-38, and that the first two bases in the codon pair with this, using standard base pairing.

The termination of peptide synthesis requires the release of a carboxyl group from its ester linkage to the final aminoacyl-tRNA (cf. p. 224). Various observations indicated that artificial polynucleotides used as messenger generally lack certain nucleotide sequences (codons) which appear to be essential for the release of polypeptide chains. It seemed plausible to assume that such "release" triplets should not code for any amino acid at all (nonsense codons). If a nonsense codon was produced by mutation, one would expect a premature release of a peptide, and this was indeed found to be the case (Sarabhai *et al.*, 1964). Thus a nonsense mutation blocks the complete synthesis of a polypeptide chain specified by the gene in which the mutation occurs. On the other hand, this effect can eventually be reversed by a phenomenon known as suppression, which enables the triplet to code again for an amino acid and thereby allows protein synthesis to proceed to completion (see p. 244). Analyses of the genetic control of suppression of nonsense mutations have contributed to the present knowledge of nonsense codons. The terms Ochre and Amber (Table 15) refer primarily to mutant strains of bacteria that can suppress certain nonsense mutations. It is not yet known whether all three nonsense codons are normally used for the process of chain termination. Although the triplet UGA has now been confirmed as a nonsense codon, it seems unlikely that its function is the termination of polypeptide chains; instead it is believed that this codon might be necessary as a "space" to separate genes in a polycistronic message (Brenner *et al.*, 1967).

Recently, insight was obtained into the mechanism of chain initiation in *E. coli*. Marcker and Sanger (1964) discovered a particular methionine tRNA upon which methionine can be formylated. Somewhat later, Adams and Capacchi (1966) and Webster *et al.* (1966), using N-formylmethionyl-tRNA containing labeled formyl groups, showed that these groups were incorporated into proteins, apparently initiating protein synthesis: In the peptides synthesized *in vitro*, the normal N-terminal alanine was covered by N-formylmethionine. Thus it seems that a start signal at a messenger leads first to the incorporation of an amino acid with a blocked amino group. Many *E. coli* proteins have either methionine, alanine or serine as their N-terminal amino acids (Waller, 1964). Therefore, it is postulated that after *in vivo* synthesis certain amino acids from the *E. coli* proteins just completed are split off by specific aminopeptidases; the number of amino acids removed probably depends on the tertiary structure of the various proteins. The discovery of this *start* signal for chain initiation now also explains why peptide synthesis cannot restart when it has been blocked due to an unsuppressed nonsense mutation. It is an open question whether chain initiation by N-formylmethionyl-tRNA as observed in *E. coli* is a general phenomenon and/or whether other amino acids might also be involved in starting protein synthesis. This principle of chain initiation should lead one to expect the 5′-terminal codon for the start signal to be recognized preferentially by the specific tRNA charged with an amino acid with a blocked amino group (in *E. coli* N-formylmethionyl-tRNA), while the same triplet in an internal position should code only for the tRNA with the normal amino acid (methionine in *E. coli*). From various observations, it is evident that the reading of a messenger proceeds in the 5′ ⟶ 3′ direction along the polynucleotide chain (e.g., Salas *et al.*, 1965; Thach *et al.*, 1965; Lamfrom *et al.*, 1966; Fig. 110). Because the process of transcription at

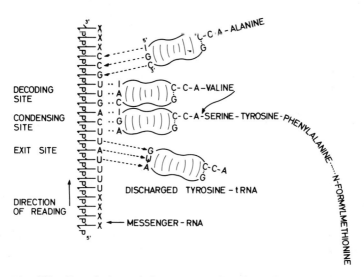

Fig. 110. Translation of the message by charged tRNA molecules at the ribosome (cf. Fig. 106).

the DNA template also takes place from 5′ to 3′ (with the same polarity; see p. 202), the consideration that protein synthesis may in some cases begin at the nascent mRNA (Goldstein *et al.*, 1965) is suggested. So far, the possibility that a few codons might be ambiguous cannot be excluded; these codons would then determine more than one amino acid. From all the available evidence it seems likely that the genetic code is universal or at least very similar in all species (cf. p. 240).

The correct reading of a message can be disturbed by various aminoglycoside antibiotics (streptomycin, neomycin, kanamycin, paromomycin, gentamycin, hygromycin B and viomycin; cf. p. 302). The extent and nature of misreading vary with the antibiotic used and with the tRNA concentration (Davies *et al.*, 1965). Both transition and transversion misreadings (see p. 000) seem to be possible. Another effect of streptomycin is to permit single-stranded DNA to be read directly as a messenger (McCarthy and Holland, 1965; cf. p. 225).

Substitution of the four standard bases in a messenger may have various effects. Poly(ribo)T was found to code for phenylalanine in the same way as polyU. However, a 5-bromocytidine polynucleotide coded for threonine in addition to proline (Grunberg-Manago and Michelson, 1964); 2-aminopurine has coding properties somewhere between adenine and guanine (Wacker *et al.*, 1966), while dihydrouridine apparently cannot be recognized at all when it is part of a codon (Rottman and Cerutti, 1966).

STUDIES OF MUTATIONS

The correctness of the code as summarized in Table 15 can be checked (not proved) by studying the effects of mutations on the amino acid sequence of a protein. In this type of experiment, protein synthesis takes place in intact cells and not in cell-free systems. Base changes in a polynucleotide chain can be produced chemically; the altered template properties then may result in an exchange of amino acids in the coded proteins. Studies of this kind were first done with TMV (see p. 261). The genetic information of a virus is contained in its nucleic acid (RNA in the case of TMV) which may maintain infectivity after careful isolation and can, under appropriate conditions, lead to the formation of new, complete viruses. As an example of base alteration, the effects of nitrous acid will be discussed briefly. By treatment with nitrous acid, it is possible to deaminate some of the bases in the TMV RNA, thus producing an altered RNA which in some cases is still infectious. This then directs the formation of "base substitution" mutants of the original virus. In many instances, the base changes in the RNA of those mutants result in a change in the amino acid composition of the protein coat. Base deamination leads only to transitions (see p. 241), i.e., changes within the existing polynucleotide purines or pyrimidines, with no exchange of purines for pyrimidines and *vice versa* (transversions).

Nitrous acid can convert cytosine to uracil (C $\longrightarrow$ U), adenine to hypoxanthine (A $\longrightarrow$ H), and guanine to xanthine (G $\longrightarrow$ X). In the course of synthesis of new TMV RNA, the "false" bases cannot be replicated, since only naturally occuring bases are available as riboside triphosphates for RNA biosynthesis. In the RNA template, hypoxanthine now acts functionally like guanine. The deamination of adenine gives rise therefore, in the subsequent generation, to a transition to guanine [A ⟶ (H) ⟶ G]. Likewise, the deamination of cytosine is a permanent change in the RNA template (C ⟶ U). It was first assumed that the xanthine produced by the deamination of guanine functions in the RNA template like guanine (Wittmann, 1961); the deamination of guanine was therefore considered to be reparable [G ⟶ (X) ⟶ G] or not to impair the template function of the viral RNA. However, the enzymatic studies of Basilio *et al.* (1962) indicated that this is not the case. Thus, it seems doubtful that xanthine is again replaced by guanine in the synthesis of viral RNA, just as xanthine cannot be utilized as its deoxyriboside triphosphate in the enzymatic synthesis of DNA (Bessman *et al.*, 1958).

Only some of the mutants of TMV produced by treatment with nitrous acid result in changes in the amino acid composition of the viral protein

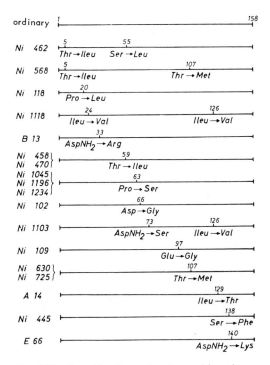

Fig. 111. Localization of amino acid replacements in the protein subunits of 18 TMV mutants. Fifteen of the mutants were produced by treatment with nitrous acid (Ni), the remaining ones arose spontaneously (from Wittmann, 1962).

(Wittmann, 1961, 1962; Tsugita and Fraenkel-Conrat, 1962). This seems quite understandable since the TMV RNA also directs the metabolic processes which lead to virus multiplication (see p. 265), and therefore only a certain part of the RNA determines the amino acid sequence in the protein coat of the completed virus. Figure 111 shows the localization of the exchanged amino acids of 18 mutants of TMV; the content of 158 amino acids per protein subunit is maintained in all mutants.

All the results with base-substituted mutants of TMV, obtained by the German group in Tübingen and the American group in Berkeley, indicate

TABLE 17.   Changes of Amino Acids in Abnormal Hemoglobins (from Braunitzer and Rudloff, 1962) together with the Appropriate Codons[a].

Type of Hemoglobin	From	Exchange	To
		5′ 3′ 5′ 3′	
C	Glutamic acid	$\begin{cases} \underline{G}AA \rightarrow \underline{A}AA \\ \underline{G}AG \rightarrow \underline{A}AG \end{cases}$	Lysine
D$_{Punjab}$	Glutamic acid	$\begin{cases} G\underline{A}A \rightarrow \underline{C}AA \\ G\underline{A}G \rightarrow \underline{C}AG \end{cases}$	Glutamine
E	Glutamic acid	$\begin{cases} \underline{G}AA \rightarrow \underline{A}AA \\ \underline{G}AG \rightarrow \underline{A}AG \end{cases}$	Lysine
G$_{Honolulu}$	Glutamic acid	$\begin{cases} G\underline{A}A \rightarrow \underline{C}AA \\ G\underline{A}G \rightarrow \underline{C}AG \end{cases}$	Glutamine
G$_{San\ José}$	Glutamic acid	$\begin{cases} GA\underline{A} \rightarrow G\underline{G}A \\ GAG \rightarrow G\underline{G}G \end{cases}$	Glycine
M$_{Boston}$	Histidine	$\begin{cases} \underline{C}AU \rightarrow \underline{U}AU \\ \underline{C}AC \rightarrow \underline{U}AC \end{cases}$	Tyrosine
M$_{Milwaukee-1}$	Valine	$\begin{cases} GU\underline{A} \rightarrow G\underline{A}A \\ GU\underline{G} \rightarrow G\underline{A}G \end{cases}$	Glutamic acid
M$_{Saskatoon}$	Histidine	$\begin{cases} \underline{C}AU \rightarrow \underline{U}AU \\ \underline{C}AC \rightarrow \underline{U}AC \end{cases}$	Tyrosine
M$_{Zürich}$	Histidine	$\begin{cases} CA\underline{U} \rightarrow C\underline{G}U \\ CA\underline{G} \rightarrow C\underline{G}C \end{cases}$	Arginine
Norfolk	Glycine	$\begin{cases} GG\underline{U} \rightarrow G\underline{A}U \\ GG\underline{C} \rightarrow G\underline{A}C \end{cases}$	Aspartic acid
S	Glutamic acid	$\begin{cases} GA\underline{A} \rightarrow G\underline{U}A \\ GAG \rightarrow G\underline{U}G \end{cases}$	Valine

[a]The changed bases are underlined. Although there are two possible ways for base changes to occur, most of them result from transition.

that the exchange of an amino acid in the coat protein is due to the. alteration of a single base in the viral RNA. In a similar way, an exchange of single amino acids was found in various abnormal hemoglobins which occur in certain hereditary human diseases. In Table 17, several such single amino acid exchanges (about 40 are now known) have been correlated with the triplets taken from Table 15. Again it can be seen that the alterations in the proteins can be ascribed to changes of single bases in the templates.

Another type of mutation results from the addition or deletion of single bases, thus causing a shift in the reading frame (cf. Fig. 109) and, as a consequence, a completely erroneous reading of a message. A second frame-shift mutation can eventually correct the error and restore the normal reading; if this correction is located close enough to the first mutation, a functionally active protein might be produced. Recently, an even more complicated mutation of this kind was analyzed (Okada et al., 1966), showing that the lysozyme primary structure from a double-mutant strain (eJ17eJ44) of phage T4 differed from that of the wild type by a sequence of five amino acids and by the insertion of an additional amino acid next to these. Using the triplets from Table 15, the probable nucleotide sequence can be deduced for the messenger which would code for the wild-type lysozyme and for that observed in the double-mutant strain (Fig. 112).

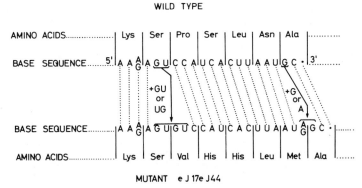

Fig. 112. Two frame shift mutations in the lysozyme produced by T4 phage. The double mutant strain eJ17eJ44 of phage T4 leads to a protein with one additional amino acid. As a consequence of the frame shift mutation at the left (insertion of two bases), the message cannot be correctly translated (exchange of five amino acids). The second frame shift mutation (insertion of one additional base, then permits again a correct reading (from Okada et al., 1966). (Reproduced with permission of National Academy of Sciences)

Thus far, all studies on the effects of mutations on amino acid changes in proteins have lead to results which are in excellent agreement with the genetic code as elucidated directly from experiments in cell-free systems. So far, this agreement has not been limited to certain species, which supports the view that the code is (at least) largely universal and has apparently remained constant over a long period of evolution. It seems plausible to assume that life might have begun with a more "primitive" code, simpler than the present one. Although there is no approach available to investigate whether such a primordial code did indeed exist, it may be stated that the origin of life is intimately connected with the origin of the genetic code.

CHAPTER X *Biochemistry of Mutations*

An essential property of the hereditary information is its stability. Gene replication generally leads to exact duplication of parent molecules. However, errors sometimes occur, causing certain alterations in the gene templates, with more or less harmful consequences. Normally such (spontaneous) alterations are relatively rare, but their frequency can be increased considerably with so-called mutagenic agents which include ultraviolet light, ionizing radiation and a variety of chemicals. Chemical mutagens and ultraviolet light have become particularly important tools in the field of molecular genetics. The preceding section illustrated the importance of mutations in the investigation of the genetic code. In this chapter, various minor mutations will be discussed in general terms. The problem of altering nucleic acid templates will be considered again in the chapters on viruses, chemotherapy and radiation effects (cf. pp. 256, 294 and 329).

The possible types of base alteration in a double-stranded DNA chain are summarized in Fig. 113. A *base pair* can be *exchanged* (*replacement*) or *lost* (*deletion*), and finally an *additional base pair* may be *inserted* during the process of replication. Two possible groups of base exchanges—*transitions* and *transversions*—can be distinguished (Freese, 1959b). In transition, a purine is converted to another purine (and, correspondingly, the complementary pyrimidine to another pyrimidine). Transversions are exchanges from purine to pyrimidine, and vice versa (Fig. 114). Thus far, among induced mutations, transversions have not been clearly demonstrated. Base exchanges, predominantly transitions, can be induced by the incorporation of base analogs (cf. pp. 271 and 310) or by chemical treatment of nonreplicating nucleic acid. A relatively well-understood example of the latter type is the deamination of nucleic acid bases by nitrous acid, whose effects on TMV RNA have already been discussed in connection with the genetic code (see p. 236). Figure 115 shows the mutagenic effect of nitrous acid on phage DNA. As a consequence of a single base transition in double-stranded DNA, half of the progeny DNA is altered, the other half unaltered; in single-stranded DNA, the effect corresponds to that in TMV RNA.

241

G	T	A	C	T	A	G	T	A		
C	A	T	G	A	T	C	A	T		ORGINAL DNA
1	2	3	4	5	6	7	8	9	10	

G	T	A	G	T	G	G	T	A		
C	A	T	C	A	C	C	A	T		REPLACEMENT
1	2	3	4	5	6	7	8	9	10	

G	T	A	C	T	G	T	A	C		
C	A	T	G	A	C	A	T	G		DELETION
1	2	3	4	5	7	8	9	10		

G	T	A	C	T	U	A	G	T	A	
C	A	T	G	A	G	T	C	A	T	INSERTION
1	2	3	4	5		6	7	8	9	

Fig. 113. Various types of base pair exchanges in DNA. Deletion as well as insertions cause frame shift mutations (cf. p. 239
and Fig. 112).

Base analogs incorporated into DNA may affect the template properties, thus causing errors in replication and transcription. 5-Bromouracil
(BU), a thymine analog, can replace a considerable part of the normal
thymine in bacterial DNA. The relatively low mutagenic effect of this
base analog is apparently caused by a shift of the tautomeric equilibrium
toward the enol state (Terzaghi *et al.*, 1962); thus the probability of a
wrong reading (BU-G instead of BU-A) is increased (Fig. 116), finally
leading to transitions from A-T to G-C. The incorporation of 5-bromoura-
cil into T4 phage DNA induces specific mutations (Benzer and Freese,
1958). The replacement of normal thymine by BU makes DNA much more

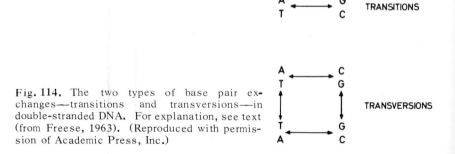

Fig. 114. The two types of base pair exchanges—transitions and transversions—in
double-stranded DNA. For explanation, see text
(from Freese, 1963). (Reproduced with permission of Academic Press, Inc.)

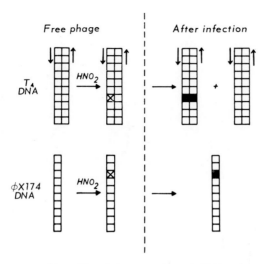

Free phage | After infection

Fig. 115. The behavior of viral DNA after mutation by *in vitro* treatment with nitrous acid: (a) T4 phages, double-stranded DNA; (b) φX174 phage, single-stranded DNA (from Tessman, 1962).

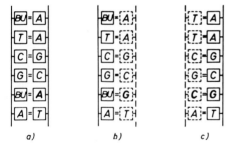

a)   b)   c)

Fig. 116. The mutagenic effect of 5-bromouracil (5-BU). 5-BU can replace part of the normal thymine in the DNA (a). During replication of this 5-BU-containing DNA, a small fraction of the analog may be present in the enol form which then pairs with guanine (b). The subsequent DNA replication leads to a permanent transition of T-A to C-G (c).

sensitive to chemical attack, ultraviolet and visible light (see p. 329), and ionizing radiation (cf. p. 334). Apparently, these effects are lethal and not mutagenic. 2-Aminopurine (AP) is a highly effective mutagen with a different mechanism. When incorporated into DNA, AP can pair in its normal tautomeric form with either thymine (by two hydrogen bonds) or cytosine (by one hydrogen bond; Fig. 117), the former being more frequent (Freese, 1959a). Thus it seems that AP induces transitions in both directions.

The effects of certain induced mutations can be reversed by other mutagenic agents. Such a reversion may only be phenotypic, if in the process of translation wrong information leads to a correct reading, i.e., the

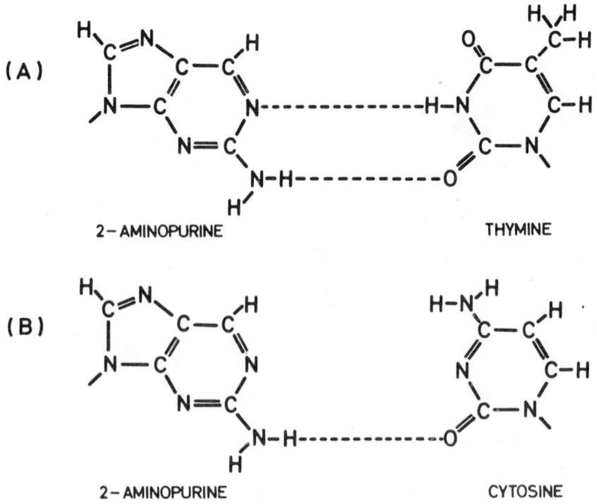

Fig. 117. More frequent (A) and rare (B) base pairing of 2-amino-purine (from Freese, 1963). (Reproduced with permission of Academic Press, Inc.)

formation of a protein with a normal amino acid sequence. 5-Fluorouracil (FU; cf. p. 312) is an analog which is incorporated exclusively into RNA. Fluorouracil may sometimes change the properties of an RNA template—as bromouracil does in DNA—resulting in a base pairing of FU-G instead of FU-A (Fig. 118). Partial phenotypic reversions due to FU have been observed in some mutants of T4 phages (Champe and Benzer, 1962) and *Neurospora crassa* (Barnett and Brockman, 1962); mutants of the latter could also be partially reversed by 8-azaguanine (see p. 318).

A mutation may also be rendered ineffective because of *phenotypic suppression*. Among suppressible mutations, the nonsense mutations have been analyzed in detail both chemically and genetically. It was concluded (Benzer and Champe, 1962; Garen and Siddiqi, 1962) that the suppression of nonsense mutations occurs at the level of translation. While unsuppressed nonsense mutations cause a premature termination of peptide chains, suppression results in the insertion of an amino acid at the position of the nonsense mutation (which is actually not read as a nonsense codon), thus permitting the growing peptide to be completed (Notani *et al.*, 1965; Stretton and Brenner, 1965; Weigert and Garen, 1965). Suppression of a nonsense mutation in an RNA phage has been achieved *in vitro* by the addition of tRNA from another bacterial strain (Engelhardt *et al.*, 1965). This indicates that the wrong reading of a nonsense mutation due to suppression results from an altered tRNA (which codes for serine in the case of the bacterial suppressor gene Su-1;

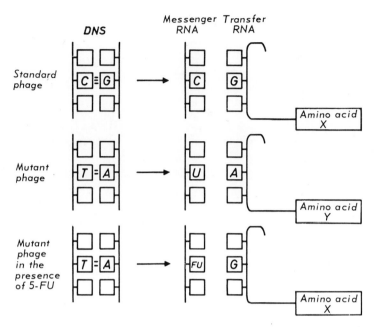

Fig. 118. A proposed mechanism for the phenotypic reversion of mutations by 5-fluorouracil.   For explanation, see text (from Champe and Benzer, 1962).

Engelhardt *et al.*, 1965).   It is not yet known whether the structure of this altered tRNA is determined directly by a suppressor gene or indirectly via an enzyme which can modify an existing tRNA.   Amber and Ochre are the names of two types of nonsense mutants which were ana-- lyzed as UAG and UAA, respectively (Brenner *et al.*, 1965; cf. Table 15). By genetic studies, it was found that Amber suppressors can suppress exclusively Amber mutants.   Ochre suppressors are defined by their suppression of nonsense mutations which are not affected by Amber suppressors.   While Ochre suppressors may suppress Amber mutants, the converse does not seem to be true (Signer *et al.*, 1965).

In most cases, the exchange of a single base in a template leads to an exchange of the one amino acid which is coded by the altered triplet. However, if the third base of a triplet is changed, this may not affect the coded protein at all (especially in the case of transitions), because of the degeneracy of the genetic code (cf. Table 15).   When all the codons remain in phase following base exchanges, deletions or insertions of single bases cause a phase shift, and thus a wrong reading (cf. Fig. 112). Such changes of a DNA template seem to occur during the process of replication if the DNA has previously interacted with certain chemical

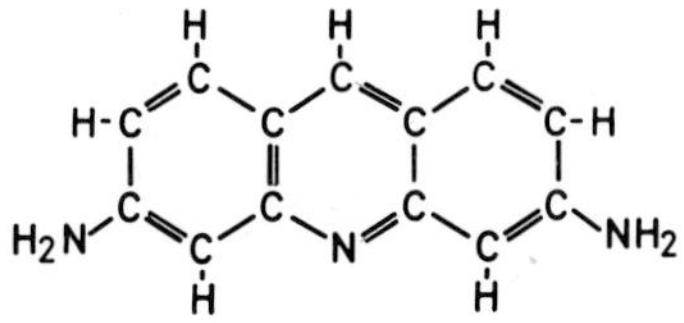

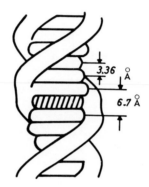

Fig. 119. Structure of proflavine.

agents. Lerman (1961) has postulated that proflavine (Fig. 119) can in-
tercalate between two adjacent purine bases of DNA (Fig. 120); this
conclusion is based from X-ray data and on the observation that proflavine greatly increases the viscosity of DNA solutions. The postulated intercalation effect of proflavine led Brenner *et al.* (1961) to pro-

Fig. 120. Intercalation of a proflavine molecule between two base pairs of the DNA helix (from Lerman,
1963).

pose that acridines induce mutations only by producing deletions or insertions of single base pairs in DNA. The mechanism of action of
alkylating agents which may also induce phase shift mutations will be
discussed in a later chapter (see p. 304).

Spontaneous mutations may be the result of a variety of events. Some
of these mutations result from mistakes in the process of replication.
The (rare) enol form of thymine may pair with guanine, or the (still less
frequent) imino form of cytosine with adenine, thus causing transitions.
Transversions would require a wrong pairing between two purines or two
pyrimidines (Freese, 1959b).

The molecular mechanisms of mutations have been reviewed in detail
by Freese (1963).

CHAPTER XI *Control of Metabolism*

Correct and cooperative functioning of all the different metabolic processes within a living cell or a tissue requires the existence of regulatory systems. The investigation of these systems is becoming an increasingly important and exciting field. Various control mechanisms are effective at different levels and with different rates; they can be triggered by many agents (e.g., substrates for specific enzymes or hormones), and in certain cases they seem to influence each other, either by amplification or damping.

DIFFERENTIATION

Although all the somatic cells of a multicellular organism contain a complete set of chromosomes (either diploid or occasionally polyploid), not all the genetic information can be used for transcription by a given cell. There is a restriction of the DNA template (cf. Paul and Gilmour, 1966), corresponding to the cell's special functions. For some time it was thought that histones might be responsible for this restriction in the chromatin. This view no longer appears to be acceptable; rather, it seems that the restriction of DNA occurs via heterochromatization (cf. pp. 89, 189 and 205). Thus in somatic cells only the DNA of euchromatin should be available for transcription. In agreement with this, it was found that the ratio of euchromatin to heterochromatin differs in various rat tissues (Harbers and Vogt, 1966); the highest ratio was observed in the liver (which also contains the largest number of enzymes); the lowest, in tumors (see p. 284). The factors that determine differentiation are not yet known. Although in the very early stages of embryologic development the full potentialities of fertilized eggs are retained, these potentialities are later curtailed as cell differentiation proceeds. These observations suggest that heterochromatization might be an irreversible process. Although this is still an arguable point, it seems useful to consider the "inactive" heterochromatin as those parts of the genetic information which have become switched off permanently at particular stages of differentiation (without completely excluding the possibility of reversibility under certain conditions), in contrast to the phenomena of repression and de-repression which will be discussed in the following

247

section. Therefore, the terms "repressed chromatin" (for heterochromatin) and "de-repressed chromatin" (for euchromatin; cf. Frenster, 1965) should be avoided.

REGULATION OF PROTEIN SYNTHESIS

In a normal cell, the rate of production of different proteins is a function of the varying requirements of intermediary metabolism. Thus a control mechanism is needed which permits a selective synthesis of specific proteins. The existence of such a regulatory device was first demonstrated with the discovery of *enzyme induction*. If a cell is provided with a substrate whose utilization requires an enzyme that either is absent or has low activity, then the substrate may rapidly induce the formation of this enzyme. The synthesis of the enzyme will be stopped when the substrate has been depleted. This phenomenon was originally called "adaptive enzyme formation."

The processes of enzyme induction have been most extensively studied in microbial systems, especially in *E. coli* β-galactosidase, an enzyme which splits lactose into glucose and galactose. It has also been shown that enzyme induction occurs in mammalian cells. Kinetic studies in the presence of various inhibitors indicated that the introduction of an appropriate substrate (inducer) is followed by the synthesis of a specific RNA necessary for subsequent enzyme formation. Under optimal conditions, the time between the addition of the inducer and the appearance of the enzyme (lag period) is quite short (about 3 minutes in *E. coli*). When the inducer is removed, enzyme formation in microorganisms rapidly ceases, in about 5 minutes for *E. coli* (Pardee and Prestidge, 1961). In mammalian tissues, enzyme synthesis appears to continue up to several hours longer (Feigelson *et al.*, 1959).

From extensive genetic studies of a variety of *E. coli* mutants, Jacob and Monod (1961a and b) developed a tentative model for the control of gene activity which was able to explain enzyme induction as well as the regulation of anabolic processes. In both cases, the control of gene activity occurs through a negative feedback system. According to the theory of Jacob and Monod, the *regulation of enzyme synthesis results from the interplay of three different kinds of genes*. The amino acid sequence of a protein is determined by the nucleotide sequence of a *structural gene* via an mRNA. The transcription of a structural gene is controlled by an adjacent specific region of the DNA, the *operator gene*. A single operator gene may control several structural genes, together with which it forms a unit termed the *operon* (Fig. 121). The function of an operator can be blocked by a *repressor* (R), which is

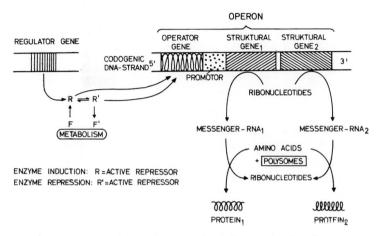

Fig. 121. Model for the control of the synthesis of gene-dependent proteins (modified from Jacob and Monod, 1961a). (Reproduced with permission of Academic Press, Inc.)

synthesized via a *regulator gene*. The repressor molecules (which appear to be proteins) can exist in active or inactive form; they interact very specifically with small molecules (specific metabolites) termed *effectors* (F). There are two distinct kinds of effectors: *inducers* and *corepressors*. The interactions of repressor and effector may be described simply:

$$R + F \rightleftharpoons R' + F'$$

In enzyme induction, only the R form of the repressor will interact with the appropriate operator gene. Combination with the effector (in this case as an inducer) then leads to the trapping (inactivation) of the repressor and the release of the operator gene and, consequently, of the processes leading to protein formation. In the control of anabolic processes, on the other hand, only the altered repressor R' has an affinity for the appropriate operator gene. Here the effector (i.e., the corepressor) blocks the activity of the operon (repression of enzyme formation); only the elimination of the corepressor, often a metabolite at the end of a series of biosynthetic reactions, permits the operon to become active again. Functionally cooperating enzymes are often located within one (polycistronic) operon (cf. p. 225); thus a single mRNA molecule might carry the messages of all the structural genes from one operon which is controlled by one regulatory gene. While it seemed at first that transcription (i.e., synthesis of mRNA) begins very close to the operator, it was observed later that an additional section of DNA (a *promotor*) must exist between the operator and the structural gene (Jacob and Monod, 1965); the promotor appears to be the starting point for mRNA synthesis.

When the Jacob-Monod model was developed, nothing was known about the nature of the postulated repressor. Recently, the β-galactosidase repressor in *E. coli* was isolated and identified as a protein with a molecular weight of about 150,000–200,000, occurring in about 10 copies per gene (Gilbert and Müller-Hill, 1966). Lactose itself is not the most efficient inducer for the synthesis of β-galactosidase; rather this seems to be a derivative, isopropylthiogalactoside. In an earlier chapter on the biosynthesis of nucleotides, the role of allosteric proteins (cf. Monod *et al.*, 1963) in feedback inhibition at the enzyme level was discussed (cf. p. 154). The binding of specific metabolites at a site distinct from the active site of the enzyme causes an allosteric transition, i.e., an altered conformation of the protein which then modifies the properties of the active site. In this way, regulation of enzyme activity occurs by negative feedback inhibition. It has been suggested that repressors are allosteric proteins whose conformation is changed by the binding of specific effectors. So far, nothing is known about the interaction between the repressor and the DNA of the operator. Because this interaction must be highly specific, one may speculate that the rare bases—or in the case of nucleohistones, certain protein moieties—might provide this specificity. Mutation of an operator gene (and also of a regulator gene) can cause both inducible and repressible enzymes to become constitutive (i.e., their synthesis is no longer under genetic control).

The model of Jacob and Monod provides a mechanism which permits the regulation of gene activities for enzyme induction as well as enzyme repression by a negative feedback system. This control device allows the cell to operate very economically. When enzymes are required for the formation of necessary cell constituents, they are synthesized; when they are not needed, they will disappear. Enzyme repression, i.e., blocking the biosynthesis of new protein, may be accompanied by an allosteric inhibition of the preexisting enzyme (cf. p. 154). In such situations, both control systems might cooperate to inhibit or even terminate a sequence of metabolic reactions (see Fig. 122). However, by de-repression (i.e., synthesis of new enzyme), it is possible to increase the total amount of the enzyme, which results in an altered enzyme level. As a consequence, the adjustment of the allosteric feedback mechanism is changed (higher concentration of the allosteric protein requires a higher level of the interacting metabolic end product). Observations by Nass and Neidhardt (1967) indicate that the rate of synthesis of some aminoacyl-tRNA synthetases is at least partially controlled by repression which can be lifted by restriction of the amino acid supply. However, an additional regulatory mechanism seems to be involved.

There is good evidence that the Jacob-Monod model also applies in the cells of multicellular organisms, where its effectiveness is probably

ENZYME   REPRESSION

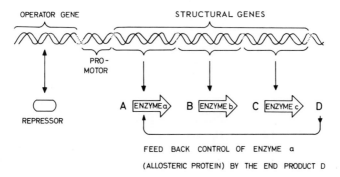

Fig. 122. Cooperative effects of enzyme repression and allosteric inhibition in the control of anabolic metabolism.

being augmented by the interaction of additional agents such as
hormones (cf. p. 252). If the concept of the role of euchromatin and
heterochromatin is correct (see pp. 89 and 189), the processes of enzyme repression and induction should be limited to the DNA of euchromatin in which the only available operons and appropriate regulator
genes would be located (cf. p. 248). One may consider as an integral
part of, or addition to, the Jacob-Monod mechanism the processes which
are alleged to modify the fine structure of nucleohistones which contain
in their histone moities acetyl, methyl (cf. p. 90) and perhaps phosphoryl groups (Stocken, 1965). In isolated lymphocytes, nucleohistone
acetylation occurs mainly in the arginine-rich fraction; the process is
independent of protein synthesis and in intact nuclei is probably limited
to the euchromatin (Allfrey *et al.*, 1966). As a consequence of the
acetylation of nucleohistones, the template activity for RNA polymerase
increases. A high ATP-dependent turnover rate of protein-bound phosphate in the chromatin, especially in the euchromatin fraction, is also
considered as part of a mechanism for modifying the structure of nucleohistones (Allfrey *et al.*, 1966). However, one must bear in mind the fact
that procedures for the isolation of nucleohistones, and especially of
their subfractions, do not yield highly purified material.

The Jacob-Monod mechanism is discussed in more detail in the following reviews: Jacob and Monod (1961a and b, 1963); Monod (1962).

OTHER CONTROL SYSTEMS

The two types of control mechanisms discussed in the preceding section do not suffice for an effective and fairly flexible regulation of cell

metabolism. Thus, additional control systems acting at different levels have been postulated. So far, there is not enough experimental evidence to prove these various hypotheses, examples of which will be mentioned briefly here.

From the results of kinetic studies with a polycistronic messenger, Ohtaka and Spiegelman (1963) concluded that a control mechanism which determines the time sequence and the frequency of translation of each cistron must exist. Because of the degeneracy of the genetic code, several types of tRNA are required for each amino acid (cf. p. 230). This suggests that tRNA may also have a regulating function in the translation process. Stent (1964) suggested that variations in the levels of some kinds of tRNA molecules relative to others might be an integral part of enzyme induction and repression. In the "adaptor modification hypothesis," Sueoka and Kano-Sueoka (1965) postulated that by modifying a specific tRNA molecule the function of some of the genes which are transcribed can be shut off, while the other genes are kept functional at the translation level. The template stability of messengers for different enzymes may vary considerably in highly differentiated cells; as suggested by Pitot and Peraino (1964), this may be due to interactions of polysomes with membranous surfaces within the cytoplasm of the cell, i.e., mainly with the endoplasmic reticulum. In this connection, it is important to note that the lifetime of mRNA can be altered in hepatomas, compared to that in normal liver (cf. p. 284).

ACTION OF HORMONES

Hormones play a key role in the coordination of cellular functions in multicellular organisms. It seemed useful to extend the Jacob-Monod concept and consider the primary action of certain hormones as a selective de-repression or repression of specific genes, thus regulating the amount of enzymes and other cell constituents. According to this view, the hormones would act as allosteric effectors. Such a hormone-gene hypothesis introduced by Karlson (1961, 1963) and Zalokar (1961) proved very stimulating. However, hormonal effects are very diverse, and certain regulatory activities of hormones seem to result from direct interaction with enzymes, membranes or allosteric proteins which are rate limiting for particular metabolic processes (cf. Tata, 1964). On the other hand, a stimulation of RNA synthesis by certain hormones in the target organs could indeed be demonstrated. This was generally considered to reflect the formation of new mRNA (cf. Table 18); however, since all the criteria necessary to characterize mRNA were not applied, it seems that much of the rapidly labeled RNA observed was of

TABLE 18. Effective Stimulation (+) of Nuclear RNA Synthesis *in vivo* or in Isolated Nuclei by Various Hormones[a]

Hormone	Species	Target Tissue	Stimulatory Effect on the Incorporation into RNA	Authors
ACTH	Rat	Adrenals	+	Branson and Chargaff (1964)
Cortisol	Rat	Liver	+ Stimulation of RNA coding for tyrosine transaminase	Jervell (1963); Kenney and Kull (1963); Sekeris and Lang (1964a); Dukes and Sekeris (1965); Lang and Sekeris (1966)
Ecdysone	Blowfly (*Calliphora*)	Epidermis and whole larvae	+ Stimulation of RNA coding for DOPA decarboxylase	Sekeris and Lang (1964b); Sekeris, Dukes and Schmid (1965); Sekeris, Lang and Karlson (1965); Karlson (1966)
Estrogen	Rat	Uterus	+ mRNA from uterus of estrogen-treated rats, applied intrauterine to ovarectomized animals caused morphological changes of the endometrium similar to those induced directly by estrogens	Hamilton (1963); Wilson (1963); Mansour and Niu (1965); Mueller (1965); Segal, Davidson and Wada (1965)
Growth hormone	Rat	Liver	+	Korner (1964); Talwar *et al.* (1964)
Insulin	Rat	Diaphragm	+	Wool and Munro (1963)
Testosterone	Rat	Prostate Seminal vesicles	+	Liao (1965) Greenman *et al.* (1965)
Thyroid hormones	Rat Tadpole Tadpole	Liver Liver Tail	+	Tata and Widnell (1966) Tata (1965, 1966a)
TSH	Sheep	Thyroid	+	Begg and Munro (1965)
Plant hormones and regulators (auxin, cytokinins, gibberellins)	Various plants	−	+	Van Overbeek (1966)

[a]Most of the references are selected examples.

the ribosomal type. Although hormone action can be inhibited by actinomycin (cf. p. 295), the interpretation of this as evidence for direct hormonal control of mRNA synthesis has frequently been criticized. The problem of the specificity of hormone action being limited to certain target tissues is intimately connected with the question of the intra-cellular localization of receptors. Several groups have reported the oc-currence of hormones in the cell nuclei (cf. Karlson *et al.*, 1964; Wilson and Loeb, 1965a) and the interaction (especially of steroids) with his-tones (cf. Sluyser, 1966). In studies on the intranuclear localization of labeled testosterone in the preen gland of the duck, Wilson and Loeb (1965b) observed that the highest uptake occurred in the euchromatin fraction (see pp. 89 and 189). Dahmus and Bonner (1965) claim a 30% increase in template activity of (total) nucleohistones isolated from livers of adrenalectomized rats which had been injected with hydrocorti-sone; this was explained by the postulated existence of a hormone-repressor (histone) interaction. However, Dingman and Sporn (1965), using labeled hydrocortisone, could detect very little hormone in liver nuclei. It is hard to derive definite conclusions from experiments of this type because an essential part of the hormone might be extracted during the isolation of nuclei or nucleohistones. Furthermore, the actual amount of hormone at the few receptor sites might be extremely small and there-fore hard to detect.

There appear to be two possible kinds of hormonal stimulation of RNA synthesis in the target tissues. First there is an increase in the forma-tion of rRNA, and later in (specific) mRNA (Tata, 1966b). The hormonal induction of certain enzymes is highly specific; in these cases, more of the DNA template—which was previously repressed—must become available for transcription. Hill *et al.* (1964) have additional evidence to suggest the existence in nuclei of pre-formed polysomes. Hiatt *et al.* (1965) observed that hydrocortisone can increase mRNA transport into the cytoplasm of liver cells. These observations led Tata (1966b) to suggest that a hormone may exert its action in the cell by coupling the synthesis of mRNA (which contains the information for specifying the en-zymes induced by the hormone) with that of rRNA (which provides "pro-grammed" polysomes for transport into the cytoplasm).

Attempts have been made to demonstrate the direct biological effects of mRNA whose synthesis had been previously induced by hormone ac-´ tion. Segal *et al.* (1965) and Mansour and Niu (1965) reported uterotrophic effects due to the application of RNA extracts isolated from the uteri of rats that had received estrogen treatment. Sekeris and Land (1964a) claimed that nuclear RNA from rat liver, extracted with hot phenol from animals pretreated with cortisone, induces the *in vitro* hormone-specific synthesis of the enzyme tyrosine transaminase. In a similar way, RNA isolated from epidermal nuclei of insects to which the hormone ecdysone has been applied causes the formation of DOPA decarboxylase in a cell-free system of rat liver (Sekeris and Lang, 1964b).

This discussion of the action of hormones has been limited here to their effects on the metabolism of RNA and proteins. An article by Tata (1966b) deals with also further biochemical effects of hormones.

CHAPTER XII *Intercellular Transfer of Genetic Information*

The primary functions of the genes (transcription and subsequent translation) are usually limited to the cell which contains the particular genome. In certain situations, however, portions of genetic material may be transferred from one cell to another and then become incorporated and eventually functionally active in the new (host) cell. Several kinds of information transfer have been studied primarily in microorganisms. In Fig. 123 the phenomena of (a) *transformation*, (b) *transduction* and (c) *bacterial conjugation* are illustrated diagrammatically. Since genetic information may also be introduced into a cell by viral infection, the action of viruses will be discussed here too.

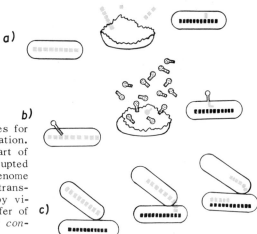

Fig. 123. Various possibilities for the transfer of genetic information. (a) *transformation*: a small part of the DNA liberated from disrupted cells is incorporated into the genome by cells of another strain; (b) transport of genetic information by viruses (*transduction*); (c) transfer of genetic material by *bacterial conjugation* (from Zinder, 1958).

VIRUSES

Viruses are submicroscopic particles which can penetrate and multiply in suitable host cells. Formation of new viruses always requires the presence of a living cell. Viruses are commonly classified according to their hosts, i.e., into *bacteriophages* which can only infect bacteria,

256

phytopathogenic viruses which attack plants, and *animal viruses* which include the viruses of human diseases. Most viruses are either spherical or rod shaped. All contain a nucleic acid (either RNA or DNA, but never both) and proteins, part of which serve as a protective envelope for the nucleic acids. Larger viruses may contain other compounds such as lipids. At the present time, viruses are often classified also as RNA or DNA viruses.

The mechanism of virus infection has been most intensively studied in bacteriophage systems. While most bacteriophages contain DNA, several RNA phages have recently been isolated (e.g., Davis *et al.*, 1961; Loeb and Zinder, 1961; Hofschneider, 1963b). DNA containing T-phages are composed of a head and a cylindrical tail (Fig. 124) which differs in length in various types of phages. The DNA is located in the head of the phage, with different proteins making up the tail and the coat. The head has the form of a bipyramidal hexagonal prism (Figs. 124 and

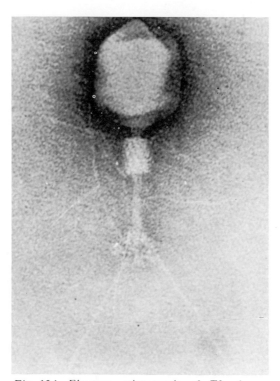

Fig. 124. Electron micrograph of T2 phage, demonstrating the head, core, sheath and tail fibers (from Brenner *et al.*, 1959). (Reproduced with permission of Academic Press, Inc.)

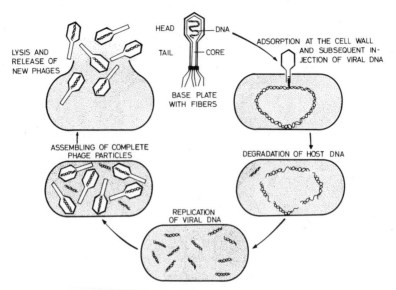

Fig. 125. The life cycle of virulent T-even phages.

125). The tail has a complex structure, with an inner core coaxial with the sheath, which at the distal end develops into several outstretched fibers. If a bacteriophage comes in contact with a suitable host cell, it is adsorbed at the cell wall by its tail (Fig. 125). After passing through several intermediate stages, the DNA and a small amount of protein are injected into the bacterium (Hershey and Chase, 1952). In a *virulent phage*, this event diverts the entire metabolism of the host cell to produce new phages and finally the cell perishes by lysis with the liberation of the completed phages. In the so-called *temperate phages* (Lwoff, 1953), the injection of viral DNA does not lead immediately to the formation of new phages and no lysis occurs. Nevertheless, the new information is not lost; instead it is passed on synchronously with the host cell genome to subsequent generations. In such *lysogenic cells*, the phage DNA—which is not part of an intact virus—is termed *prophage*. A prophage can become virulent either spontaneously or as a result of external factors such as irradiation with ultraviolet light or treatment with certain chemicals. Then normal phage multiplication ensues with subsequent lysis.

It seemed reasonable to assume that the genes of a prophage cannot function because of the presence of a repressor (cf. p. 248), made via a gene of the prophage. Inactivation of this repressor should then trigger the production of new phages and subsequent lysis. Recently, such a postulated repressor (of the λ phage) has been isolated from *E. coli* cells (Ptashne, 1967a). The repressor was found to be an acidic protein with a molecular weight of approximately 30,000 (see p. 250). It binds specifically and with high affinity to the phage DNA (Ptashne, 1967b). Thus, the transcription of viral DNA can apparently be blocked by this type of repressor. The same repressor seems to be responsible for the immunity of lysogenic cells to superinfection by virulent phages similar to the prophage. Superinfecting phages may inject their DNA into host cells; however, this will not be used for either transcription or replication.

In earlier investigations, it had been observed that a phage vanishes immediately after infection and does not reappear until shortly before lysis when newly formed virus particles can be found (see Fig. 125). The period of time during which no active virus can be detected in the infected cell is called the *eclipse*; the virus is present then in a *vegetative form* (cf. p. 266). Hershey and Chase (1952), who were the first to show that only the DNA of T phages enters the cell during infection, were also able to follow the course of the subsequent DNA synthesis during phage multiplication by the use of labeled precursors (Fig. 126). One of the most important fields in modern virus research is the in-

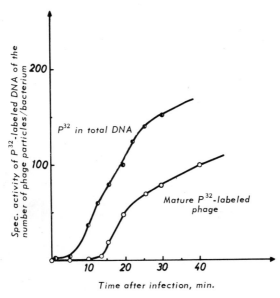

Fig. 126. Time course of P^{32} incorporation into total DNA and into DNA of newly formed viruses following infection of *E. coli* cells with T2 phages (from Hershey, 1954).

creasingly detailed analysis of the various processes that take place during the eclipse (see pp. 265 and 270).

Properties of Viral Nucleic Acids

The results of older investigations on the molecular weight of phage DNA seemed to indicate that the DNA of each virus particle was composed of several subunits. Careful study of T2 phage, showed that the DNA of the virus exists as a single molecule with a molecular weight of 130×10^6 (Davison *et al.*, 1961; Rubinstein *et al.*, 1961). However, in certain other DNA viruses, nucleic acid subunits appear to occur. The DNA of some very small bacteriophages lacks the usual double helix with two polynucleotide strands. The best-studied example is the ϕX174 phage, which contains one single-stranded molecule of DNA (molecular weight 1.7×10^6). In contrast with the Watson-Crick model, its molar base ratios for adenine:thymine (0.76) and guanine:cytosine (1.29) deviate from unity. Another characteristic of this single-stranded DNA is its sensitivity to the action of formaldehyde. Furthermore, the optical density at 260 mμ in the 20–60° C temperature range is markedly dependent on temperature (cf. Fig. 45) as well as ionic strength (Sinsheimer, 1959). Fiers and Sinsheimer (1962) have shown that native single-stranded DNA from ϕX174 phages is circular; later Marvin and Schaller (1966) obtained evidence that the phage fd DNA has a ring structure. Meanwhile, the DNA of other viruses, as for example phage S13 (Tessman, 1959) and apparently the vaccinia virus (Pfau and McCrea, 1962), had also been found to be single stranded.

The single strandedness of viral DNA requires a special mechanism for DNA replication. As an intermediate, a double-stranded replicative form (RF DNA) must be synthesized (cf. p. 267). The RF DNA of phage ϕX174 as well as that of phage M13 was also found to be circular (Sinsheimer, 1961; Hofschneider, 1963a; Kleinschmidt *et al.*, 1963; Ray *et al.*, 1966). Such circular RF DNA is resistant to thermal denaturation and cannot be attacked by exonucleases (Burton and Sinsheimer, 1963). Preparations of ϕX174 RF DNA usually contain a mixture of at least two components in variable amounts (Burton and Sinsheimer, 1965), which can be distinguished by differences in sedimentation and denaturation behavior. A more detailed analysis revealed that the tertiary structure of component I is a highly twisted circle and of component II an extended circle (Jaenisch *et al.*, 1966). Component II is a double strand with at least one single-strand break.

The DNA of T-even phages differs from normal DNA in that it contains 5-hydroxymethylcytosine instead of cytosine (Wyatt and Cohen,

1953), part of which has some glucose or gentiobiose bound to it (Sinsheimer, 1954; Volkin, 1954; Jesaitis, 1956). The molar ratio of 5-hydroxymethylcytosine to glucose and the type of glycosidic bond differ in the various T phages (Table 19). In the DNA of phage SP8

TABLE 19. 5-Hydroxymethylcytosine Content of T2, T4 and T6
Phage DNA, with Glucose in Either an α- or a
β-Glycosidic Linkage or with α-Gentiobioside
(from Lehman and Pratt, 1960)

	T2	T4	T6
Free 5-hydroxymethylcytosine	25%	0	25%
α-Glucoside	70%	70%	3%
β-Glucoside	0	30%	0
α-Gentiobioside	5%	0	72%

(host cell *B. subtilis*), thymine is replaced by 5-hydroxymethyluracil (Kallen *et al.*, 1962); this DNA also contains a high amount of glucose (0.98 mole/P), while in the mutant SP8* T_s, the hexose is D-mannose (Rosenberg, 1965). In other phages, thymine is replaced by uracil (Takahashi and Marmur, 1963). 6-Methylaminopurine is found in small amounts (1 mole per 200 moles of adenine) in the DNA of T2 and *Salmonella* phages (Dunn and Smith, 1958). In addition to high molecular weight DNA, some phages contain small amounts of nucleoside triphosphates (ATP and another unknown nucleotide) (Wahl and Kozloff, 1962). With the hybridization technique (see p. 276), it has been shown that the nucleotide sequences in the DNA of the various T-even phages are largely homologous; similar relationships were detected between T3 and T7 phages. However, no evidence for hybrid formation between DNA from even-numbered and odd-numbered T phages or between these and the DNA from the *E. coli* host cells has been found (Schildkraut *et al.*, 1962). On the other hand, Cowie and McCarthy (1963) established that the DNA of λ phages is partially homologous with the DNA from *E. coli* (cf. pp. 276 and 279).

Among the various RNA-containing viruses, the most widely used for experimental purposes is the rod-shaped tobacco mosaic virus (TMV) which was the first virus obtained in crystalline form (Stanley, 1935). According to present knowledge, plant viruses always contain RNA (Schuster, 1960). TMV particles have a length of about 3000 Å and a maximal diameter of 170 Å. The RNA is present as a single-stranded polynucleotide arranged in a helical manner within the walls of the screw-like rods (Fig. 127). The RNA, which constitutes only about 5%

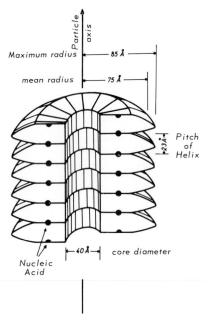

Fig. 127. Longitudinal section of a short length of a TMV particle. The picture shows the helical arrangement of the protein subunits and the hollow axial core in the interior of the particle. The viral RNA is located in a helical groove (from Franklin *et al.*, 1957).

of the total virus (Knight and Woody, 1958), is completely enclosed by a protein envelope. The bases lie parallel to the long axis in the RNA spiral, which has a diameter of 80 Å; there are about 50 nucleotides per turn with a pitch of 23 Å (Fig. 128). The molecular weight of the co-herent RNA of a TMV particle, which is composed of 6600 nucleotides, is about 2×10^6 (Schuster, 1960). While detailed amino acid sequence analyses are available for the protein portion of TMV, which consists of many subunits, probably homologous (molecular weight 17,500 each; cf. p. 238), nothing is as yet known of the base sequence of its RNA. Aden-osine is liberated by alkaline hydrolysis of TMV RNA as the terminal nucleotide in a 5′-ester linkage (Sugiyama and Fraenkel-Conrat, 1961). In one of the smallest plant viruses hitherto detected ("brome-grass mosaic virus"), whose total molecular weight amounts to only 4.6×10^6, the RNA has a molecular weight of 1×10^6 (Bockstahler and Kaesberg, 1962). Finally, the satellite tobacco necrosis virus (STNV) contains an RNA with a molecular weight of 400,000. The genome of STNV (ap-proximately 1200 nucleotides) contains just enough information to code

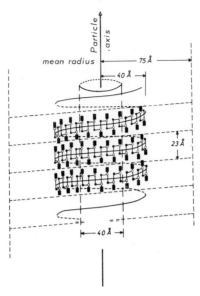

Fig. 128. Helical arrangement of the single-stranded RNA in TMV particles. The symbols represent: ● = phosphate, o = ribose, ■ = bases (from Ginoza, 1958).

its coat protein (about 400 amino acid residues) and thus appears to represent a monocistronic messenger (Clark *et al.*, 1965; cf. p. 225).

In general, viral RNA was found to be single stranded. More recently, however, double-stranded RNA has been detected in several cases, first in the Reo virus (Gomatos and Tamm, 1963) and in the wound tumor virus (Black and Markham, 1963), then also in the rice dwarf virus (RDV) (Miura *et al.*, 1966). The various physical and chemical characteristics of RDV RNA are similar to those of double-stranded DNA (Sato *et al.*, 1966); thus the molar ratios A/U and G/C are both equal to one. Partially helical structures resulting from hydrogen bonding can apparently also exist in single-stranded RNA to permit a denser packing within the protein coat of some smaller viruses (Mitra *et al.*, 1963).

As in the case of phages with single-stranded DNA, the synthesis of new single-stranded viral RNA takes place via a replicative form. Such an RF RNA was first detected after infection with RNA phages (Weissmann and Borst, 1963; Ammann *et al.*, 1964; Kaerner and Hoffmann-Berling, 1964) and was then generally found to be an intermediate during the multiplication of RNA viruses, e.g., TMV (Burdon et al., 1964; Shipp and Haselkorn, 1964), polio virus (Baltimore, 1964), and the encephalomyocarditis virus (Horton *et al.*, 1964). The RF RNA—as it is isolated from the host cell—is double stranded and therefore partially resistant to pancreatic ribonuclease (but see p. 269); the replicative form of phage M12 RNA is biologically active (cf. the following part of this section) only after melting to single strands (Ammann *et al.*, 1964).

The successful infection and subsequent production of complete virus particles by a pure nucleic acid preparation were first accomplished with TMV RNA extracted by mild isolation procedures which permitted the preparation of infectious native RNA, with the exclusion of possible contamination with intact virus (Gierer and Schramm, 1956a). Since then, infection with other isolated viral nucleic acids (e.g., Cheng, 1957; Colter *et al.*, 1957; Wecker, 1959; Sanders, 1960), including RNA-containing polio virus (Alexander *et al.*, 1958; Koch *et al.*, 1960) and influenza viruses (Maassab, 1959) as well as DNA from various phages (Fraser *et al.*, 1957; Guthrie and Sinsheimer, 1960; Hofschneider, 1960), has been accomplished. Generally, a much higher amount of pure nucleic acid is required, compared to the nucleic acid of intact viruses, to bring about infection. This may be attributed to the fact that much of the free nucleic acid is inactivated by cellular nucleases. In certain cases, it seems possible to stabilize infectious RNA by coupling with basic proteins (Smull *et al.*, 1961). On the other hand, the isolated nucleic acid of cucumber mosaic virus was found to be more infectious than the complete virus (Schlegel, 1960). It is an interesting observation that in some cases, cells resistant to the intact virus can be infected with pure nucleic acids (Fraser *et al.*, 1957; Holland *et al.*, 1959; Gordon and Smith, 1960; Davis *et al.*, 1961). The successful infection of bacteria with isolated DNA from an animal virus represents an extreme case (Abel and Trautner, 1964). An infection with pure nucleic acids of some of the tumor viruses (cf. p. 288) has also been achieved. Thus, isolated DNA from polyoma virus showed tumor-inducing activity as does the complete virus (DiMayorca *et al.*, 1959). Similarly, the DNA from the Shope papilloma virus was also reported to be effective (Hodes *et al.*, 1962). On the other hand, it has not yet been possible to isolate an infectious nucleic acid from the Rous sarcoma virus (Weil *et al.*, 1960). In various cases, a successful induction of animal tumors by RNA from neoplastic tissues was observed (e.g., De Carvalho *et al.*, 1960; Lacour *et al.*, 1960; Huppert *et al.*, 1961). However, experimental results of this kind cannot be considered proof that the donor tumor was originally caused by a virus.

In the antibody reaction against a virus, only its protein moiety is involved; thus the antibodies may leave the viral nucleic acid unaffected. Infectious TMV RNA can be recovered from antibody-inactivated material by phenol extraction (Rappaport, 1959).

Metabolic Events after Virus Infection

The first effect of viral nucleic acid after it has been introduced into a host cell is the initiation of the production of virus-specific enzymes which are required for virus multiplication; in addition, enzymes are often synthesized to inhibit normal cellular metabolism and/or to degrade host cell metabolites needed for the formation of new viral precursors. The genetic information for all these enzymes is provided by the viral nucleic acid. In the case of DNA viruses, a specific RNA must first be synthesized at the viral DNA template (e.g., Nomura et al., 1960; Volkin, 1960), while the RNA of RNA-containing viruses is used directly for translation. All of these processes utilize the apparatus of the host cell, which after infection is more or less devoted to virus multiplication. The host cell provides the necessary ribosomes for protein synthesis (Scharff et al., 1963), as well as tRNA, activating enzymes, etc. (Penman et al., 1963); however, it was recently observed that the DNA of the herpes simplex virus apparently also contains the information for several types of tRNA (Subak-Sharpe and Hay, 1965).

The most important enzymes are those for the replication of viral nucleic acids (cf. p. 266). In addition, many enzymes have been discovered which are present in a cell only after virus infection; such enzymes are different from the corresponding enzymes of the host cells and therefore may be separated (e.g., Bello et al., 1961a; Greenberg et al., 1962). The DNA polymerase, isolated after infection with T2 phages, differs from that of normal E. coli cells (Aposhian and Kornberg, 1961). The task of other enzymes synthesized via the DNA of T-even phages appears to be to prevent cytosine from being utilized for DNA synthesis. This is effected by a deoxycytidine triphosphatase (Koerner et al., 1960; Zimmerman et al., 1961), which splits off pyrophosphate and converts dCTP to dCMP, thus removing dCTP from the pool of deoxyribonucleoside triphosphates (cf. p. 130). A phage-induced deoxycytidine hydroxymethylase (Flaks and Cohen, 1959) and hydroxymethyl deoxycytidylate kinase (Somerville et al., 1959) are responsible for 5-hydroxymethylcytosine being incorporated into phage DNA instead of cytosine. Infection with T5 phages causes the formation of a new deoxyribonucleotide kinase, whose action is not limited to only one of the four DNA bases (Bessman et al., 1965). Thus, in contrast to the corresponding enzymes of the host cell, this kinase lacks base specificity; another difference is its very low activity with ribonucleotides as substrates. The supply of precursors for viral nucleic acids is usually assured by the enzymatic degradation of the host cell DNA (Kunker and Pardee, 1956; Pfefferkorn and Amos, 1958; Crawford, 1959; cf. Fig. 125). The scheme in Fig. 129

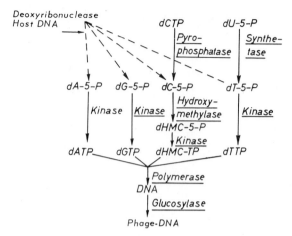

Fig. 129. Enhanced enzyme activities (or additional enzyme synthesis, respectively) in *E. coli* cells after infection with T2 phages. Infection does not increase the conversion of 5'-dAMP to dATP (from Kozloff, 1960).

summarizes the most important of the known changes in the enzyme pattern following T2 phage infection. The glucosylation—with uridine diphosphate glucose as donor for hydroxymethylcytosine bases in the T-even phages (cf. Table 19)—is mediated by specific enzymes, whose formation is initiated by phage infection (Kornberg *et al.*, 1961; Josse and Kornberg, 1962); this glucose remains as a DNA component during DNA replication (Cocito and Hershey, 1960).

RNA-containing viruses cause metabolic effects in their host cells similar to those of DNA viruses. Thus, infection with TMV results in the rapid degradation of the RNA in the infected leaves, with the released nucleotides available for the synthesis of TMV RNA (Reddi, 1963). Holland (1962a and b) was able to show that the synthesis of DNA-dependent RNA is blocked in polio virus infected cells, whose nucleic acid metabolism is routed entirely to the production of RNA-dependent viral RNA. After infecting Ehrlich ascites tumor cells with an RNA-containing virus, Holoubek and Rueckert (1964) observed intensive formation of histones which may have inhibitory effects on the DNA-dependent RNA synthesis of the host cell (cf. p. 204).

Synthesis of Viral Nucleic Acids

Because of the various kinds of viral nucleic acids (DNA or RNA, single or double stranded), different mechanisms for nucleic acid synthesis are required. The replication of double-helical viral DNA occurs in the same semiconservative manner as that of cellular DNA (cf. p. 178). Although a virus-specific DNA polymerase has been found (Aposhian and Kornberg, 1961), it seems that viral DNA is often synthesized via the DNA polymerase of the host cell. In the early phase of phage T4 DNA synthesis, a circular and folded tertiary structure of the replicative DNA is postulated (Kozinski and Kozinski, 1965). Similarly, a circular duplex DNA seems to be formed after lysogenic induction with λ phages; the appearance of this circular λ DNA coincides with the onset of DNA synthesis in the induced calls (Lipton and Weissbach, 1966). There are contradictory reports about the effects of animal DNA viruses on DNA synthesis of the host cell. With the polyoma virus (which may induce tumor formation; see p. 288), an inhibition of cell DNA synthesis (Birnie and Fox, 1965) preceding the replication of viral DNA (Sheinin, 1966), as well as a stimulation (Dulbecco et al., 1965; Weil et al., 1965), has been observed. The lack of uniformity in experimental conditions may be responsible for the difference in results.

Double-stranded viral RNA (cf. p. 263) is assumed to be replicated in a semiconservative manner like DNA. The Reo virus RNA was reported to be an effective template in vitro for DNA synthesis catalyzed by the DNA polymerase from E. coli (Gomatos et al., 1965). However, with ribonucleoside triphosphates as precursors no RNA synthesis, or only a very slight amount, was observed in the Reo virus RNA-primed DNA polymerase reaction. In addition, it had been claimed that double-stranded Reo virus RNA was an active primer for E. coli RNA polymerase (Gomatos et al., 1964). Both RNA-primed reactions and the multiplication of total Reo virus were found to be inhibited by the presence of actinomycin (see p. 294). Somewhat later, Shatkin (1965) reported that highly purified Reo virus RNA does not direct polynucleotide synthesis in E. coli at all—either by RNA polymerase or by DNA polymerase; the positive results of Gomatos et al. (1964, 1965) were ascribed to the presence of some host cell DNA in the RNA preparations. These differing experimental observations both indicate that Reo virus RNA does not serve as a template for host cell polymerases; thus a new Reo virus-specific replication system appears to be produced in the course of infection (Shatkin, 1965).

The synthesis of single-stranded viral DNA has been found to be a more complicated process that takes place in several steps. Most studies have been done with φX174 phage. Immediately after infection, the circular single-stranded DNA is converted into a double-stranded RF DNA by synthesizing a complementary strand from low molecular weight precursors (Denhardt and Sinsheimer, 1965a). This step usually requires about 3 minutes; the process is apparently catalyzed by a host cell enzyme. Subsequently this RF DNA, containing the one (primarily infecting) parental strand, is replicated semiconservatively, thereby leading to an accumulation of progeny RF molecules. During the eclipse period, the rate of increase in the amount of total RF DNA is fairly constant. It appears then that the progeny RF DNA does not replicate; a possible reason for this is that it is used exclusively as a template for mRNA synthesis (Denhardt and Sinsheimer, 1965b). About 10 minutes after infection, the formation of progeny RF molecules is completed and synthesis of progeny single-stranded circular DNA begins. The results of P^{32}-decay inactivation experiments suggest that only the one RF DNA molecule containing the original parental single strand can be used as a template for this process; apparently this template remains as a double-stranded DNA ring, at which the single-stranded progeny DNA is synthesized in a conservative manner (Denhardt and Sinsheimer, 1965b). The various steps in this replication mechanism are summarized in Fig. 130.

The multiplication of single-stranded viral RNA also proceeds via an intermediate replicating form; the parental plus strand then codes for a complementary minus strand. When isolated from infected cells, this RF RNA was found to contain double-stranded regions as well as a number of single-stranded tails, corresponding to partially synthesized nascent plus strands of different lengths (Fenwick *et al.*, 1964; Weissman *et al.*, 1964). Francke and Hofschneider (1966) observed, after in-

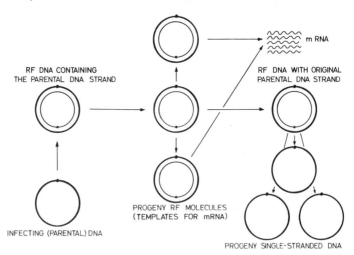

RF DNA CONTAINING
THE PARENTAL DNA STRAND

m RNA

RF DNA WITH ORIGINAL
PARENTAL DNA STRAND

INFECTING (PARENTAL) DNA

PROGENY RF MOLECULES
(TEMPLATES FOR mRNA)

PROGENY SINGLE-STRANDED DNA

Fig. 130. Suggested model for the replication of ϕX174 phage's single-stranded DNA. For explanation, see text (modified from Denhardt and Sinsheimer, 1965b). (Reproduced with permission of Academic Press, Inc.)

fection of *E. coli* with M-12 phage, two different classes of infectious RF RNA (both of which should contain an intact plus strand). It has been concluded that one class (15S, RNase resistant to 86%) is composed of complementary double-stranded molecules with twice the molecular weight of mature phage RNA, while the other class (22S) has, in addition, single-stranded tails. Thus, intact plus and minus strands are very likely the building blocks of infectious 15S RF RNA of phage M-12.

A clearer picture of the process of viral RNA replication has come from enzymatic studies. Haruna and Spiegelman (1965a) succeeded in isolating an RNA-dependent RNA polymerase from *E. coli* cells infected with bacteriophages MS-2 or Q_β (the enzyme was termed replicase; cf. p. 206). With respect to the mechanism of RNA multiplication in this system, contradictory observations were at first reported. Haruna and Spiegelman (1966) claimed that the purified Q_β RNA replicase should produce *in vitro* unlimited amounts of complete and biologically competent replicas of Q_β RNA (plus strands), without any detectable minus strands. Weissmann and Feix (1966), however, using the same kind of enzyme preparation, found in the early phase of synthesis only minus strands, preceding the later production of plus strands. More recent studies indicate that in the case of Q_β RNA the replication starts with the (relatively slow) formation of minus strands at the viral (plus) RNA template, followed by the (more rapid) synthesis of new viral RNA, presumably with the single-stranded minus strands as templates (Feix *et al.*, 1967). Some of the minus strands appear in a 40S complex associated with plus strands. However, in this complex, the plus and minus strands are not hydrogen bonded and are, therefore, sensitive to RNase. By treatment with phenol the complex can be converted into an RNase-resistant double-helical form. These observations by Feix *et al.* (1967) indicate that the RF RNA isolated from infected cells is—at least in the case of Q_β and MS-2—an artificial product due to the removal of proteins by phenol or a detergent. In agreement with these conclusions, it was found that the RNase-resistant double-stranded Q_β RNA was completely ineffective as a primer for the RNA replicase. As yet the mechanism of action of the Q_β replicase is not completely understood. Because plus and minus strands are complementary, it seems that a spontaneous formation of a double helix is prevented by the enzyme (or further proteins); on the other hand, the template function requires hydrogen bonds which might immediately be disrupted by the replicase when it advances along the template. In Fig. 131, the various steps of the replication mechanism for viral RNA are summarized. Brown and Martin (1965) postulated that the synthesis of the plus strands takes

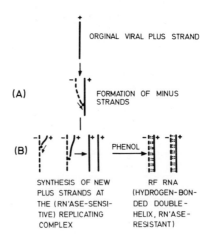

(A) ORGINAL VIRAL PLUS STRAND

FORMATION OF MINUS STRANDS

(B)

PHENOL

SYNTHESIS OF NEW
PLUS STRANDS AT
THE (RN'ASE-SENSI-
TIVE) REPLICATING
COMPLEX

RF RNA
(HYDROGEN-BON-
DED DOUBLE-
HELIX, RN'ASE-
RESISTANT)

Fig. 131. Single-stranded viral RNA replication. It is postulated that a minus strand
must be synthesized as a first step (A).
The resulting replicative forms (whose
RNA is not an RNase-resistant double
helix, at least in the case of phage Q_β) permits the formation of new plus strands (B).

place at a circular template with the RNA-forming enzyme rotating
around the cyclic primer, as is similarly suggested for the synthesis of
single-stranded viral DNA (see Fig. 130).

The problem of phage RNA replication has been discussed in detail
by Weissmann and Ochoa (1967).

Morphopoiesis of Virus Particles

The formation of a new virus requires that all of its building blocks be
synthesized in the infected host cell and subsequently assembled to
form a complete virus particle. The viral nucleic acid normally contains
the information for its own replication as well as for its protein moieties
and various specific enzymes. In addition, a control is necessary for the
morphopoiesis of virus particles which may be defined as the process
which leads from simple subunits to a complex morphologically defined
end product. The more simple viruses such as TMV contain only one
type of protein as a subunit. By the addition of such subunits in identical positions, the protein coat of a virus might be formed in a kind of
self-assembly process. In the case of TMV, Fraenkel-Conrat and Williams
(1955) succeeded in converting isolated TMV proteins (a homogeneous
fraction) into virus-like rods under appropriate conditions; these rods,
which did not contain viral RNA and were almost indistinguishable from

the intact TMV particles, seem to have been formed spontaneously. However, the information contained in the protein subunits alone is not sufficient for the helical structure of the rod (cf. Fig. 127 and 128); the
TMV RNA contributes additional information and probably also determines the length of the completed TMV particles.

For the investigation of viruses as self-assembling systems, the results of X-ray diffraction studies and the theoretical considerations by
Caspar and Klug (1962) have become fundamental background material.
The basic assumption of the Caspar and Klug hypothesis is that a viral
protein coat is determined by the specific bondings between its identical
subunits. Under appropriate environmental conditions, a self-assembly
process (i.e., the formation of specific structures with the subunits as
building blocks) can take place in a regular manner. The other important point of their theory is that it postulates in general an icosahedral
symmetry of the shell which had previously been established for quite
unrelated viruses. Triangulating a closed surface and thus producing a
network with icosahedral symmetry offers the optimum conditions for a
state of minimum energy for the total shell. The results of structural
studies on a number of different viruses have shown that their protein
shells are built up according to designs as predicted by the theory of
Caspar and Klug, including viruses with two shells [i.e., having an inner
protein core (Klug *et al.*, 1966)]. According to this picture, a self-
assembling of larger isometric viruses would require the formation of an
icosahedral protein core followed by a condensation of the viral nucleic
acid with the outer protein coat. The assembling of more complex structures like that of T phages is a more complicated process, the details of
which can be investigated by *in vitro* model experiments (Edgar and
Wood, 1966; Takai, 1966) and by identifying the so-called morphopoietic
genes (Kellenberger, 1966). The mutation of such genes causes disturbances in the assembly mechanisms.

Morphopoiesis of virus particles is of considerable general interest as
a relatively simple example of structural biological organization at the
molecular level. These very briefly considered problems of virus assembling were discussed in more detail at a CIBA Foundation Symposium
by Caspar (1966), Kellenberger (1966), Klug *et al.* (1966), and Pirie (1966).

Effects of Certain Drugs on Viral Nuclei Acids

Various drugs which interfere with nucleic acid metabolism can also
change the properties or influence the replication of viruses. For example, 5-fluorouracil (5-FU) and 5-fluorodeoxyuridine (5-FUDR) inhibit
DNA synthesis by blocking the methylation reaction which gives rise to
thymine (see p. 313); both compounds can also cause an inhibition of the

multiplication of DNA-containing viruses by the same mechanism (e.g., Goldé and Vigier, 1961; Simon, 1961; Volkin and Ruffili, 1962). 5-FU is incorporated into RNA, whose coding properties may thus be changed ("fraudulent" RNA; cf. Fig. 118, pp. 245 and 314). In TMV RNA, up to 50% of the normal uracil can be replaced by 5-fluorouracil (Gordon and Straehlin, 1959). Such viruses still have normal infectivity, and the progeny RNA again show a normal base composition. Very high doses of 5-FU may cause some inhibition of TMV multiplication (Davern and Bonner, 1958) and also an increase of the normal mutation rate (Kramer et al., 1964). Munyon and Salzman (1962) observed an inhibition of polio virus production due to 5-FU; although up to 35% of the uracil in the viral RNA had been replaced, no changes in the properties of the virus could be detected. In contrast, Cooper (1964) reported a fairly high rate of mutations for polio viruses that had replicated in the presence of 5-FU. In RNA phages, 5-fluorouracil can cause a loss of infectivity (Gomatos et al., 1962). 2-Thiouracil leads to a considerable decrease in the infectivity of intact TMV but has less effect on the virulence of isolated TMV RNA (Francki and Matthews, 1962). Thymine analogs like 5-chlorouracil, 5-bromouracil and 5-iodouracil can partially replace the normal thymine in DNA (see pp. 242 and 310). The resulting changes in the template properties cause either mutations or complete inactivation in T phages (e.g., Dunn and Smith, 1954; Terzaghi et al., 1962). Actinomycins (cf. p. 294) preferentially inhibit the multiplication of DNA viruses (Reich et al., 1961). However, the RNA-containing influenza virus and two other RNA viruses were also found to be inhibited by actinomycin. Mitomycin C, an antibiotic which interacts with DNA (see p. 299), has an inhibitory effect on the DNA-containing vaccinia virus (Reich and Franklin, 1961). (These effects of certain drugs on the properties of viruses are just a few selected examples from the vast literature.) Unfortunately, none of these drugs specifically affects only the viruses, i.e., without harming the host organism; thus, they cannot be used for therapeutic purposes to cure viral infections.

Interferon

Cells infected by a virus can often prevent or inhibit the various viral effects due to the intracellular production of a compound termed *interferon*, which has been identified as a protein; chicken interferon has a molecular weight of 27,000–35,000 (Burke and Ross, 1965). The synthesis of interferon takes place via transcription and translation like other proteins; thus interferon formation can be inhibited by actinomycin (Gifford and Heller, 1963) as well as by puromycin (Wagner and Huang, 1965; cf. pp. 294 and 301). Apparently the production of new interferon is initiated by the viral nucleic acid or, in general, by nucleic acids foreign to a particular cell (Rotem *et al.*, 1963). According to recent observations, the antiviral effect of interferon is due to an interaction with cellular ribosomes. Ribosomes prepared from cells which had been exposed to homologous interferon could still use cellular mRNA, while their ability to interact with viral mRNA was markedly reduced. Thus it seems that the ribosomes from cells exposed to interferon are altered selectively to permit only certain (cell-specific) messengers to be bound and translated (Carter and Levy, 1967). The observation that avirulent viruses are generally more effective inducers of interferon synthesis than virulent viruses suggests that this difference might be an essential basis for virulence (Wagner, 1965). Recent experimental results by Wagner and Huang (1966) support the hypothesis that viral virulence is at least partially determined by the inhibitory effect of a virus on cellular RNA synthesis, a prerequisite for interferon formation. In contrast to antibodies, the intracellular antiviral effect of interferon is a fairly fast-acting mechanism; another difference from the action of antibodies is that interferon does not appear to have a specificity for particular viruses. In the blood of mice, extracellular interferon has been detected (Baron and Buckler, 1963); however, no interferon, or at most a very small fraction, was found to be taken up by cells during the period of induction of antiviral activity (Buckler *et al.*, 1966). Thus the antiviral effect of interferon seems to be limited to the cells by which it is produced.

TRANSFORMATION

Transformation is the transfer of genetic information by means of pure extracellular DNA, i.e., the introduction of purified chromosomal material. This phenomenon, at first thought to occur only in bacteria, was originally observed *in vivo* by Griffith in 1928 and was finally correctly interpreted when confirmed *in vitro* (Dawson and Sia, 1931; Alloway, 1932). In 1944, Avery *et al.* showed that the agent responsible for transformation is highly polymerized DNA. In transformation studies, DNA must be carefully isolated from donor cells and added to a culture of competent recipient cells. The earlier experiments were done primarily with *Diplococcus pneumoniae*, of which two forms exist, one with a carbohydrate capsule (S-form) and another one without this capsule (R-form). The ability to synthesize a carbohydrate capsule can be transferred to R-cells by DNA extracted from S-cells. The acquired property is retained in subsequent generations of the transformed cells.

During the last decade, many genetically determined properties have been successfully transferred from one bacterial strain to another by transformation. Among these, the transfer of relative resistance against certain drugs has become an important type of experiment. At first, the transformation of penicillin-sensitive pneumococci to resistant strains was obtained (Hotchkiss, 1951). Subsequently, other sensitive bacterial strains were transformed to resistant forms (part of these listed by Wacker, 1959). Finally, the converse was also found to be possible, i.e., transfer of increased sensitivity to a particular antibiotic (e.g., Hotchkiss and Marmur, 1954; Wacker, 1959; Nathanson, 1962).

Hotchkiss (1957) developed methods which made it possible to study the transformation process on a quantitative basis and thus permitted a more detailed analysis. The uptake of transforming DNA by competent recipient cells occurs very rapidly and is followed by a delay of several hours before cell division can be observed in the transformed cells (Fig. 132). On the other hand, the growth of the culture as a whole

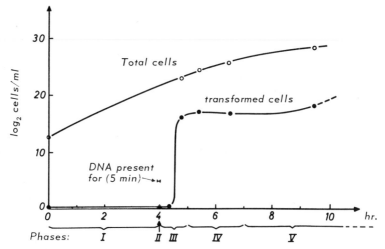

Fig. 132. Kinetics of the transformation of streptomycin resistance in pneumococci (from Hotchkiss, 1957).

(upper curve, Fig. 132) is hardly affected by the process of transformation. The number of transformed cells is dependent on the amount and quality of DNA supplied (cf. Fig. 133). By using P^{32}-labeled DNA, it could be shown that the uptake by bacterial cells is not restricted to genetically active or homologous DNA. However, RNA or highly degraded DNA is not incorporated by bacterial cells (Fox, 1957). The mechanism by which DNA may penetrate the cell wall is not yet known. Pretreatment with deoxyribonuclease decreases both the penetrability and the biological activity of transforming DNA. The incorporation of

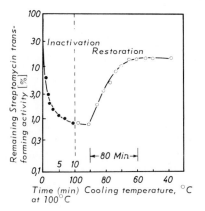

Fig. 133. Transforming activity of native, thermally denatured and renatured pneumococcal DNA in relation to the DNA concentration,(from Marmur and Lane, 1960).

transforming DNA into the genome occurs shortly after uptake by the bacterial cell. The introduction of new genetic material does not appear to be associated with DNA replication (Voll and Goodgall, 1961; Fox, 1962). For a DNA fragment to be active, it should contain at least 900 nucleotides, but it is generally much longer (Litt et al., 1958; Fox, 1962).

Transformation experiments provide means for testing the effects of various treatments on the genetic activity of DNA (see p. 172). Thermal denaturation of DNA normally leads to a loss of most of the transforming activity. However, by cooling the DNA slowly, a partial renaturation is possible (cf. p. 85), accompanied by restoration of biological activity (Fig. 134). With rapid cooling, strand separation is largely maintained.

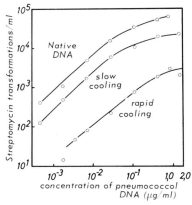

Fig. 134. Thermal inactivation and subsequent reactivation of the transforming ability of pneumococcal DNA (from Marmur and Lane, 1960).

The degree of reactivation after slow cooling is concentration dependent; thus at concentrations below 3 $\mu g/ml$, less than 5% of the initial activity is restored (Marmur and Lane, 1960). The optimum temperature for re-naturation is about 25°C below the T_m value (Marmur et al., 1961). Completely denatured DNA may still show slight transforming activity (Ginoza and Zimm, 1961; Guild, 1961). The heat sensitivity of different genes need not be uniform and thus a preferential inactivation of certain genes seems possible (Roger and Hotchkiss, 1961).

When a solution of DNA molecules from two different, but related, species is heated and then slowly cooled, double-helical hybrid mole-cules can be formed by the combination of strands which have more or less complementary base sequences, but are not necessarily exactly complementary. The capacity for such hybridization can be used to test the degree of genetic relatedness of microorganisms on a molecular basis (Marmur et al., 1961; Schildkraut et al., 1961); similarly, hybridiza-tion between labeled mRNA and (template) DNA is possible (McCarthy and Bolton, 1963). Hoyer et al. (1964) developed a technique for DNA-DNA hybridization that could also be applied to the detection of common base sequences in the genes of multicellular organisms. The method for the formation of DNA-RNA hybrids (Gillespie and Spiegelman, 1965) as well as for DNA-DNA hybrids (Warnaar and Cohen, 1966) has been im-proved considerably by immobilizing the denatured DNA on nitrocellu-lose membranes. It should be noted that these hybridization techniques are becoming increasingly important in the qualitative characterization of the information content of a particular DNA, especially in comparative studies.

At first, information transfer by pure DNA seemed to be limited to bacterial cells, although it had been shown that mammalian cells can in-corporate high molecular weight DNA (e.g., Gartler, 1959; Sirotnak and Hutchison, 1959). However, there was no evidence for a functional role of the incorporated DNA until Japanese workers succeeded in transform-ing Yoshida ascites tumor cells, previously sensitive to a nitrogen mustard derivative (Kurita, 1960) and to mitomycin C (Hoshino et al., 1960), into resistant cells. Finally, the Szybalskis (1962) demonstrated the transformation of a defective mutant in human tissue culture cells under clearly defined conditions. It now appears, therefore, that the transformation process is a more generally possible phenomenon. The question of whether in vivo transformations can be achieved, as postu-lated in the much disputed duck experiments of Benoit et al. (1958), is still unresolved. As in the case of bacteria, the extent of DNA uptake by mammalian cells depends on its double strandedness and molecular weight (Wilczok, 1962); coupling to basic proteins stabilizes DNA (Tabor,

1961) and RNA, and also leads to increased incorporation (Amos, 1961; Cocito *et al.*, 1962).

While bacterial transformation occurs exclusively by way of DNA, RNA being unable to penetrate the bacterial cell wall, high molecular weight RNA can be incorporated by mammalian cells and may eventually have some metabolic effects (Schwarz and Rieke, 1962). Thus, with immature erythrocytes from patients with sickle cell anemia (cf. p. 239), the formation of normal hemoglobin could be induced *in vitro* by RNA from normal red blood corpuscles (Weisberger, 1963), whereas DNA was found to be inactive. In this special case, the added RNA was apparently functioning as a messenger. Some other metabolic changes initiated by RNA in mammalian cells can persist through many generations (Niu *et al.*, 1962; Niu, 1963). This suggests that the incorporated RNA has a permanent effect on the genome, perhaps by changing the availability of the genetic information stored in the DNA (cf. pp. 89 and 189). It is tempting to explain in this way too the observation that pure RNA can induce tumor formation (see p. 290).

The problems of transformation have been reviewed more extensively by Ravin (1961) and Schaeffer (1964).

TRANSDUCTION

The phenomenon of transduction was first observed by Zinder and Lederberg in 1952. After mixing cell suspensions of two defective mutants of *Salmonella typhimurium*, these authors found wild-type cells, whose formation could not be explained by back mutation, transformation or conjugation. A more detailed study revealed that a bacteriophage had mediated the transfer of genetic material (Fig. 135). During eclipse, a certain portion of the bacterial DNA can be incorporated into the phage and then carried to new host cells (recipient cells) at the next infection. Thus, genetic material is transferred from one cell to another using the phage's coat for transportation. There are two types of transduction which differ in the nature and amount of bacterial DNA in the phages. In one type, only a certain portion of the phage genome is exchanged for the bacterial genome, leaving the phage defective in terms of its ability to direct the production of new phages ("restricted transduction"). In the other, the DNA from the donor cell corresponds in size to the entire phage genome (Luria, 1959), thus permitting a coupled transduction of adjacent gene loci ("generalized transduction"). The frequency of transduction is about 10^{-5}–10^{-6} per bacterial cell. The transferred information is incorporated into the genome of the recipient cell and then transmitted to all daughter cells. In other cases, the for-

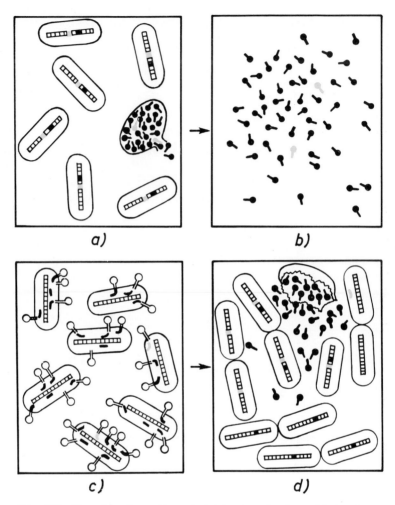

Fig. 135. Transfer of genetic material by transduction. (a) A culture of bacterial donor cells is infected with phages. Some of the newly formed phages (b) have incorporated part of the information of the host cells (gray). Upon infection of another bacterial strain by these phages (c), virus replication and lysis can occur (d, top center); some of the cells, however, incorporate genetic material from the donor cells (first bacterial strain, red squares), from latent viruses (black squares), or from both. Occasionally a cell may take up genetic material that is not subsequently incorporated into the genome (d, top right); in this case, the new information is passed on during cell division to only one daughter cell (abortive transduction; from Zinder, 1958).

eign DNA may be functionally active in the recipient cell, but it is evidently not incorporated and therefore not replicated with the recipient cell genome ("abortive transduction"; see Fig. 135). Thus the inheritance during abortive transduction is nonlinear; with each cell division, the abortive fragment of DNA will remain in one or the other of the two daughter cells. According to present knowledge, relatively few bacteriophages have the ability to undergo transduction.

In *B. subtilis*, transducing bacteriophages that contain a glucosylated DNA have been found (Takahashi and Marmur, 1963a and b). In some of these phages, uracil replaces thymine in the DNA. DNA from *B. subtilis* and from these phages may be separated by Cs_2SO_4 density gradient centrifugation. With this method, it has been shown that transducing SP10 phages contain two kinds of DNA—one has transforming activity (see the preceding section), while the other represents the genuine viral DNA; the former corresponds in density and T_m value to the DNA from *B. subtilis* (Okubo *et al.*, 1963).

A phenomenon somewhat similar to transduction was recently observed by Rogers (1966). The Shope papilloma virus which induces tumors in rabbits (cf. p. 288) has been found to cause the formation of a new type of arginase, which does not normally occur in rabbit epithelium. There are several indications that the information for the synthesis of the enzyme is not part of the rabbit genome, but instead must be derived from the viral genome. As one of the consequences of the virus action, infected rabbits develop a low concentration of arginine in their blood. It was startling to observe the same decrease in arginine level in people who had worked with the Shope virus. This indicates that these research workers are apparently carrying "virus information," fortunately without any pathological consequences. It is very interesting to speculate whether it might be possible in the future to tie specific DNA information onto the genome of passenger viruses to supplement the genetic information of patients with diseases caused by genetic deletions (cf. p. 293).

The phenomenon of bacterial transduction has been discussed in more detail in several reviews (Stent, 1963; Campbell, 1964; Hayes, 1964).

BACTERIAL CONJUGATION

In 1946, Lederberg and Tatum discovered a sexual differentiation in *E. coli* cells. The ability for "mating" or conjugation is determined by an extrachromosomal DNA-containing sex factor F (for fertility). Thus *E. coli* cells without this factor (F^-) are females that can be converted to F^+ males by conjugation; the size of the DNA in the sex factor corresponds to that of a bacteriophage. The conjugation systems which have been found so far in only a few species of bacteria are fairly complicated (for reviews, see Jacob and Wollman, 1961; Hayes, 1964, 1967). Here, only the process of conjugation itself shall be briefly discussed.

A prerequisite for conjugation is physical contact between the two cells involved. It is not known whether this contact occurs via a cytoplasmic bridge or whether hair-like fimbriae of the male cells (Brinton *et al.*, 1964) form the bridge. The sex factor F seems to be attached to the cell membrane at the site where contact is formed with a recipient cell. The transfer is limited to DNA; no measurable transfer of cytoplasmic material was detected (Silver, 1963). For the transfer process to occur, the DNA must be replicated. According to more recent observations (Gross and Caro, 1966), replication takes place in a semiconservative way during conjugation. Thus the transferred double-helical DNA contains one old strand and its newly synthesized complement. These results agree with a model proposed by Jacob and Brenner (1963) which postulates that the transferred DNA is a copy containing one preformed and one new strand, transferred as it is synthesized.

CHAPTER XIII   *Nucleic Acids in Tumor Cells*

The distinguishing characteristic of tumor cells compared with the normal cells from which they are derived is that their growth is no longer subject to regulation.   Tumors (also termed neoplasms) grow autonomously so that the tumor tissue continues to increase in mass without restraint. Loss of growth control does not necessarily imply that tumors always grow rapidly (like most of the transplantable neoplasms).   Tumors can in fact grow more slowly than the normal tissues from which they are derived.   However, the equilibrium between cell formation and cell death may be disturbed by an extended survival time of cells, and consequently the amount of such neoplastic tissue will increase.   Malignant tumor cells have additional features: They show invasive growth and often metastasize, i.e., they can be transferred via the blood or lymphatic tissues to other organs and form secondary tumors.   These phenomena may depend largely on changes in the chemical structure of the cell membranes.   It is suggestive to consider such alterations as partially responsible for the loss of growth control, particularly the control of cell division (Pardee, 1964).   A central question of cancer research is: what kinds of primary lesions cause the loss of growth control whose normal mechanisms are unfortunately still not understood (cf. p. 185)?   There are more and more indications that changes in the functions of the genetic apparatus of the cell play an essential role in the primary steps of carcinogenesis.

In the following sections, some problems of carcinogenesis will be briefly discussed from the viewpoint of the involvement of nucleic acids and their metabolism.

NUCLEIC ACID METABOLISM IN NORMAL AND NEOPLASTIC TISSUES

Much of the biochemical investigation of neoplastic cells has been directed toward the discovery of metabolic and compositional differences between normal and tumor tissues, with the hope that these differences might explain the phenomenon of carcinogenesis and/or provide a rational basis for therapy, especially with certain drugs. Considering the immense number of such studies, the results obtained seem fairly disappointing. Thus far, only quantitative, but no qualitative, differences have been observed to exist between normal and tumor cells. Moreover, most of the metabolic changes that occur during carcinogenesis appear to be secondary phenomena which eventually permit a preferential attack against tumor cells by appropriate chemotherapeutic agents, but which have nothing to do with the process of tumor formation itself.

As one would expect, in rapidly growing tumors an intensive synthesis of RNA and DNA takes place (e.g., Barnum *et al.*, 1953; Tyner *et al.*, 1953); however, this is also true for normal proliferating tissues and therefore is not specific for neoplastic growth. It has frequently been observed that tumor cells lose their ability to degrade certain metabolites which can then no longer be withdrawn and are thus available for various anabolic processes, especially those involved in growth (Fig. 136). Thus

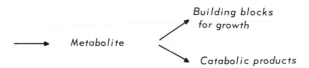

Fig. 136. Intermediary metabolites are either utilized for anabolic processes (including growth) or removed by degradation (alternate pathways).

in neoplastic tissues, the normal equilibrium between anabolism and catabolism may be shifted in favor of the formation of metabolites for growth. For example, only minute amounts of uracil can be utilized by normal rat liver for nucleic acid synthesis; most of it is rapidly degraded (Canellakis, 1956, 1957c; cf. p. 136). In hepatomas induced by 2-acetyl-aminofluorene (2-AAF), this ability to break down uracil is gradually lost, and the pyrimidine is utilized at an increasing rate for nucleic acid synthesis (Rutman *et al.*, 1954). In several transplantable tumors, an increased incorporation of labeled uracil into RNA and DNA as compared with most normal tissues was observed (Heidelberger *et al.*, 1957). Similarly, neoplastic tissues often show a decreased capacity for degradation and, as a consequence, an increased utilization of purines (Wheeler and Alexander, 1961). Xanthine oxidase (cf. p. 120) whose activity is usually diminished in tumor cells (Bergel *et al.*, 1957) probably plays an essential role in this metabolic alteration, since intraperitoneal administration of purified xanthine oxidase has been shown to inhibit the growth of experimental tumors to varying, but significant, degrees (Haddow *et al.*, 1958). The chemotherapeutic effectiveness of certain purine and pyrimidine analogs in neoplastic diseases is largely due to metabolic alterations of this kind (cf. pp. 312 and 318). However, all these compounds are capable of only preferentially inhibiting certain tumor cells; they do not attack them specifically, and so far no drug has been found that causes a real cure of cancer. Because the purine and pyrimidine analogs also affect the functioning of normal tissues, their therapeutic index is necessarily small.

Most investigators consider cancer to be a genetic disease; i.e., the normal genetic information is assumed to be altered—in an as yet unknown way. One of the main arguments for the hypothesis that carcinogenesis is initiated by somatic mutations (Boveri, 1914) was the observation that the conversion to tumor cells is generally permanent. Intensive cytological and cytochemical investigations showed that most neoplasms, at least as primary tumors, did not exhibit visible chromosomal alterations (e.g., Bayreuther, 1960; Hauschka, 1963). A convincing exception is the abnormal "Philadelphia" chromosome which is observed fairly consistently in myeloid cells of patients with chronic myelocytic leukemia (Newell and Hungerford, 1961; Sandberg, 1962; Jacobs et al., 1966). In experimental hepatomas induced by azo dyes (cf. p. 285), aneuploidy and mitotic irregularities have been found (Stich, 1963). The question of whether aneuploidy, which in some cases occurs in precancerous phases (Stich, 1960), may be directly related to the process of tumor formation, is still open (cf. Stich and Steele, 1962).

Comparative biochemical investigations on the properties of DNA isolated from normal and tumor tissues at first revealed no significant differences between them (Kit, 1960a and b). Later, Fiel et al. (1965), in similar studies, did observe differences in the properties of DNA from human neoplasms and the tissues of their origin. However, the possibility cannot be excluded that the observed differences (lower molecular weight and higher viscosity of tumor DNA) were due to the effects of the isolation procedures. The same group (Bardoz et al., 1965) found that DNA isolated from human neoplasms was a more efficient primer for DNA polymerase than DNA from the corresponding normal tissues. When fractionating histones from various tissues, Davis and Busch (1959, 1960) isolated a protein which was claimed to be specific for tumor cells; in an earlier publication, Cruft et al. (1954) had reported an abnormal histone fraction isolated from neoplastic cells.

Although alterations of the properties of DNA and the amount per nucleus (at least as far as they can be recognized with the available methods) do not appear to be prerequisites for carcinogenesis, it seems that disarrangements within the chromatin play an essential role. The results of more recent investigations indicate that in several transplantable tumors, the ratio of DNA in euchromatin and heterochromatin (see pp. 89 and 189) is lower than that in normal tissues (Harbers and Vogt, 1966; Sandritter *et al.*, 1968). In addition, with the induction of hepatomas in the rat by diethylnitrosamine or N-nitrosomorpholine (cf. Fig. 137), a decrease in the percentage of DNA in the euchromatin fraction of the parenchyma cells of rat liver was observed (Harbers *et al.*, 1968). This indicates that in these cases, the arrangement of the genetic material had apparently been altered with a concomitant loss of information compared to that in normal tissues (Harbers, 1966). The decrease in the availability of genetic information might explain why with tumor formation various enzymes are lost or no longer inducible (cf. pp. 282 and 292). It is not yet known whether the nuclear DNA which has become part of heterochromatin during carcinogenesis is altered in its properties or whether this part of the genome is intact but simply not available for cellular functions (see p. 293).

DNA has been found to be a building block of mitochondria (cf. p. 91). In mitochondria of rat hepatomas, the amount of DNA was found to be higher than in mitochondria of normal rat liver; however, partial hepatectomy also caused some increase in the DNA content per mitochondrion (Neubert, 1966b). Apparently the average lifetime of mitochondria in rat hepatomas is longer than in normal liver (Neubert, 1966b). The size distribution of polysomes in rat liver cytoplasm is changed by neoplasia and also by physiological stress; in addition, hepatomas showed a difference in the degree of association of polysomes with the endoplasmic reticulum (Webb *et al.*, 1964, 1965; Webb and Morris, 1966). These *in vivo* observations were confirmed by *in vitro* studies which then led to the hypothesis that membrane alterations might be responsible for changes in the stability of mRNA templates (Pitot *et al.*, 1965; Pitot, 1966a). According to a model suggested by Pitot, environmental stimuli (including carcinogenic agents) may alter the molecular arrangement of the mosaic patterns of the membranes in the endoplasmic reticulum; it is postulated that these changes are maintained without any further genetic alterations (cf. p. 293).

CHEMICAL CARCINOGENESIS

A variety of agents (chemicals, ionizing radiation, ultraviolet light and certain viruses) can induce tumor formation. This permits the experimental study of carcinogenesis under controlled conditions which may reveal processes either responsible for, or substantially involved in, tumor induction. In view of the multiplicity and variety of compounds that can lead to cancer, it is unlikely that tumor formation is due to a single, uniform mechanism.   Rather, the mode of action of a particular carcinogen can be considered merely an example of one possible way to induce tumors.   In this section, the biochemical effects of several more intensively studied chemical carcinogens will be discussed; viral carcinogenesis and radiation effects will be dealt with later (see pp. 288 and 329).

The formation of rat liver hepatomas as a result of prolonged feeding with azo dyes [especially 4-dimethylaminoazobenzene (DAB), "butter yellow"] has been carefully investigated by the Millers.   These azo dyes are specifically bound by the liver cell proteins (Miller and Miller, 1947).   A number of observations support the postulated causal relationship between protein binding and tumor formation: (1) Azo dyes with low carcinogenic activity are less strongly bound than DAB.   (2) A protective action of certain ingredients of the food causes a decreased binding of DAB.   (3) In mature hepatomas induced by DAB, the dye is no longer bound.   These results suggested that hepatoma cells lack specific proteins which in normal liver cells are responsible for binding the dye. Using immunological methods, Weiler (1956) was subsequently able to demonstrate directly the loss of certain proteins during DAB-induced carcinogenesis.

This observation of protein binding by carcinogenic azo dyes was later also shown to be true for carcinogenic hydrocarbons (Miller, 1951; Wiest and Heidelberger, 1953) and for AAF (Weisburger and Weisburger, 1958).   As with the azo dyes, the extent of binding was generally found to parallel the biological activity.   However, there were exceptions (Miller and Miller, 1953; Heidelberger and Moldenhauer, 1956).   Moreover, whereas hyphosectomy results in a marked increase in the resistance to the effect of carcinogens, the binding of azo dyes by liver proteins in these animals was not impaired (Ward and Spain, 1957).   Thus, protein binding of certain carcinogens seems to be at most a necessary, but not a sufficient, prerequisite for tumor formation.

More recently, numerous attempts have been made to investigate the possibility of carcinogens interacting with nucleic acids.   Such studies involve great technical difficulties because it is not easy to exclude contamination by small amounts of protein residues or artificial binding that occurs during the preparative steps.   Heidelberger and Davenport (1961) were the first to demonstrate the binding of a C^{14}-containing polycyclic hydrocarbon [1,2,5,6-dibenzanthracene (DBA)] to DNA and RNA of mouse epidermis cells.   Subsequently, Brookes and Lawley (1964a and b) observed the binding of various labeled aromatic hydrocarbons to DNA and RNA under similar conditions.   The binding to the nucleic acids was found to persist after extensive purification, and (within the limited series of hydrocarbons studied) there was a significant positive correlation between extent of binding to DNA and carcinogenic activity. No such correlation could be detected in the binding to RNA or proteins. Later, Brookes and Heidelberger (1966) found a considerably higher amount of hydrocarbons bound to DNA in normal cells than in various tumor cells. It is tempting to speculate that the higher heterochromatization of tumor cells may contribute to differences in the binding capacity of nuclear DNA (cf. pp. 284 and 296). From the results of model experiments, Boyland and Green (1960, 1962) suggested that cancer-producing polycyclic hydrocarbons interact with DNA by intercalation between two base pairs in the double helix; this intercalation hypothesis was derived from observations on the interaction of acridines with DNA (see p. 246). However, Giovanella et al. (1964) showed that the experiments of Boyland and Green did not permit the postulated conclusions; in addition, Dannenberg (1966) also found no evidence for intercalation of hydrocarbons. Model experiments in which carcinogenic hydrocarbons and isolated nucleic acids (or proteins) are brought together in appropriate solvents are of limited value because in living cells the carcinogens appear to undergo metabolic transformations before they can be bound (cf. Heidelberger, 1964).

In rat liver, metabolites of labeled 4-dimethylaminoazobenzene (with high specific activity) were covalently bound to rRNA and DNA, as well as to proteins (Roberts and Warwick, 1966; Warwick and Roberts, 1967). While the amount of radioactivity bound to rRNA and proteins paralleled the turnover of these constituents, the level of DNA-bound material remained almost constant for three months. C^{14}-labeled 2-AAF administered in vivo was bound to rat liver protein and RNA (Marroquin and Farber, 1962); interaction of 2-AAF with liver RNA was also demonstrated in in vitro experiments (Brown and Ashton, 1962).

Several alkylating agents (which attack nucleic acids as well as proteins) are used for cancer chemotherapy (cf. p. 304); however, various alkylating substances can also induce cancer. A series of nitroso compounds has been found to be powerfully carcinogenic. The tumor-inducing activity of the simplest compound, dimethylnitrosamine (cf. Fig. 137) was discovered in 1956 by Magee and Barnes, and since then many other nitrosamines have been synthesized by Druckrey's group and found to be carcinogenic. A surprising phenomenon, which cannot yet be explained, is that some of these substances cause tumor formation only in certain tissues and in a very specific way (e.g., Druckrey et al., 1967). The most reactive sites for alkylation of nucleic acids are the ring nitrogen atoms of the bases, especially nitrogen 7 of guanine and nitrogen 3 of adenine (see p. 304). While alkylating effects of this kind had initially

(A) 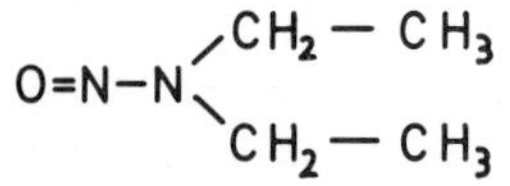   O=N−N⟨ CH₂ — CH₃ / CH₂ — CH₃

(B) 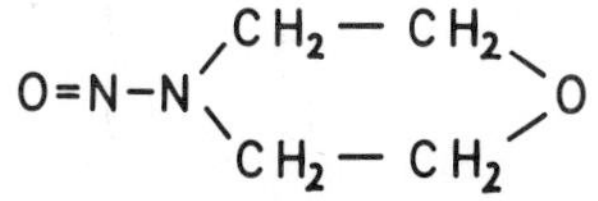   O=N−N⟨ CH₂ — CH₂ \ O / CH₂ — CH₂

Fig. 137. Structure of diethylnitrosamine (A) and of N-nitrosomorpholine (B).

been demonstrated only *in vitro*, Magee and Farber (1962) showed by *in vivo* experiments that dimethylnitrosamine (which induces hepatomas in the rat) led to the formation of 7-methylguanine in both types of nucleic acids. 7-Methylguanine is a normal component of human urine (Krüger and Salomon, 1898) and also of the urine of the rat (Mandel *et al.*, 1966). When rats had been treated with C^{14}-dimethylnitrosamine of high specific activity, the animals excreted C^{14}-containing 7-methylguanine (Magee, 1966); this had apparently been derived from the methylated nucleic acids (and not from free nucleotides) because no labeled 7-methylguanine could be detected in the acid-soluble fraction of rat liver. It seems that the demethylation of the azo dye 4-dimethylaminoazobenzene can also lead to a methylation of nucleic acid purines (Miller and Miller, 1961); in agreement with this, Berenbom (1962) found that after administration of the methyl-labeled azo dye to rats, the liver nucleic acids contained some of the C^{14}-methyl groups as purine constituents. Among the further effects of alkylating agents, Brookes and Lawley (1964a) have discussed two possibilities for inducing mutations: anomalous base pairing or depurination with 7-methylguanine being split off from the DNA chain (i.e., deletions; see p. 307).

From the experimental data presently available, it is impossible to decide whether the interaction of chemical carcinogens with proteins or with the nucleic acids (in the sense of possible mutations) is the essential one. The main argument against the mutation theory of cancer (cf. p. 293) is that the originally postulated correlation between mutagenic and carcinogenic activities has not been demonstrated so far in a convincing way. However, from a more modern point of view, it seems unreasonable to expect such a simple correlation to be apparent. Freese and Bautz-Freese (1966) give the following reasons: (1) Some carcinogens, such as viruses (see the following section) and hormones, are not generally mutagenic. (2) Many carcinogens have such a low solubility in water that their effect is hard to test. (3) Various carcinogens are converted inside the cell into the active compound. The required enzymes may not be present in the organism employed for mutagenic studies.

These arguments however do not exclude the importance of interactions with proteins. In contrast to DNA, in more simple test systems like microorganisms or phages the nucleohistones in somatic cells also contain proteins, and it seems conceivable that the alteration of such proteins by the interaction of carcinogenic agents may disturb the functioning of the genetic apparatus—and consequently result in a loss of growth control.

Because carcinogenic hydrocarbons inhibit the formation of interferon (cf. p. 272), it has been suggested that latent tumor viruses may be activated when interferon synthesis fails (de Maeyer and de Maeyer-Guignard, 1963). The possibility of such a mechanism receives some support from the reports of a cell-free transmission of leukemia induced by methylcholanthrene (Irino et al., 1963; Toth, 1963). At present it seems unlikely that these observations are of any more general importance.

TUMOR INDUCTION BY VIRUSES

It has been known for more than 50 years that viruses can induce tumor formation. However, until about a decade ago, the observed phenomena were believed to be of little importance because of their rarity. This attitude has now changed appreciably since many virus-like agents have been discovered which are capable of tumor induction. Some of these contain DNA, others, RNA; furthermore, these viruses can differ considerably in size and morphology. As with other viruses, infection (in the sense of tumor induction) is possible with viral nucleic acids (see p. 264) or with nucleic acids isolated from neoplastic tissues (e.g., Latarjet et al., 1958; Bielka and Graffi, 1959; Ito and Evans, 1961; Cantarow et al., 1966).

The DNA tumor virus most widely used experimentally is the polyoma virus which contains a double-stranded DNA as a single circular molecule (Dulbecco and Vogt, 1963). Other DNA tumor viruses are the Shope papilloma virus (which induces papillomas in rabbits) and the papovavirus (usually termed SV 40 = simian virus 40), found in simian kidneys. Under appropriate conditions, viruses which normally appear to be harmless can be carcinogenic. Thus, the adenoviruses of type 12 and 18, which frequently occur in human subjects, can induce tumors in hamsters (Huebner et al., 1962; Trentin et al., 1962). More detailed investigations of the mechanism of viral carcinogenesis became possible with the use of tissue cultures which permit the observation of neoplastic transformation under controlled conditions in vitro. With respect to the response of a host cell to a virus, one can distinguish a productive infection (i.e., the virus multiplies and finally kills the host; cf. p. 258) from a transformation which leads to conversion into a cancer cell. For example, the DNA-containing polyoma virus rapidly kills embryonic mouse cells in tissue culture; this is accompanied by the liberation of large amounts of new viruses. In contrast, hamster cells are not killed by the virus *in vitro*; instead the virus initiates proliferation of the cells, accompanied by the formation of aberrant cell forms—a phenomenon never observed *in vitro* with mouse cells (Dulbecco, 1960). The major effect of polyoma virus—*in vitro* and *in vivo*—is usually cell proliferation in the hamster, whereas it is virus multiplication in the mouse. One possible explanation is that for the process of transformation, the polyoma DNA must be incorporated by the genome of the host cell (Vogt and Dulbecco, 1962). The question of whether all or only a certain part of the viral DNA has to be incorporated cannot be answered as yet; but, in any case, only a part of the viral genome is effective in the host cell. By immunological studies, it has been shown that in transformed cells no viral coat protein is synthesized; thus the part of the viral DNA which contains the information for this particular protein is not functioning. As a consequence, no complete viruses can be produced in transformed cells. On the other hand, other parts of polyoma DNA are utilized in the host cell for transcription, as demonstrated by DNA-RNA hybridization studies (Benjamin, 1966); the transformed cells were found to produce a polyoma-specific mRNA. Thus there is a continuous effectiveness of genetic material derived from the virus, which is apparently also responsible for the morphology of the transformed cell. All the functions of viral DNA in the host cell are not known in detail. One such function, however, is the production of a virus-specific antigen which is localized at the cell surface (Habel, 1961; Sjögren *et al.*, 1961), and another is the initiation of DNA replication and the formation of the required precursors, i.e., loss of growth control. According to a hypothesis by Dulbecco (1967), transformation by DNA-containing viruses occurs in two steps: At first, after the viral DNA has entered the nucleus of the host cell, only the "early genes" of the viral information begin to function (this includes an activation of those cellular genes that control the synthesis of host DNA). During this phase, the cell is only transiently transformed. The transformation is a permanent one when the viral DNA has become part of the cellular structure. This seems to require that the "late genes" of the virus (which include those containing the information for the viral coat proteins) be inactivated. The nonfunctioning of these late genes prevents a productive infection, and thus cell death. Some details of this hypothesis are still in need of experimental verification.

The DNAs of the two adenoviruses, types 12 and 18 (carcinogenic for hamsters), which occur in man, and of the polyoma and the Shope papilloma virus, have a lower GC content (48–49%) than the other, noncarcinogenic adenoviruses (56–57%), and also differ in T_m values (89.0 °C versus 92.5 °C; cf. p. 78). Furthermore, it has been shown by hybridization experiments that these two adenoviruses are closely related genetically (Green and Pina, 1963).

The observed carcinogenic activity of type 12 and 18 adenoviruses in hamsters raises the question of whether these or other viruses may play a role in the origin of human neoplasms. A decisive answer still cannot be given. From the present knowledge, it is quite evident that the absence of a virus in a human cancer can no longer be considered proof against its original induction by a virus. While the results of epidemiological studies did not offer much information (with the exception of the "Burkitt" tumor), it seems theoretically possible to detect certain footprints of a DNA virus in transformed cells by analyzing virus-specific mRNA (cf. Benjamin, 1966) or antigens (cf. Habel, 1966). However, this type of analysis by DNA-RNA hybridization or by antigen detection requires that the suspected virus be known and available in a highly purified form.

In contrast to the behavior of DNA viruses, it is apparent that in RNA tumor viruses, replication and release of new virus particles can continue along with survival and multiplication of transformed cells. A somewhat special case is the avian Rous sarcoma virus (RSV) with which extremely detailed studies have been carried out. RSV induces malignant transformation with high speed and high frequency in appropriately infected cells. Each RSV particle contains one single-stranded molecule of RNA (Robinson et al., 1965). While RSV alone is capable of causing a permanent malignant transformation, it cannot replicate and form new viruses unless a helper virus RAV (Rous avian virus) is present (Rubin, 1964); RAV is a virus of the avian leukosis group which is required for the synthesis of the viral coat protein of RSV. The transformation of cells infected by RSV is accompanied by the activation of certain genes, as a consequence of which various properties of the cell are changed (e.g., increased rate of glycolysis and increased rate of synthesis of acid mucopolysaccharides; Temin, 1966). Temin (1964) suggested that these activated genes are not originally present in the host cell but are newly formed via the viral RNA (transcription in the opposite direction); another possible explanation is a change in the pattern of functionally available DNA in the cell nucleus (cf. pp. 89 and 189). This last theory also seems to explain the observations that isolated (nonviral) RNA can initiate permanently the synthesis of new enzymes (Niu et al., 1962) or intervene in morphogenesis (Hillman and Niu, 1963). Again it appears that this RNA does not introduce new information but affects the availability of the information already present in the cell.

The report by de Carvalho and Rand (1961) suggests that with appropriate RNA preparations, *in vitro* metabolic and morphological changes may be induced in tumor cells that appear as a beginning of reversion of carcinogenesis (cf. p. 293). The authors incubated Novikoff hepatoma cells (transplantable rat ascites tumor) for 15 hours in isotonic salt solutions at 2°C; RNA preparations from different sources had previously been added to portions of the tumor cells. The low temperature was used to minimize enzymatic degradation of the added RNA. After incubation, the cells were transplanted. Their subsequent behavior varied, depending on the origin of the added RNA. RNA from the Novikoff hepatoma increased the growth rate, while RNA from normal tissues (except liver) had no effect on growth—the cells being essentially the same as the controls. Liver RNA, on the other hand, decreased the growth rate considerably. At the same time, the tumor cells developed a tendency to resemble morphologically normal parenchyma cells of the liver. Apparently this effect is not species specific, because rabbit liver RNA had the same effect. Other authors have reported similar results (Aksenova *et al.*, 1962; O'Neal *et al.*, 1962).

HYPOTHESES ON THE ORIGIN OF CANCER

The still unresolved cancer problem has led to many hypotheses attempting to explain the transformation from normal to malignant tissues. A few of these concepts will be briefly discussed here. More than four decades ago, Warburg (1924) found a high rate of glycolysis in various tumors. From these classical observations he concluded that the transformation from normal to tumor cells is due to an irreversible injury to respiration (which is then compensated by an increased glycolysis; cf. Warburg, 1930). There is still dispute as to whether the claimed relationship between glycolytic rate and growth rate of tumors really exists, and whether the latter always have a higher rate of glycolysis than normal tissues (cf. Pitot, 1966b). The main argument against the *Warburg hypothesis* is that metabolic changes are considered a direct cause for carcinogenesis and not—as one would postulate from our growing knowledge of cellular control systems—a consequence of certain primary lesions. Thus, Potter (1958) when discussing the Warburg hypothesis emphasized that "any difference [between normal cells and cancer cells] must be identified with a *mechanism* before it can be claimed as a defect" (in the sense of a respiratory defect). The *Greenstein hypothesis* (cf. Greenstein, 1956), which can be considered an extension of the

Warburg concept, postulated a general tendency of tumor cells toward biochemical uniformity; however, as Greenstein himself pointed out, several exceptions were soon observed which disagreed with his theory of convergence.

The binding of carcinogenic azo dyes to proteins of rat liver (cf. p. 285) led Miller and Miller (1947) to the *deletion hypothesis*. In its initial form it postulated a permanent alteration or loss of specific proteins which are essential for the control of growth.   Potter (1948) modified this concept to the *catabolic deletion hypothesis*, according to which deletions during carcinogenesis should be associated mainly with enzymes involved in catabolic reactions (cf. p. 282 and Fig. 138).   This hypothesis was initially supported by the results of many metabolic studies.   Sub-

Fig. 138. Model of the deletion hypothesis of carcinogenesis (from V. R. Potter, 1962).

sequently, however, in the so-called minimum deviation hepatomas, substantial amounts of catabolic enzymes were detected.   As a consequence, the original ideas have now been newly formulated in terms of molecular biology as the *feedback deletion hypothesis* (Potter, 1964) which postulates that in tumor cells "one or more connecting links in feedback loops have been deleted or altered to the point of functional ineffectiveness."   Finally, the *virus theory* of carcinogenesis is well accepted now; however, it still cannot be decided to what extent viral tumor induction is of general importance and whether viruses may play a role in the formation of human neoplasms.

The mechanism of growth control is still not really known, but there are indications that it operates on the principle of a negative feedback systems made up of several connected regulatory circuits in which specific genes are also involved.   It is plausible that the normal functioning of such circuits can be disturbed in different ways.   Not knowing the parts of the circuits, one can merely distinguish in a simplified fashion two kinds of possible interactions—with DNA or with non-DNA building blocks.   If various carcinogenic agents are compared from this point of view, certain general statements can be made.   In the case of DNA-containing tumor viruses, neoplastic transformation requires the incorporation of viral DNA into the genome of the host cell.   Although the total amount of DNA in such cells ought to be somewhat increased, the newly added genes appear to cause a rearrangement of the chromatin, thus leading to the inactivation (heterochromatization?) of other parts of the cellular genome (including those responsible for growth control). In contrast to DNA viruses, certain chemical carcinogens that interact with cellular DNA may cause mutations. If genes which are involved in growth control are mutated, the appropriate regulatory circuit becomes ineffective, and autonomous growth might begin; at the same time, mutations of other genes could cause deletions of various enzymes. Other ways to switch off the regulatory circuits are the interaction with non-DNA building blocks (protein binding?) or the inactivation (not alteration) of the involved genes. The observation that in tumor cells the amount of functionally available DNA in the euchromatin is decreased (cf. p. 284) suggests that an inactivation of DNA via heterochromatization might indeed be a very common reason for loss of growth control as well as for deletion of enzymes; the last effect would then be an accompanying one without much importance for carcinogenesis itself. Perhaps this picture can be extended to include the hypothesis that a decrease in functionally available DNA can cause a great variety of deletions which might finally lead to either cell death or—for the surviving cells—neoplastic transformation (Harbers and Sandritter, 1968). One conclusion from these as yet fairly speculative considerations is that for the process of carcinogenesis, an alteration of DNA does not always appear to be necessary, although in certain cases (e.g., DNA viruses) changes in the cellular DNA seem to be essential for tumor induction. Carcinogenesis without DNA alteration might offer the hope for a possible reversion of neoplastic transformation. Based on the Jacob-Monod hypothesis (cf. p. 248), Pitot and Heidelberger (1963) proposed a regulatory circuitry in which a short exposure to a carcinogen could cause a permanent metabolic disturbance. If this or similar models of the mechanism of tumor formation without direct involvement of cellular DNA prove to be correct, then "a reversion from the malignant to the nonmalignant state is well within reason," as the authors pointed out. A further encouragement in this direction is the clinical observation that spontaneous regressions of malignant tumors do occasionally occur in humans (e.g., Brunschwig, 1963; Legier, 1964).

The biochemistry of cancer has been treated in many recent reviews and symposia (e.g., Emmelot and Mühlbock, 1964; Holzer and Holldorf, 1966; Pitot, 1966b). In addition, specific topics are discussed in three monograph series (Haddow and Weinhouse; Homburger; Rentchnick).

Effect of Various Drugs on Nucleic Acid Metabolism

CHAPTER XIV

A wide variety of drugs can interfere with the function or metabolism of nucleic acids. Some, such as the analogs of nucleic acid bases, have been synthesized by chemical methods; others are natural products whose biological activities have been determined empirically. Few of them have clinical importance. Apart from some antibiotics used in the treatment of infections, these drugs have been utilized primarily in the chemotherapy of malignant tumors and, recently, in attempts to treat certain virus diseases. Many of these drugs have proved useful in the analysis of metabolic processes and the induction of certain types of mutations.

On the following pages, the action of some drugs that interfere with nucleic acid metabolism will be briefly discussed. Many such compounds have been described; these have usually resulted from attempts to find more effective anticancer agents. This chapter will deal with only a small fraction of these drugs, and will generally be concerned with those that are clinically important or whose mechanism of action is reasonably well understood.

ANTIBIOTICS

Among the ever-increasing number of antibiotics, the actinomycins are of special interest because of their mode of action. At first considered unsuitable for therapeutic use because of their extremely high toxicity, these compounds were later found to inhibit the growth of some experimental tumors (Hackmann, 1953) and were therefore introduced briefly into clinical cancer therapy. Brockmann (1960) and his co-workers determined the complete structure of actinomycin, which consists of a peptide and a chromophore (Fig. 139), and finally confirmed this by synthesis.

All actinomycins investigated thus far have the same chromophore and differ only in the amino acid composition of their peptide moiety. Figure 139 shows the structure of the most frequently used actinomycin D.

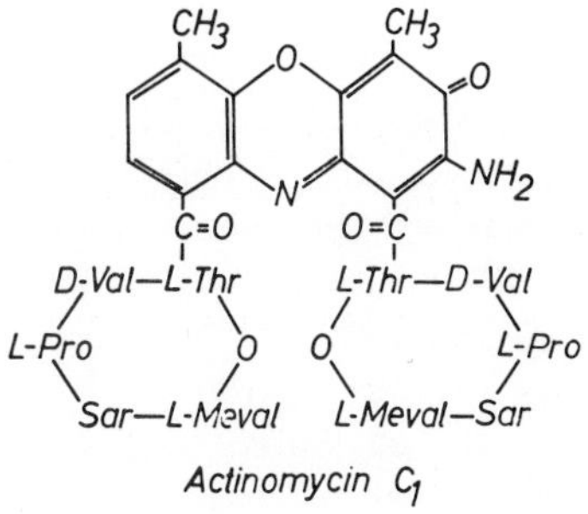

Fig. 139. Actinomycin D(= C₁, in the German literature).

The observation that the effect of actinomycins on microorganisms can be reversed by the addition of DNA to the culture medium provided the first clue to the mechanism of action of these antibiotics (Kersten *et al.*, 1960; Kirk, 1960); RNA was found to be much less effective. Actinomycins form complexes with DNA (Rauen *et al.*, 1960), with binding occurring at the deoxyguanosine (Kersten, 1961). In comparing DNA binding *in vitro* and the biological effects of various actinomycins and synthetic derivatives, only the bacteriostatically active compounds were found to form complexes with DNA (Müller, 1962). Subsequently DNA binding of radioactively labeled actinomycin was demonstrated in living cells (Harbers and Müller, 1962; Harbers *et al.*, 1963). Metabolic experiments showed that even low concentrations of actinomycin inhibit RNA synthesis, while DNA synthesis was found to be less affected. An obvious interpretation of these results was that actinomycin binding to the DNA template prevents the synthesis of DNA-dependent RNA (especially mRNA). Later investigations showed that the sensitivity of different types of RNA toward actinomycin varies, depending on their molecular weight. The *in vivo* incorporation of labeled precursors into rat liver rRNA is inhibited at extremely low doses; higher doses are required for the inhibition of mRNA formation, while the synthesis of low molecular weight tRNA is scarcely affected by actinomycin. These *in vivo* observations were later confirmed with DNA-dependent RNA polymerase systems (Goldberg and Rabinowitz, 1962; Goldberg *et al.*, 1962; Hartmann and Coy, 1962; Hurwitz *et al.*, 1962). Because of the specific effect of actinomycins on DNA-dependent RNA synthesis, they have become an important experimental tool, permitting the direct inhibition or blocking of the expression of genes and the metabolic processes they control.

DNA binding by actinomycins preferentially inhibits the multiplication of DNA viruses (Reich *et al.*, 1961). However, several RNA viruses have also been found to be inhibited, e.g., the influenza virus (Barry *et al.*, 1962), the avian Rous sarcoma virus (Temin, 1963), and the Reo virus with its double-stranded RNA (Gomatos *et al.*, 1962).

No details are known as yet about the nature of the actinomycin binding to DNA. Hamilton *et al.* (1963) have suggested that the chromophore is bound to deoxyguanosine via hydrogen bonds. More recent studies have shown that actinomycins form very stable dimers ($k_0 > 10^5$ liters/mole) in aqueous solutions (Müller and Emme, 1965) with two independent binding sites for deoxyguanosine (Müller and Spatz, 1965). When these dimers are bound to DNA, a lattice is formed which with small doses of the antibiotic is mainly intramolecular but becomes increasingly intermolecular as the dosage increases (Müller, in preparation). While unchanged DNA can be recovered *in vitro* from the actinomycin-DNA complex, the antibiotic causes *in vivo* a dose-dependent decrease in the viscosity of the DNA, isolated later, apparently due to some depolymerization (Müller *et al.*, 1963). Isolated nucleohistones have only half the binding capacity for actinomycin of pure DNA. Furthermore, in somatic cells actinomycin is bound mainly to the DNA of the euchromatin (Table 20). When pure DNA in solution complexes with

TABLE 20.   DNA-bound H^3-labeled Actinomycin C_1 in Three Fractions of Rat Liver Cell Nuclei after *in vitro* Incubation with the Labeled Antibiotic. Quantities in Parenthesis Indicate the Conditions of Centrifugation Needed to Sediment the Three Nuclear Fractions (from Harbers and Vogt, 1966)

Nuclear Fraction	DNA-bound Actinomycin $\dfrac{\mu g\ Actinomycin}{mg\ DNA}$	Relative Values
Heterochromatin (1000g; 10 min)	0.34	16
Intermediate fraction (3000g; 30 min)	1.6	76.5
Euchromatin (78,000g; 60 min)	2.1	100

actinomycin, an equilibrium between excess and DNA-bound actinomycin is established after a while. Thus, by means of sedimentation analysis and equilibrium dialysis the binding of the various actinomycins to pure DNA or nucleohistones can be studied. However, when isolating the DNA-actinomycin complex from tissues, the equilibrium becomes disturbed, and the recovery of bound actinomycin is thus always incomplete.

In mitochondria, RNA formation is not inhibited at all by actinomycin *in vivo*. This appears to be due to the impermeability of the mitochondrial membrane to the antibiotic, because *in vitro*, especially after pretreatment of the isolated mitochondria with digitonin, RNA synthesis is inhibited by actinomycin in the same way as in cell nuclei (Neubert, 1966a). Even in extremely high concentrations, actinomycin generally has very little effect on DNA synthesis (e.g., Reich et al., 1961; Harbers and Müller, 1962). If cells whose DNA is complexed with actinomycin do replicate, the antibiotic seems to be "slipped off" during DNA duplication. In autoradiographic studies with H^3-labeled actinomycin, it was observed that in mitotic cells the antibiotic either could not be found at all or was present only in relatively minute amounts (Backmann, 1963); moreover, when DNA replication was blocked by X-rays, actinomycin remained in the cells for a longer time (Harbers et al., 1963; cf. p. 349). In addition to inhibition of RNA synthesis, some depolymerization of RNA by actinomycin has been observed in B. subtilis (Acs et al., 1963).

Growth inhibition by actinomycin seems to be caused primarily by blocking RNA synthesis in the G_1 period (cf. p. 182), thus preventing adequate production of DNA polymerase and thereby inhibiting DNA replication (Lieberman et al., 1963). On the other hand, a direct inhibition of DNA synthesis may be greater than appears in most studies because the end group reactions at the primer DNA are much less sensitive to actinomycin than real DNA replication (Keir et al., 1963; cf. p. 167). Using C^{14}-labeled actinomycin D_1 it was shown that the antibiotic is concentrated most intensively in the spleen of mammals; very high concentrations were also found in the liver. Actinomycin is excreted in the bile and then partially reabsorbed in the small intestine; apparently it cannot be metabolized (or only to a very low degree) by the mouse, as demonstrated by carrier studies (Harbers et al., 1964). This, together with a very slow excretion (e.g., over a period of 48 hours, the mouse can excrete only about 10% of intraperitoneally administered C^{14}-labeled actinomycin in the urine and feces), contributes to the markedly cumulative effect of the actinomycins. It is surprising that the concentration of actinomycin in experimental tumors is very small, even in those that respond to the antibiotic. Further factors, therefore, must play a role in the preferential effect on certain tumors. An important one seems to be the low euchromatization in tumor cells (cf. p. 284); the small amount of functionally available DNA requires lower amounts of actinomycin to be blocked. The lack of sensitivity of some microorganisms to actinomycin seems to be due to its inability to penetrate the cell wall (Haywood and Sinsheimer, 1963).

Because of its ability to specifically block gene function, actinomycin is now widely used in experimental research to investigate the gene dependence of metabolic processes, e.g., the effect of hormones or antibody formation. However, all metabolic processes are ultimately gene dependent. Thus an inhibitory effect of actinomycin (e.g. on the action of a hormone) is not necessarily an indication of direct involvement of this particular hormone at the gene level. To avoid pitfalls, one must be very careful in analyzing the effects of actinomycin, especially under *in vivo* conditions.

Results of studies with *Drosophila* suggest that the actinomycins have no mutagenic effect; on the contrary, they seem to provide some protection against the mutagenic effect of X-rays (Burdette, 1961). On the other hand, combined actinomycin and radiation treatment have been found to increase the biological effectiveness of X-rays in cancer therapy (see p. 349). At present, these contradictory observations cannot be explained.

Reich and Goldberg (1964) have reviewed a considerable amount of literature on the mechanism of action of the actinomycins and their various applications.

After the mechanism of action of the actinomycins had been clarified, other compounds were detected which also interact with DNA and inhibit the synthesis of DNA-dependent RNA. In contrast to actinomycin, some of these compounds also have an inhibitory effect on DNA replication or can interact with RNA.

Pluramycin A has properties similar to those of actinomycin in several respects (Tanaka *et al.*, 1965). It is bound by DNA, thus causing an increase in the T_m value; RNA polymerase is inhibited more than DNA polymerase. The antibiotic chromomycin A_3 was found to selectively inhibit RNA synthesis in mammalian cells (Wakisaka *et al.*, 1963); an interaction with DNA (and also with tRNA) was demonstrated by Behr and Hartmann (1965). More recently, several anthracyclins (daunomycin, cinerubin A and B, isoquinocycline and rhodomycin) were reported to inhibit the enzymatic synthesis of RNA and DNA to almost the same extent (Koschel *et al.*, 1966). When testing the effects of various antimicrobial agents on RNA polymerase activity, Ward *et al.* (1965) and Waring (1965) observed considerable inhibition with echinomycin, mithramycin, neomycin, olivomycin, suramin, ethidium bromide, proflavin sulfate and prothidium bromide; in several cases, the DNA primer was precipitated by these agents.

Phleomycin, an antibiotic that inhibits the growth of experimental tumors, is also bound by DNA. In contrast to the actinomycins, this compound preferentially blocks the action of DNA polymerase; only with very high concentrations can the RNA polymerase be inhibited (Falaschi and Kornberg, 1964). The inhibitory effects parallel the AT content of the DNA primer; thus it seems that (again in contrast to actinomycins) the AT bases are involved in the binding process. Similar effects have been observed with negalomycin (Bhuyand and Smith, 1965).

Aflatoxin B_1, a metabolite produced by some strains of the mold *Aspergillus flavus*, is an extremely toxic compound that binds to DNA and strongly inhibits RNA synthesis (Rees *et al.*, 1966; Sporn *et al.*, 1966). In contrast to the majority of the actinomycins, aflatoxin B_1 is a very potent carcinogen for rat liver. As yet, this difference in the carcinogenic activity of the two agents, which seem to have very similar mechanism of action, cannot be explained. For the acridines, which are bound by DNA as well as RNA, an intercalation between the bases of double-helical DNA has been postulated (cf. p. 246).

Basic dyes, used for the histological staining of cell nuclei, are bound to the DNA via the phosphate groups. In this case, the degree of binding depends, among other things, on the extent of histone masking; higher salt concentrations lead to a dissociation of the DNA-histone complex and hence to the release of additional phosphate groups, which then become available for reaction with the dye (cf. Klein and Szirmai, 1963).

Another antibiotic that interferes with nuclear metabolism is mitomycin C (Fig. 140). This compound, discovered in Japan, has been employed

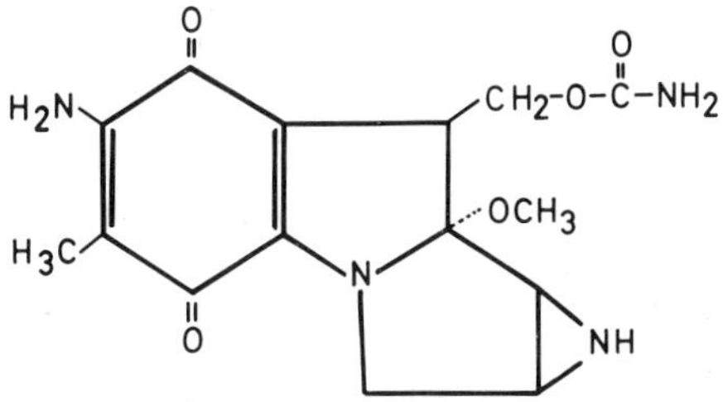

Fig. 140. Mitomycin C.

in cancer chemotherapy. Shortly after administration, mitomycin C prevents the incorporation of labeled thymidine into DNA (Schwartz, 1962), without essentially affecting RNA and protein synthesis. This is followed by disintegration of the nucleus with partial depolymerization of the DNA, which then appears as droplets in the cytoplasm (Shatkin *et al.*, 1962). While H. Kersten *et al.* (1964) claimed that the degradation of DNA is due to an activation of a cellular deoxyribonuclease, findings by Szybalski's group showed that mitomycin (which is not active in its natural form) becomes converted into a reactive alkylating product that causes cross-linking of DNA (cf. p. 304). At first, the alkylating effect could only be demonstrated *in vivo* (Iyer and Szybalski, 1963). Later, it was observed that the active form of mitomycin, a bifunctional alkylating agent, is obtained by a reduction which can be performed either chemically or enzymatically (Szybalski and Iyer, 1964). The extent of cross-linking seems to be proportional to the GC content of the DNA; the cross-linking favors a renaturation of DNA after thermal denaturation (cf. p. 83). The biological effects of mitomycin C were found to be enhanced in combination with either 5-fluorouracil or 6-thioguanine (cf. pp. 312 and 320; Sartorelli and Booth, 1965). Compounds related to mitomycin C (mitomycin A and B, porfiromycin, 7-oxyporfiromycin) apparently have the same mechanism of action.

The antibiotics discussed so far act at the genetic level, thus mainly disturbing either transcription or DNA replication, or both. In contrast, the compounds treated in the following paragraphs interfere in various ways with the translation mechanism, thus inhibiting or modifying the process of protein synthesis.

Chloramphenicol (Fig. 141), a drug of considerable clinical importance, causes an intensive inhibition of protein formation in bacteria;

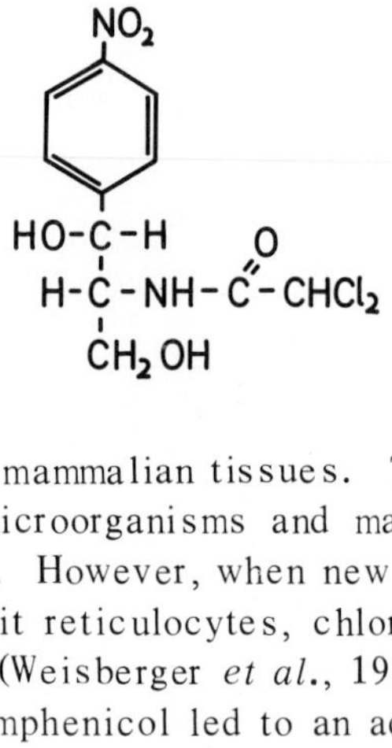

Fig. 141. Chloramphenicol.

the effect is much less pronounced in mammalian tissues. This difference in the response of cells from microorganisms and mammals was also demonstrated in cell-free systems. However, when new mRNA was added to a system prepared from rabbit reticulocytes, chloramphenicol was found to be an effective inhibitor (Weisberger et al., 1963). Rendi and Ochoa (1962) observed that chloramphenicol led to an accumulation of RNA (probably mRNA) in bacterial cells. These experimental results suggested that the antibiotic interferes with the binding of mRNA to ribosomes, while it should have no effect on polysomes (in which this binding has already occurred). The specificity of the effect of chloramphenicol on bacteria was at first considered an indication of structural differences in mammalian and microbial ribosomes. Now, it seems more probable that the much longer lifetime of mRNA in mammalian cells plays an essential role in their low sensitivity to this antibiotic (Weisberger and Wolfe, 1964). Chloramphenicol appears to inhibit only processes which require the formation of new polysomes. In growing bacteria, new polysomes must be formed continuously; thus the antibiotic is rapidly and highly effective. In contrast, polysomes of mammalian cells have a much longer lifetime; therefore, in these cells chloramphenicol should disturb only metabolic events which are connected with polysome formation.

Puromycin (Fig. 142) is another antibiotic that interferes with protein synthesis, leading to the premature release of incomplete peptides

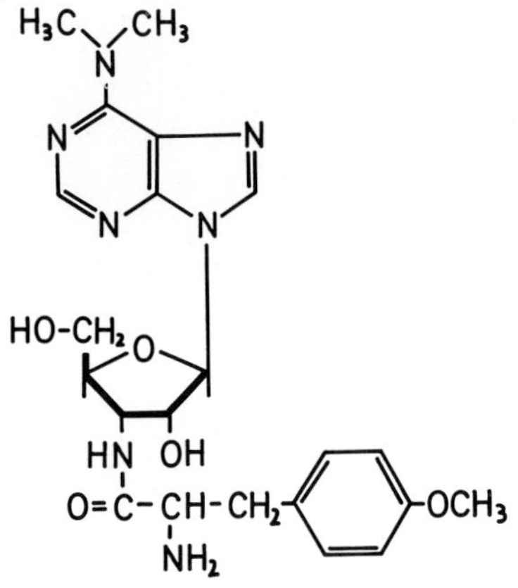

Fig. 142. Puromycin.

(Rabinowitz and Fisher, 1962). Apparently, puromycin competes with normal tRNA by acting as a phenylalanyl-tRNA analog (Yarmolinski and de la Haba, 1959). This concept was finally confirmed by results of experiments with C^{14}-labeled puromycin (Allen and Zamecnik, 1962) which, when supplied to a cell-free system from rabbit reticulocytes, was found to be bound to the carboxyl end of the released peptides; under saturation conditions, 1 mole of the antibiotic was bound to 1 mole of polypeptide. A detailed analysis of the various steps of protein synthesis (Noll, 1965) indicates that puromycin reacts only with ribosomes in state I (cf. p. 225 and Fig. 106) during which the antibiotic competes with the incoming phenylalanyl-tRNA for peptide bond formation. Thus, ribosomes in state II (peptide-linked tRNA in the decoding site) should be puromycin resistant. In contrast to the effect of chloramphenicol, the binding of mRNA to ribosomes is not impaired, but the average spacing between ribosomes is increased in the presence of puromycin, apparently due to faster readout and premature detachment of ribosomes (Noll, 1965). The opposite effect was observed with "Acti-dione" (cyclohexi-mide) which reduces the readout rate, thus causing a closer spacing of the ribosomes at the messenger. At very high concentrations of "Acti-dione," the mRNA transport in the polysomes becomes completely blocked. The essential primary action of "Acti-dione" seems to be a specific inhibition of the translocase reaction (Noll, 1965; cf. p. 225).

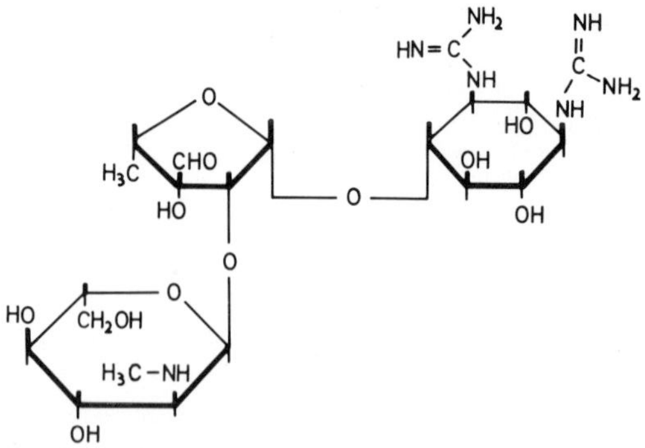

Fig. 143.   Streptomycin.

Streptomycin (Fig. 143) is an antibiotic with various effects. Because of its basic properties, it can be used to precipitate nucleic acids or nucleohistones; however, there does not appear to be any correlation between this ability and its biological effects. Streptomycin is bound to the 30S particles of the ribosomes; this binding takes place only in cells that are sensitive to or dependent on streptomycin. As a consequence of this binding, protein synthesis may be inhibited and a breakdown of rRNA can occur (Dubin and Davis, 1962). In addition, alterations of the cell membrane have been observed, causing a leakage of nucleotides (especially 5′-AMP and 5′-GMP). This nucleotide release does not play an important role in the mode of action of streptomycin, as initially assumed, because it does not take place at low, although still lethal, concentrations of the antibiotic (Tzagoloff and Umbreit, 1963). The results of more recent investigations showed that the binding of streptomycin to the 30S particles of bacterial ribosomes disturbs the reading of the message for polypeptide synthesis (Davies *et al.*, 1964); the same type of disturbance was found with other aminoglycide antibiotics like neomycin, kanamycin, paromomycin, gentamicin, hydromycin B and viomycin (Davies *et al.*, 1965). By the use of cell-free systems, the extent and kind of misreading were found to differ for the various antibiotics tested. An analysis of the results of studies done with the four homopolymers polyU, polyC, polyA and polyI suggests that the antibiotics of this group cause misreading of only one base at a time (in either the 5′- or the internal position) in the pyrimidine triplets UUU and CCC, but of all three bases in the purine triplets AAA and III; both transition and transversion (cf. p. 241) misreadings were observed (Davies *et al.*, 1965). Misreading the message is now considered to be the essential effect of streptomycin, as a result of which proteins with faulty amino acid sequences are synthesized. The dependence of certain mutant cells on the presence of streptomycin seems to be a kind of suppression at the ribosome level (see p. 244). Studies with C^{14}-labeled streptomycin indicate that the antibiotic binds noncovalently to ribosomes of both sensitive and resistant strains to a level of one molecule per ribosome (Leon and Brock, 1967). The presence of streptomycin permits single-stranded DNA to be used as a template for protein synthesis (McCarthy and Holland, 1965; cf. p. 225).

Another group of antibiotics are nucleoside analogs in which the structure of the base (cf. p. 322) or that of the pentose moiety has been altered. Cordycepin, an antibiotic produced by the fungus *Cordyceps militaris*, proved to be 3′-deoxyadenosine. In Ehrlich ascites tumor cells, the deoxyribonucleoside analog was found to be converted to its triphosphate (Klenow, 1963a) and to inhibit the incorporation of labeled precursors into RNA as well as DNA (Klenow, 1963b). Cory *et al.* (1965) observed that the analog can be utilized for the synthesis of both RNA and DNA. The 3′-deoxyribose is preserved after incorporation into the nucleic acids; therefore, it is postulated that a 2′,5′-phosphodiester bond must be formed. A principal effect seems to be a blocking of the formation of PRPP (Overgaard-Hansen, 1964; Rottmann and Guarino, 1964) (cf. p. 150); as a consequence of which many metabolic processes are disturbed, including the *de novo* synthesis of purine nucleotides. The metabolic effects of further analogs in this group [e.g., psicofuranine (adenine-β-D-psicofuranoside), an antibiotic that inhibits growth of bacteria and several transplantable tumors] have as yet not been studied in detail. Among various purine and pyrimidine arabinosyl nucleosides, 9-β-D-arabinofuranosyl-adenine is converted to the triphosphate which is a noncompetitive inhibitor of DNA polymerase (York and LePage, 1966). 9-β-D-Xylofuranosyladenine, another adenosine analog, acts by blocking the formation of PRPP, which is supposed to be responsible for the inhibition of the synthesis of purines as well as of RNA and DNA (Ellis and LePage, 1965). The analog controls the *de novo* synthesis of purines by feedback in the same way as natural adenosine does (cf. p. 156).

1-β-D-Arabinofuranosylcytosine was found to inhibit mitosis and DNA replication in mouse fibroblasts (Silagi, 1965). The analog is incorporated into DNA and, to a lesser extent, into RNA. Configurational changes in small regions of helical DNA are postulated to be the essential effects.

Nucleic acid base analogs will be treated in a later section (see p. 310). A general review on the mechanism of action of antibiotics has been written by Newton (1965); the clinical aspects of antibiotics in cancer chemotherapy have been discussed by C. G. Schmidt (1963, 1966).

The nucleoside antibiotics (Fox *et al.*, 1966) and the D-arabinosyl nucleosides (Cohen, 1966) have been treated in more special reviews.

ALKYLATING AGENTS

Many of the drugs used for clinical cancer chemotherapy are alkylating substances. These highly reactive compounds have widely varying structures and exhibit a broad range of effects in living cells since they can react with many cell components. It is postulated that the interaction with DNA plays the major role in their biological activity. The structure of the basic molecule is less important for the biological effects than is the presence of functional groups like β-chloroethylamino, ethylenimino or methyl sulfonic acid residues. Bifunctional substances have proved to be especially active. Figure 144 shows, as an example

Fig. 144. Nitrogen mustard [methyl-bis(β-chlorethyl)-amine].

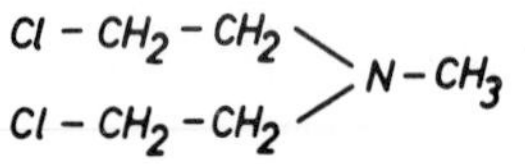

$$Cl-CH_2-CH_2 \searrow$$
$$\qquad\qquad\qquad N-CH_3$$
$$Cl-CH_2-CH_2 \nearrow$$

of such compounds, the structure of nitrogen mustard, one of the first drugs used for cancer chemotherapy. The suggestion that DNA—and probably also some nuclear proteins—constitutes the essential site of action of the alkylating substances was based primarily on the detection of mutagenic effects (Auerbach, 1946) and chromosomal alterations. Most investigations of the mode of action of alkylating agents are performed *in vitro* under conditions which do not always permit a simple extrapolation to living systems. The work of Lawley (1957) on the action of alkylating agents on DNA has indicated that the most sensitive point of attack is nitrogen 7 of guanine, whose alkylation results in a shift of electrons leading to cleavage of the glycosidic bond at nitrogen 9 (Fig. 145); subsequently, the phosphate is cleaved from the deoxyribose [reaction (a) in Fig. 145 below] or a cyclic phosphate is formed [reaction (b) in Fig. 145 below]. Bifunctional substances thus cause cross-linking of the two DNA strands of a double helix (Fig. 146), with the result that either the molecule breaks apart or in the next DNA replication deleted daughter molecules are produced. Figure 147 illustrates schematically the various possible consequences of the action of mono- and bifunctional agents. As can be seen, the bifunctional mode leads to a more severe damage of the DNA than does the monofunctional one (Lawley, 1962). In model experiments, at least, an interstrand cross-linking may also be produced by bifunctional agents (Kohn *et al.*, 1966). It seems likely that the mechanism of action described applies in principle to all

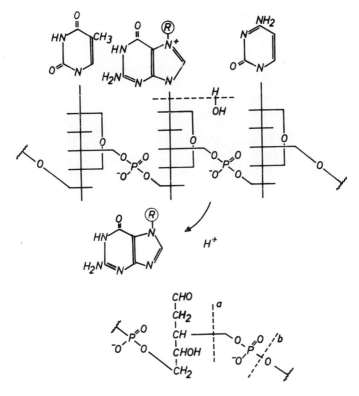

Fig. 145. The hydrolysis of alkylated DNA (R = alkyl group).
For explanation, see text (from Lawley, 1962).

alkylating compounds. In addition to their interaction with DNA, cross-
linking between DNA purines and the carboxyl groups of aspartic and
glutamic acids in the acidic proteins of nucleohistones (Steele, 1962)
may be of considerable importance for the biological activity of these
drugs; there does not appear to be any interaction with basic histones.

DNA alterations have been found in leucocytes of patients with
chronic myelocytic leukemia after treatment with "Myleran" (Bendich
et al., 1957). The interaction of alkylating agents with DNA is not
limited to guanine, but under conditions which approximate those of
clinical therapy, of pH and concentration in the cell reactions
with other bases are negligible. RNA is also attacked by alkylating
drugs (Fraenkel-Conrat, 1961; Lawley, 1962). Alterations in RNA may
contribute to the biological effects, but they appear to be less severe
since they can be repaired as long as the correct DNA template is
available.

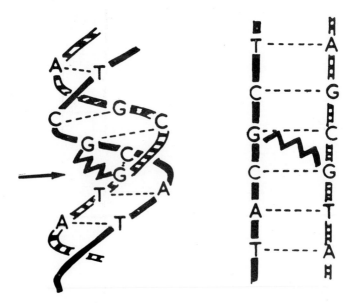

Fig. 146. Postulated mechanism of action of bifunctional alkylating agents. The alkyl chain is indicated by the zig-zag line (from Lawley, 1962).

Alkylating drugs inhibit to a greater or lesser extent the incorporation of labeled precursors into DNA and RNA. Studies of the mitotic cycle have indicated that cells in the G_2 period are not blocked by nitrogen mustard, but are prevented from entering mitosis during the next cycle. It seems that under these particular conditions, all the G_2 cells can pass at least once through the S period and are then blocked afterward (Levis *et al.*, 1965). When radioactive alkylating substances were administered *in vivo*, their concentration in tumor tissues was found to be quite low (e.g., Maller and Heidelberger, 1957; Rutman *et al.*, 1961). Thus, the preferential effect of some of these drugs against certain tumors cannot be explained as yet in a convincing way. Perhaps the lower euchromatization in tumor cells may contribute to their sensitivity as seems to be true with the actinomycins (cf. p. 296).

Just as DNA damage induced by ultraviolet light can be repaired in microorganisms (cf. p. 343), alterations due to the action of alkylating agents can also be repaired (Hanawalt and Haynes, 1965; Lawley and Brookes, 1965; Searaski and Strauss, 1965). Many reviews have been published on the chemistry, mode of action and general pharmacology of alkylating drugs (e.g., Philips, 1950; Ross, 1958; Stacey *et al.*, 1958; Lawley, 1966).

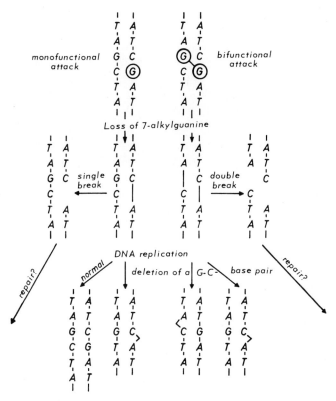

Fig. 147. Effects of hydrolysis of alkylated DNA on the base sequences of newly formed DNA, (from Lawley, 1962).

ANALOGS OF NATURAL METABOLIC PRODUCTS

In the past decade a great many so-called analogs have been added to the armamentarium of drugs utilized in the treatment of neoplasms. The effect of such analogs, most of which are synthetic, depends widely on their ability to "fool" the cell because of their similarity to natural metabolites and thus to replace the normal intermediates in metabolism. The structural differences of these analogs cause metabolic disturbances that range from trivial inhibitions to complete blocking of a reaction (hence the term "antimetabolite"). Nucleic acid base analogs, when incorporated into nucleic acids, can also change the template properties.

The first such analogs used in therapy were the sulfonamides introduced by Domagk in 1935. In 1940, Woods was able to demonstrate that the bacteriostatic effect of the sulfonamides could be overcome by *p*-aminobenzoic acid—later shown to be a precursor for the biosynthesis of folic acid (cf. p. 107); hence, the sulfonamides block the utilization of an intermediate necessary for folic acid formation in bacteria. Numerous other antagonistic drugs have since been tested on microorganisms in attempts to block growth by interfering with the utilization of natural metabolites (Woolley, 1952). The mechanism of action of various analogs has been clarified in more detail by using labeled precursors and cell-free systems. In the following sections, folic acid analogs will be discussed very briefly, while some of the more important pyrimidine and purine analogs will then be dealt with in somewhat greater detail.

Folic Acid Analogs

The importance of folic acid for mammalian cell growth became clear as a result of observations on patients with chronic myelocytic leukemia. Administration of folic acid intensified the disease, whereas remissions occurred if patients were given a diet low in folic acid together with a poorly absorbed sulfonamide (which inhibited folic acid formation in the intestinal flora; Heinle and Welch, 1948). Aminopterin and especially amethopterin (usually termed methotrexate) proved to be the most active folic acid analogs therapeutically. In contrast to normal folic acid, the two drugs have an amino group substituted on carbon 4; amethopterin also contains a methyl group on the nitrogen in position 10 (Fig. 148). Both compounds have been used since 1948 for the chemotherapy of acute leukemia in children (Farber *et al.*, 1948) and more re-

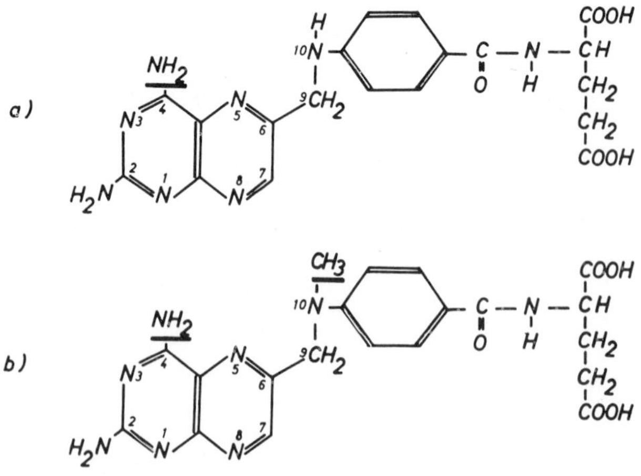

Fig. 148. The two folic acid analogs most frequently used clinically: (a) aminopterin and (b) amethopterin (methotrexate).

cently for the treatment of inoperable chorion epithelioma (Hertz *et al.*, 1958). The biological action of folic acid analogs is due to an inhibition of the reduction of folic acid to FH_2 and of FH_2 to FH_4, the effective coenzyme (cf. p. 151 and Fig. 75). In this way, the transfer of C_1 units is prevented, thereby blocking the *de novo* synthesis of purines (see p. 112) as well as the methylation step in thymine synthesis (cf. p. 129); as a consequence, DNA replication and growth are inhibited (Fig. 149). These effects result from the complex formation of analogs

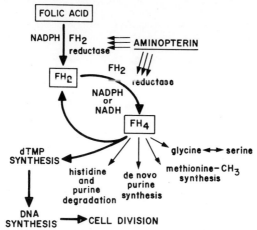

Fig. 149. Sites of attack of folic acid analogs in proliferating tissues. The broad arrows indicate the preferential effect on dTMP synthesis—and hence on DNA replication (from O'Brien, 1962). (Reproduced with permission of American Association for Cancer Research, Inc.)

and enzyme (Greenspan *et al.*, 1950). The affinity of aminopterin and amethopterin for folic acid reductase is 10^5 times greater than that of the natural coenzyme (Werkheiser, 1961). The resulting firm binding has a prolonged effect; the analog cannot be removed *in vivo* by the administration of even very high doses of folic acid. Only the end product in the sequence of reactions, tetrahydrofolic acid (FH_4), is able to reverse the inhibitory effect (Schoenbach *et al.*, 1950; Burchenal and Babcock, 1951).

The precursor usually required for the formation of FH_4 in the mammalian cell is dihydrofolic acid (FH_2); however, since the two reduction steps are catalyzed by the same enzyme (cf. p. 151), the folic acid analogs can block the entire reaction. Zakrzewski (1963) has studied in detail the binding of numerous folic acid analogs to folic acid reductase.

Prolonged therapy with folic acid analogs results in the development of drug resistance (cf. p. 325), caused primarily by a severalfold increase in folic acid reductase activity (Fischer, 1961; Misra et al., 1961). Although the affinity of the folic acid analogs for the enzyme is unchanged, it is no longer possible to "saturate" the enzyme with doses that can be tolerated in vivo.

Pyrimidine Analogs

In a number of analogs of this group, the hydrogen atom on carbon 5 of the pyrimidine ring is replaced by a halogen. Substitution with chlorine, bromine or iodine is relatively simple to carry out, while the introduction of fluorine in place of hydrogen initially presented great difficulties. Uracil, with chlorine, bromine or iodine in the 5-position, behaves biologically like a thymine analog; the van der Waals diameter of these halogens comes close to that of a methyl group (cf. Fig. 151). As early as 1945, Hitchings et al. (1945) reported growth inhibition of L. casei by uracil halogenated in the 5-position; thymine reversed this effect. It has also been observed that these pyrimidine analogs are able, in some cases, to stimulate growth. Since thymine cannot be utilized by mammalian cells (cf. p. 136), compounds like 5-iodo- or 5-bromouracil are also inactive; only the deoxyribosides of these analogs have certain biological activity. Microorganisms can incorporate 5-bromouracil (5-BU) into DNA at a high rate and thus replace a considerable part of the natural thymine (as a maximum, about 50%; Dunn and Smith, 1954; Wacker et al., 1954). In the T2r$^+$ phage, nearly 100% of the thymine can be replaced by 5-BU (Litman and Pardee, 1956), with 9% of the infectivity retained. The distribution pattern of 5-BU in DNA is not uniform (Bendich et al., 1957). Mammalian cells also incorporate 5-BU into DNA if the analog is supplied as 5-bromodeoxyuridine (5-BUDR), i.e., as a thymidine analog. 5-BUDR has some inhibitory action on DNA synthesis. The incorporation of 5-BU into DNA causes an increase in radiation sensitivity (cf. pp. 77 and 346) and also in the T_m value (Kit and Hsu, 1961).

5-Iododeoxyuridine (5-IUDR) is another thymidine analog which is incorporated into DNA (Prusoff, 1960). The biological effects of 5-BUDR and 5-IUDR can be reduced or reversed by the simultaneous administration of thymidine. Thymidine also reverses the action of 5-fluorodeoxyuridine (5-FUDR), a compound that blocks thymine synthesis (see p. 314). When thymidine is replaced by 5-BUDR or 5-IUDR and the methylation reaction for thymine formation is blocked by 5-FUDR, an increased incorporation of 5-BUDR and 5-IUDR into DNA can be obtained (Lorkiewicz and Szybalski, 1960). The practical application of these

two thymidine analogs in mammals is limited by their rapid dehalogenation. The corresponding deoxycytidine analogs are less readily dehalogenated and are therefore more efficiently incorporated into DNA than
the thymidine analogs (Cramer et al., 1962). 5-IUDR was found to inhibit the carcinogenic action of adenovirus, type 12, in the hamster
(Huebner et al., 1962; cf. p. 289).

5-Iododeoxycytidine (5-ICDR) and 5-bromodeoxycytidine (5-BCDR) are
almost completely deaminated before incorporation into DNA and, therefore, like 5-BUDR and 5-IUDR, mainly displace DNA thymine (Cramer
et al., 1962). The incorporation rate of the deoxycytidine analogs is,
however, distinctly higher. Deamination of the two analogs appears to
occur at the nucleoside level (Creasey, 1963). In studies on the distribution of the deoxycytidine analogs in mammals, 5-ICDR was found to
be degraded more slowly than 5-BCDR (Kriss and Révész, 1962; Kriss
et al., 1962).

Rapid incorporation of 5-BU leads to an increase of 6-methylamino-
purine in the DNA of E. coli (Dunn and Smith, 1957); 5-amino-uracil has
the same effect in other bacteria (Theil and Zamenhof, 1963). The reason for this base exchange is not yet understood. Methionine serves as
donor of the —CH_3 group. The breakdown of these halogenated pyrimidine derivatives apparently takes place in the same way as with normal thymine or uracil. In contrast to 5-fluorouracil (cf. p. 312), the formation of the halogenated dihydropyrimidines is followed by cleavage of
the hydrogen halide and subsequent normal pyrimidine degradation (see
p. 139).

Recently, an interesting new compound has been added to the list of
thymine and thymidine analogs. Heidelberger et al. (1962) succeeded in
replacing all three hydrogen atoms in the methyl group of thymine
by fluorine. The two synthesized compounds, 5-trifluoromethyluracil
(5-trifluorothymine, F_3T) and the deoxyribonucleoside 5-trifluoromethyl-
2'-deoxyuridine (5-trifluorothymidine, F_3TDR; Fig. 150) were found to
be incorporated into the DNA in microbial and mammalian cells, respectively, and also to be effective inhibitors of DNA synthesis. This inhibition results from the interaction of 5-trifluorothymidine-5'-monophos-
phate with the thymidylate synthetase; the analog combines with the
enzyme in an irreversible fashion, thus causing a noncompetitive inhibition (Reyes and Heidelberger, 1965). The incorporation of F_3TDR into
DNA induces mutations at about the same rate as does 5-BUDR (Gottschling and Heidelberger, 1963). As a consequence of the replacement
of thymine by F_3T, the physical properties of DNA are changed; T4
phage DNA, containing 11% F_3T instead of thymine, showed a T_m value
about 5°C lower than that of normal DNA of the same origin (Gottschling

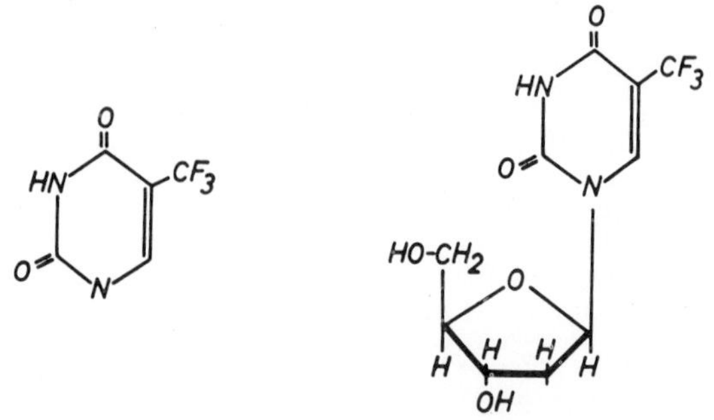

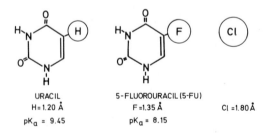

Fig. 150. 5-Trifluoromethyluracil and 5-trifluoromethyl-
2′-deoxyuridine, analogs of thymine and thymidine, respectively (from Heidelberger *et al.*, 1962). (Reproduced
with permission of Academic Press, Inc.)

and Heidelberger, 1963). Like 5-BUDR, the incorporation of F_3TDR into
the DNA of mammalian cells increases their sensitivity to X-irradiation
(Szybalski *et al.*, 1963); in contrast to 5-BUDR and 5-IUDR, F_3TDR
considerably   inhibits   the   growth   of various transplantable tumors
(Heidelberger and Anderson, 1964).   These observations indicate that
this drug should be of special interest for combined radiation and chemotherapy (cf. p. 348).   Furthermore, in tests on antiviral activity, F_3TDR
was found to be 10–100 times more effective than 5-IUDR when locally
applied to rabbit eyes infected with herpes simplex virus (Kaufman and
Heidelberger, 1964).

   5-Ethyluracil is a thymine analog which can replace up to 66% of normal thymine in T3 phage DNA (Pietrzykowska and Shugar, 1966); the incorporation of 5-ethyluracil into DNA causes a decrease in its T_m
value.

   5-Fluorouracil (5-FU) is also biologically a uracil analog; the van der
Waals radius of fluorine comes close to that of hydrogen (Fig. 151).   In
contrast to thymine, uracil can readily be utilized by mammalian tissues

URACIL

H= 1.20 Å

pK$_a$ = 9.45

5-FLUOROURACIL (5-FU)

F =1.35 Å

pK$_a$ = 8.15

Cl =1.80 Å

Fig. 151. Normal uracil and 5-fluorouracil
(5-FU) (from Heidelberger, 1965). (Reproduced with permission of Academic Press,
Inc.)

for nucleic acid synthesis; thus 5-FU is a very effective drug in mammals. Its main effect is blocking the methylation in thymine biosynthesis (Fig. 152), leading to severe inhibition of DNA replication (Bosch *et al.*, 1958; Harbers *et al.*, 1959); RNA formation, on the other hand, is only slightly diminished. Experiments with C^{14}-labeled 5-FU have shown that there is extensive incorporation into RNA, but not into DNA (Chaudhuri *et al.*, 1958); this then eventually leads to the formation of a "fraudulent" RNA with altered template properties (cf. p. 244).

The stimulus for the synthesis of 5-FU, a compound introduced to cancer chemotherapy by C. Heidelberger, was the observation that normal uracil is incorporated into the DNA of transplantation tumors at a higher rate than in normal tissues (Heidelberger *et al.*, 1957; cf. p. 282). This phenomenon is due to the fact that in tumor cells, there is often a greatly decreased capacity for the breakdown of uracil, which then becomes available for anabolic processes. The uracil analog 5-FU undergoes the same reactions and therefore can be found in relatively high concentrations in certain tumors (Chaudhuri *et al.*, 1958). Since the rate of RNA and protein turnover is less affected than DNA replication, in Ehrlich ascites tumors 5-FU led to the formation of giant cells with increased ratios of RNA and protein to DNA (Lindner, 1959); a similar phenomenon has been observed in *E. coli* treated with 5-FU (Horowitz *et al.*, 1958). The behavior of 5-FU in intermediary metabolism corresponds closely to that of normal uracil (Fig. 152). After the formation of 5-fluorodeoxyuridine monophosphate, however, the normal methylation

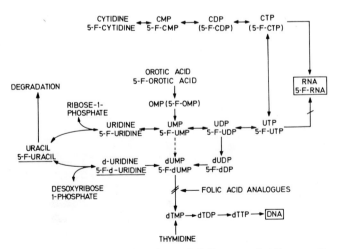

Fig. 152. The mode of action of 5-fluoro-pyrimidines. For explanation, see text (modified from Harbers, Chaudhuri and Heidelberger, 1959). (Reproduced with permission of The American Society of Biological Chemists, Inc.)

step in the synthesis of thymidine monophosphate is impossible, and thus the analog cannot be incorporated into DNA. The associated inhibition of thymidylate synthetase is competitive (Hartmann and Heidelberger, 1961). As its riboside, whose formation takes place via uridinephosphorylase, 5-FU is phosphorylated up to the triphosphate level and then converted in small amounts to 5-fluorocytidine phosphates; part of the triphosphate is then utilized for RNA synthesis (Fig. 152). The observed effects of 5-FU (as an RNA constituent) on coding properties are not uniform. In general, even extensive incorporation of 5-FU into viral RNA does not cause alterations of the properties of the virus. However, in certain cases, mutations of RNA viruses have been found (cf. p. 272). The fact that 5-FU inhibits or completely blocks enzyme induction in rat liver (Gaetani and Spadoni, 1961; Kröger and Greuer, 1965) or in microorganisms (Horowitz et al., 1960; Nakada and Magasanik, 1964) suggests that in the presence of 5-FU an altered ineffective protein is produced. 5-FU can apparently also cause alterations of the ribosomes; in *Staphylococcus aureus* the analog leads to the formation of structurelly different, nonfunctional ribosomes (Hignett, 1966).

The initial degradation of 5-FU is analogous to that of normal uracil (Chaudhuri et al., 1958). After reduction to 5,6-dihydrofluorouracil, the ring opens; this is followed by cleavage to α-fluoro-β-ureidopropionic acid and urea (cf. p. 139). During its utilization for nucleotide synthesis, 5-FU may have more or less inhibitory effects on uridine phosphorylase (Sköld, 1958). This inhibition can be extremely strong in 5-FU-resistant tumors. As a consequence, practically no formation of 5-fluorouridine (5-FUR) occurs (Reichard et al., 1959); thus, the conversion of 5-FU to a form effective in blocking the thymidylate synthetase is prevented. The uridine phosphorylase reaction and interfering side reactions such as the utilization for RNA synthesis are avoided if the drug is supplied as the deoxyribonucleoside 5-fluorodeoxyuridine (5-FUDR) (Fig. 153); the essential product for the inhibition of methylation can then be formed directly by phosphorylation (Fig. 153). After *in vivo* application of C^{14}-labeled 5-FUDR to tumor-bearing mice, the highest accumulation of the drug was found in the tumor cells (sarcoma 180) and much smaller amounts in various normal tissues (Harbers et al., 1959). In similar studies, when patients with carcinoma of the colon received labeled 5-FUDR prior to surgery, in all cases considerably more 5-FU nucleotides were present in the tumor tissue than in the normal intestinal mucosa (Heidelberger, 1966). Such a tumor-specific accumulation does not occur with the riboside of 5-FU, 5-fluorouridine (5-FUR; Fig. 153). 5-FUDR is readily hydrolyzed to 5-FU and deoxyriboside-1-phosphate. This undesirable effect can be partially pre-

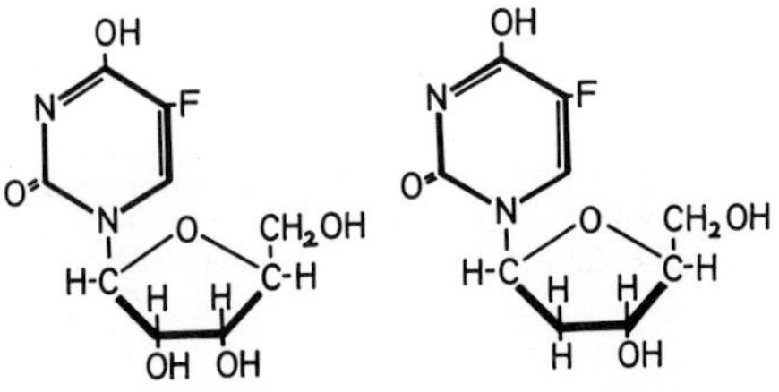

Fig. 153. 5-Fluorouridine (5-FUR) and 5-fluoro-deoxyuridine (5-FUDR).

vented by the simultaneous application of 5′-deoxyglucopyranosylthymine, an analog that inhibits the uridine-deoxyuridinephosphorylase (Langen, 1966). 5-Amino-deoxyuridine enhances the biological effects of 5-FUDR (Friedland and Visser, 1961) as does F_3TDR (see p. 311), while normal thymidine causes a reversion.

All base analogs that have been useful therapeutically are incorporated into nucleotides via the preformed pathway (cf. p. 118). It was therefore believed that the *de novo* synthesis of pyrimidines and purines would not be affected by such analogs and thus the cells would be able to compensate for an analog inhibition. However, once the principle of allosteric inhibition had been recognized (see p. 107), it was found to apply also to the effect of certain analogs, thereby greatly enhancing their chemotherapeutic effect. Biosynthetically produced nucleotides with unnatural bases can often block the initial steps of *de novo* synthesis in the same manner as natural cell products ("pseudo-feedback inhibition") (McCollister *et al.*, 1962). In this way, the cell's ability to eliminate an analog is reduced, and the biological effects of base analogs are not (or not essentially) increased by the simultaneous application of drugs that inhibit the *de novo* synthesis of nucleotides. This pseudo-feedback inhibition is also effective with 5-fluoropyrimidine nucleotides (Bresnick and Hitchings, 1961; Bresnick, 1962).

In microorganisms, the symptoms of 5-FU administration correspond closely to those occurring in thymine-dependent mutants (Cohen and Barner, 1956; Mennigmann and Szybalski, 1962), where the lack of thymine in the nutrient medium ultimately leads to the so-called thymineless death. In *Staphylococcus aureus*, 5-FU also inhibits mucopeptide synthesis required for cell wall formation (Rogers and Perkins, 1960).

5-Fluoroorotic acid (5-FO; Fig. 152) interferes with the last step in the *de novo* synthesis of pyrimidines. In a cell-free system from rat liver, 5-FO inhibited the normal OMP synthesis from orotic acid and PRPP (Stone and Potter, 1957; cf. p. 127); at the same time, 5-fluorouridine phosphates were found in the nucleotide fraction. As expected, 5-fluorocytosine exhibits only very slight activity in mammalian cells, because normal cytosine is not readily utilized. Of the nucleosides, only 5-fluorodeoxycytidine is of any interest. Most of this analog is de-aminated and then has effects similar to those of 5-FUDR (Harbers *et al.*, 1959). The ribotides of 5-FUDR, their polymers and other deriva-tives of 5-FU have shown no or only slight inhibition of the *in vitro* in-corporation of labeled formate into DNA thymine of ascites tumor cells (Mukherjee and Heidelberger, 1962).

The various effects and applications of 5-fluoropyrimidines have been extensively reviewed by Heidelberger (1965); their importance as drugs for cancer chemotherapy is discussed by the same author (1966).

4-Azauracil is another interesting uracil analog in which carbon 4 is replaced by a nitrogen atom (Fig. 154). This compound inhibits the

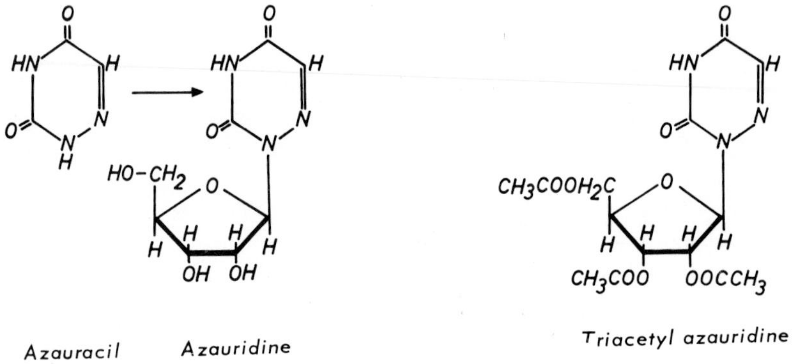

Azauracil   Azauridine   Triacetyl azauridine

Fig. 154. 4-Azauracil, 4-azauridine and the acetylated form of 4-azauridine (from Creasey *et al.*, 1963). In the literature, this uracil analog is fre-quently designated as 6-azauracil, but it is 4-azauracil according to the nomenclature used in this book (cf. p. 5).

growth of microorganisms and various experimental tumors   In clinical trials, very unpleasant neurotoxic effects were observed (Shnider *et al.*, 1960) that did not occur with the ribonucleoside, 4-azauridine (AzUR; Fig. 154); the latter has since been applied in the clinical chemotherapy of neoplasms. AzUR is converted in the cell to 4-azauridine-5′-mono-phosphate, which interferes with the *de novo* synthesis of pyrimidines by blocking the decarboxylation of OMP (Pasternak and Handschumacher, 1959; Fig. 155); as a consequence, OMP accumulates and the utilization of added uridine for nucleic acid synthesis is stimulated. The inhibition of orotidine phosphate decarboxylase by AzUR is competitive. AzUR is also converted to the riboside di- and triphosphates; the triphosphates,

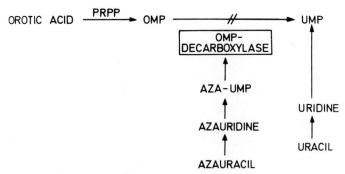

Fig. 155. Mode of action of 4-azauracil and 4-azauridine. 4-Aza-UMP, formed in the cell, blocks the decarboxylation of OMP and thus prevents the *de novo* synthesis of pyrimidine.

however, are not utilized effectively for RNA synthesis. As its riboside diphosphate, AzUR inhibits the polynucleotide phosphorylase reaction (Škoda *et al.*, 1959). Similarly, the DNA-dependent RNA polymerase is inhibited by the triphosphate of AzUR (Goldberg and Rabinowitz, 1963), but the necessary concentrations are higher than those required for the inhibition of OMP decarboxylation. As expected, AzUR is not incorporated into DNA. Chloramphenicol (cf. p. 300) enhances the therapeutic effects of AzUR in experimental tumors (Cardoso and Jaffe, 1961).

For clinical application, AzUR must be injected intravenously, as it is poorly absorbed when given orally. In addition, there is the danger with oral administration that the intestinal flora will attack it and cause liberation of the neurotoxic 4-azauracil (see above). A triply acetylated form of AzUR has been developed by Creasey *et al.* (1963) (Fig. 154) which is claimed to remain active after oral administration and to permit a ready maintenance of the necessary blood level of AzUR. AzUR leads to temporary remission of some human leukemias and shows a low toxicity in comparison with other drugs (Welch *et al.*, 1960; Handschumacher *et al.*, 1962).

4-Azathymine inhibits the growth of various microorganisms; the competitive effect may be reversed by natural thymine or thymidine (Elion *et al.*, 1954; Prusoff *et al.*, 1954). In bacteria, 4-azathymine and its deoxyriboside, 4-azathymidine, give rise to symptoms which correspond to thymineless death (cf. p. 315).

Handschumacher (1963) has reported experiments with the triazine derivative, 5-azaorotic acid, a compound that also interferes with the *de novo* synthesis of pyrimidines by inhibiting the OMP pyrophosphorylase reaction (cf. p. 127) and therefore blocks nucleotide formation from

orotic acid.   5-Azauracil noncompetitively inhibits uridine and deoxyuridine phosphorylase (Chihak and Sorm, 1964; see p. 116) and may thus
also increase the biological effect of 5-FUDR by preventing its enzymatic degradation.   5-Azacytidine is an analog which inhibits the
growth of certain mouse leukemias and microorganisms.   It is incorporated into nucleic acids (apparently only into RNA; Cihak *et al.*, 1965).
2-Thiouracil (XLIII) is primarily applied as a thyrostatic drug, although
it also inhibits the growth of experimental tumors.   Little is known of
its mechanism of action.   The analog is incorporated into RNA (including viral RNA; cf. p. 272) but not into DNA (Mandel *et al.*, 1957); it has
very little effect on the utilization of normal uracil.   The incorporation
of 2-thiouracil into RNA appears to change the template properties,
leading in *E. coli* to the formation of an altered β-galactosidase (Hamers
and Hammers-Casterman, 1961).

XLIII   XLIV   XLV

Purine Analogs

The first compound of this group to be studied in detail was 8-azagua-
nine (XLIV). The growth-inhibitory effect of this purine analog in microorganisms can be reversed by the addition of guanine or xanthine.
Though also observed in experimental tumors (Law, 1950; Sugiura
et al., 1950), growth inhibition by 8-azaguanine has not been found to
occur in clinical trials.   The compound is incorporated into RNA and,
in smaller amounts, into DNA (Smith and Matthews, 1957), replacing the
natural guanine. There seems to be no conversion to 8-azaadenine. The
formation of the riboside di- and triphosphates required for incorporation
into RNA has been confirmed by direct analysis (Brockman *et al.*, 1959).
Mammals deaminate a large part of the guanine analog to form the relatively inactive 8-azaxanthine (Carlo and Mandel, 1954).   The low guanase activity of various transplantable tumors contributes appreciably to
the tumor-specific action of 8-azaguanine (cf. p. 120).   Aminoimidazole-
carboxamide (see p. 107) leads to an inhibition of guanase activity,
thereby increasing the rate of incorporation of 8-azaguanine into nucleic
acids (Bennett and Skipper, 1957). Furthermore, 8-azaguanine was found
to inhibit protein formation in microorganisms (Chantrenne and Devreux,
1958) as well as in mammalian cells (Zimmerman and Greenberg, 1965);

RNA and DNA synthesis are inhibited to a lesser degree and at later times.

It is still not quite clear which metabolic disturbances are essential for the biological effects of 8-azaguanine. The incorporation into RNA (Otaka *et al.*, 1961), together with the inhibition of protein synthesis, suggests that structural alterations of the mRNA template play a major role by interfering with the translation process. The observations that 8-azaguanosine inhibits enzyme induction in the liver (Kröger, 1963) and that 8-azaguanine, like 5-FU, may cause phenotypic reversions of certain *Neurospora crassa* mutations (Barnett and Brockman, 1962; cf. p. 244) support these conclusions.

2,6-Diaminopurine (XLV) also inhibits the growth of microorganisms and of some experimental tumors, an effect that can be reversed by adenine (Elion and Hitchings, 1950). Diaminopurine is converted via its riboside to nucleotides and is ultimately used for RNA synthesis. C^{14}-labeled diaminopurine appeared in microbial systems as both guanine and adenine in RNA (Balis *et al.*, 1952), but only as guanine in the RNA of mammalian cells (Bennett *et al.*, 1952). Adenine reverses this effect, apparently by competing with diaminopurine in the purine nucleoside phosphorylase reaction.

An important purine derivative for practical cancer chemotherapy is 6-mercaptopurine (6-MP; XLVI), synthesized by Elion *et al.* in 1952.

XLVI

6-MP interferes with purine metabolism and as its riboside monophosphate inhibits the conversion of IMP to AMP and GMP (Salser *et al.*, 1960; Rich *et al.*, 1962; Simpson *et al.*, 1962) by occupying the active sites of the IMP dehydrogenase (cf. p. 113), probably through the formation of thioether or disulfide bridges between the 6-mercapto and the —SH groups of the enzyme (Hampton, 1963). In this way, it diminishes the amount of available purine nucleotides. At the same time, *de novo* synthesis of purines is blocked (Fernandes *et al.*, 1956; Greenlees and LePage, 1956), apparently due to allosteric inhibition by the 6-MP nucleotides of the synthesis of phosphoribosylamine (see p. 109; McCollister *et al.*, 1962, 1964). Some instances of the incorporation of S^{35}- or C^{14}-labeled 6-MP into nucleic acids have been reported (Elion *et al.*, 1954; Hamilton and Elion, 1954). However, because 6-MP cannot be phosphorylated, to the di- and triphosphate state (Paterson, 1959; Brockman, 1960), this incorporation is explained by nonspecific binding via the mercapto group (Hansen and Nadler, 1961). Another effect of 6-MP which might enhance its chemotherapeutic effectiveness is its ability to decrease NAD synthesis (Atkinson *et al.*, 1961). In mammals, especially in mammalian tumors, 6-MP nucleotides are synthesized primarily by way of a purine nucleotide pyrophosphorylase reaction (Lukens and Herrington, 1957; cf. p. 116). The results of more detailed studies by Atkinson and Murphy (1965a) showed that 6-MP is a competitive inhibitor of phosphoribosyl transfer to guanine and hypoxanthine, while there is no significant inhibition of this type of reaction with adenine. Ehrlich ascites tumor cells rapidly split the free base from the 6-MP nucleoside (Paterson, 1960), whose administration seems to have no advantage over that of 6-MP (Montgomery *et al.*, 1962). The anticancer activity of 6-MP in some forms of leukemia, as well as its toxicity, is probably enhanced by the inability of normal bone marrow to synthesize purines (Lajtha and Vane, 1958; cf. p. 118).

Microorganisms rapidly convert 6-MP to hypoxanthine, which can then be used for normal nucleotide and nucleic acid synthesis (Balis *et al.*, 1951); this detoxification process occurs to only a limited extent in mammals (Elion *et al.*, 1954). Since hypoxanthine reverses the growth inhibition of 6-MP in bacteria, the biological effect is spontaneously diminished by this mechanism (Brockman, 1963).

Polynucleotide phosphorylase can utilize 6-MP nucleoside diphosphate as substrate only in the presence of CDP, leading to the formation of a copolymer (Carbon, 1962). The enzymatic formation of polyA is inhibited by 6-MP nucleoside diphosphate. 6-MP can be used as a substrate by xanthine oxidase, which it competitively inhibits (Silberman and Wyngaarden, 1961); therefore, degradation to thiouric acid may occur (Elion *et al.*, 1954). This inactivation of 6-MP can be prevented by the simultaneous application of 4-hydroxypyrazolo-(3,4-*d*)-pyrimidine, an analog of hypoxanthine (cf. XLVIII, the corresponding adenine analog), which was found to be a powerful competitive inhibitor of xanthine oxidase (Yu and Gutman, 1964). When S^{35}-labeled 6-MP is administered to humans, radioactive material is excreted mainly as the sulfate (Hamilton and Elion, 1954).

6-Thioguanine (6-TG; XLVII), also synthesized by Elion and Hitchings (1954), differs from 6-MP only in the amino group on carbon 2. This purine derivative inhibits the growth of both microorganisms and experimental tumors (Clarke *et al.*, 1953). Its metabolism differs from that of 6-MP in that it can reach the triphosphate stage (Moore and LePage, 1958) and therefore be utilized for nucleic acid synthesis (LePage, 1960;

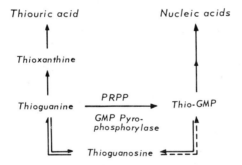

XLVII

Fig. 156). 6-TG inhibits the *de novo* synthesis of purines—probably in-directly by means of allosteric inhibition (LePage and Jones, 1961). Like 6-MP it also diminishes the incorporation of labeled guanine into acid-soluble nucleotides and nucleic acid purines (Sartorelli and LePage, 1958a); moreover, the conversion of IMP to AMP and its incorporation into the adenine moieties of nucleic acids is inhibited by 6-TG, while on the other hand the utilization of labeled adenine is not appreciably affected. The decreased turnover of C^{14}-labeled guanine in the presence of 6-TG is probably due to substrate competition in the GMP-pyrophosphorylase reaction (Fig. 156). According to LePage and Jones (1961), the most important of the various effects of 6-TG is its incorporation into DNA, as a result of which the mechanism of DNA replication is disturbed.

Fig. 156. Metabolic behavior of 6-thio-guanine (from Brockman, 1963).

The degradation of 6-TG, as in the case of 6-MP, terminates in the formation of thiouric acid (Moore and LePage, 1958); since 6-TG cannot react directly with xanthine oxidase, presumably a deamination by gua-nine deaminase first occurs. In some experimental tumors, the combined application of 6-TG and 6-MP enhanced the chemotherapeutic effect (Henderson and Junga, 1960); in the mouse sarcoma 180, this could also be achieved by combining 6-TG with a uracil mustard derivative (Booth and Sartorelli, 1962).

6-Chloropurine (6-CP) is a purine derivative which in mammals is partially oxidized to 6-chlorouric acid (Duggan and Titus, 1960); labeled 6-CP was found to be incorporated into nucleic acids in the form of adenine and guanine. 6-CP apparently inhibits normal GMP formation and thus stimulates the utilization of added adenine which is then incorporated into both adenine and guanine moieties of nucleic acids (Sartorelli and Booth, 1962).

6-Amino-4,5-pyrazolylpyrimidine (XLVIII) is an adenine analog that is converted to ribonucleoside phosphates (Henderson and Junga, 1961) and

$$NH_2$$

XLVIII

becomes incorporated into nucleic acids in small amounts (Bennett *et al.*, 1962). It appears to inhibit primarily the *de novo* synthesis of purines (Booth and Sartorelli, 1961); blocking of xanthine oxidase (Pomales *et al.*, 1963) leads to an increased utilization of hypoxanthine for nucleic acid synthesis. Allopurinol (6-hydroxy-4,5-pyrazolylpyrimidine) is the corresponding analog of hypoxanthine and a potent inhibitor of xanthine oxidase (Elion *et al.*, 1963). The compound has been found to be very effective in the treatment of gout (hyperuricemia) in man (Rundles *et al.*, 1966). In mammals, a considerable part of the drug is oxidized to the xanthine analog oxoallopurinol (Elion *et al.*, 1966). From the urine of patients treated with allopurinol, two ribonucleosides of the analog have been isolated, 9-ribosylallopurinol and the 3-ribosyl-derivative of oxoallopurinol (Krenitsky *et al.*, 1967). These observations indicate that oxoallopurinol is metabolized as a pyrimidine analog.

Formycin [7-amino-3-(β-D-ribofuranosyl)pyrazolo-(4,3-*d*)-pyrimidine; IL] is an adenosine analog with an altered base (Komaya *et al.*, 1966). The antibiotic was found to have various effects on purine metabolism. In chemotherapeutically sensitive Ehrlich ascites tumor cells, the analog seems to block primarily the *de novo* synthesis of purines by inhibiting PRPP formation (Henderson *et al.*, 1967). In addition, nucleotide and nucleic acid synthesis appears to be affected.

Many purine nucleoside analogs with abnormal sugar moieties have been found among the more recently discovered antibiotics (cf. p. 303; for a more detailed review, see Fox *et al.*, 1966). The allosteric feedback effects of purine analogs have been discussed by Henderson (1963), who observed that not all analogs are able to inhibit the *de novo* synthe-

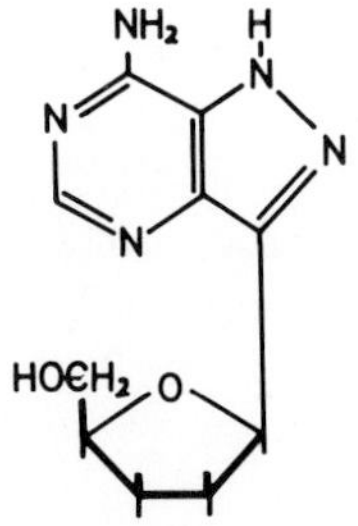

IL

sis of purines; even the most effective analogs were found to be less active than the natural purines in feedback inhibition.

Amino Acid Analogs

Azaserine (L) and 6-diazo-5-oxo-L-norleucine (DON; LI) are glutamine analogs and therefore, interfere with *de novo* purine synthesis (cf. Fig. 57), thus inhibiting nucleic acid formation. Both compounds were first discovered by virtue of their antibiotic properties. DON inhibits

$$L \qquad N^- = N^+ = CH - \overset{O}{\underset{||}{C}} - O - CH_2 - \overset{NH_2}{\underset{|}{CH}} - COOH$$

$$LI \qquad N^- = N^+ = CH - \overset{O}{\underset{||}{C}} - CH_2 - CH_2 - \overset{NH_2}{\underset{|}{CH}} - COOH$$

IIL

the phosphoribosylpyrophosphate amidotransferase reaction (cf. p. 109), interacting with the enzyme in a two-step reaction (Hartman, 1963). It also inhibits FGAR formation (Levenberg *et al.*, 1957); *in vitro* the inhibitory effect of DON is much more pronounced than that of azaserine. DON also interferes with the amination reaction in the formation of CTP from UTP (Hurlbert and Kammen, 1960), but only at a tenfold higher concentration than that required for the inhibition of purine synthesis (Moore and Hurlbert, 1961). Since glutamine is needed for the amination of XMP in the synthesis of GMP (see p. 113), this reaction too is inhibited by azaserine (Abrams and Bentley, 1959). In mouse liver, azaserine is degraded with the release of pyruvic acid and NH_3 (Jacquez and Sherman, 1962).

In the treatment of acute leukemia in children, 6-MP combined with DON may give better results than when used alone (Sullivan *et al.*, 1962). In addition to azaserine and DON, other antibiotics of similar structures with a diazo group, e.g., alazopeptin and duazomycin A (Brockman, 1963), also appear to interfere with purine metabolism.

β-Methylaspartic acid is a natural metabolite formed by the isomerization of glutamic acid with the participation of a vitamin B_{12}-containing coenzyme. This compound is an active antagonist of normal aspartic acid and therefore inhibits *de novo* pyrimidine synthesis in *E. coli* cells in a noncompetitive manner (Woolley, 1960).

DEVELOPMENT OF DRUG RESISTANCE

The practical application of the drugs whose mechanisms of action have been discussed in the foregoing sections is limited if the target cells do not respond sufficiently. This relative resistance may either be present initially in some cells or develop gradually in the course of treatment, finally making its continuance impossible. In transplantation tumors, a resistance can often be induced when the drugs are administered at low dosages over fairly long periods. Tumors of this kind, along with resistant bacterial strains, have become important tools for studying the phenomenon of resistance. Knowledge of the mechanisms by which a cell develops resistance, as well as knowledge of possible cross-resistances, is of considerable importance in decisions regarding therapeutic action if a tumor or an infectious microorganisms no longer responds to a previously effective drug.

From a theoretical point of view, drug resistance may develop for the following reasons:

(1) Decreased permeability of cellular membranes to the inhibitor;

(2) Increased activity of an inhibited enzyme;

(3) Alteration in the properties of the inhibited enzyme, so that it no longer interacts, or interacts only weakly, with the inhibitor;

(4) Increased inhibitor breakdown or some other type of blockage so that the inhibitor can no longer participate in the reactions that produce the biological effect;

(5) Decreased conversion of the inhibitor into its active form.

Point (5) appears to be the most frequent reason for resistance. However, the specific causes may be quite different in various biological systems. Moreover, several mechanisms can cooperate. Similarly, various possibilities may exist for the mechanism by which the development of resistant cells take place (cf. Klein, 1959; Law, 1952).

The biological action of actinomycins depends on their complex formation with the DNA in the cell (cf. p. 295). In a relatively resistant strain of Ehrlich ascites tumor cells that had been cultivated in the presence of actinomycin C_1, the amount of C^{14}-labeled antibiotic bound to the DNA was found to be decreased (Harbers *et al.*, 1963); whether

this was due to intracellular alterations or to a decreased permeability
to actinomycin is not known. *In vivo* studies of transplantable mouse
tumors, which differed in their actinomycin sensitivity, indicated that in
addition to the tissue concentration of the antibiotic (Harbers *et al.*,
1964), other factors, particularly the extent of euchromatization (Harbers
and Vogt, 1966), seem to play an important role in the response of tumor
cells.

It is interesting that a 5-FU-resistant strain of L1210 mouse leukemia
is highly sensitive to actinomycin (Hutchison, 1963). Insensitivity to
mitomycin C is associated with cross-resistance to alkylating substances
(Merker *et al.*, 1962); because these compounds, like mitomycin C, lead
to DNA cross-linking, this observation gives further support to the
postulated mechanism of action of mitomycin C (cf. p. 299).

The rapid development of resistance to alkylating substances by experimental tumors has been studied by Hirono. In addition to changes in
the permeability of the cell membrane (Hirono *et al.*, 1962), the content
of non-protein-bound SH-containing cell constituents seems to play a role
in causing resistance; in the case of resistant transplantable tumors,
the content of free SH-containing material was found to be definitely in-
increased (Hirono, 1960). The well-known observation that cysteine reduces the biological activity of nitrogen mustard supports this postulated
correlation (Brandt and Griffin, 1951). However, the concentration of
SH-containing constituents was not found to be increased in tumors that
were already relatively insensitive to alkylating agents. There is general cross-resistance between the various alkylating agents, but not
against folic acid, purine and pyrimidine analogs (Skipper and Schabel,
1962).

Initially, various mechanisms were suggested for resistance against
folic acid analogs; for the most part, these explanations were neither
confirmed nor supported experimentally. Finally, a considerable rise in
folic acid reductase activity was observed in mouse tumors that had become resistant to folic acid analogs (Fischer, 1961; Misra *et al.*, 1961).
This appears to be the key to understanding the alteration in metabolic
behavior (cf. p. 309). Investigations by Hakala *et al.* (1961) have shown
that the increase in resistance is proportional to the rise in folic acid
reductase activity. Similarly, in leucocytes from leukemia patients who
had gradually become resistant to amethopterin, a five- to twentyfold increase in enzyme activity was observed (Bertino *et al.*, 1962). Results
of kinetic studies have shown that the development of resistance is not
associated with changes in the properties of folic acid reductase but
simply reflects a considerable increase in the amount of the enzyme.
Resistance in microorganisms may involve other, as yet unknown, factors
(cf. Hakala *et al.*, 1961). Treatment of a substrain of mouse leukemia

L1210 with dichloroamethopterin led first to slight resistance, without increase in folic acid reductase activity; the latter appeared only when full resistance had developed (Friedkin *et al.*, 1962). Misra *et al.* (1961) have proposed breaking the resistance by means of previously reduced folic acid analogs. These should bypass folic acid reductase and continue their action at the level of the coenzyme FH_4.

Prolonged administration of the thymine analog 5-bromouracil to bacteria did not lead to the development of resistance; the initial cell damage disappeared gradually, even though a considerable part of the thymine in the DNA had been replaced by halogenated uracil. Mammalian cells, on the other hand, can develop resistance to 5-bromodeoxyuridine, resulting from a decrease in thymidine kinase activity (Kit *et al.*, 1963).

Induced resistance to 4-azauracil in *S. faecalis* evidently resulted from a decrease of uridine phosphorylase activity, because the mutated strain was no longer able to form nucleotides from the azauracil and therefore was not cross-resistant to 4-azauridine (Handschumacher, 1957). Mouse leukemia cells resistant to 4-azauridine were also unable to form nucleotides from the analog (Pasternak *et al.*, 1961) because of insufficient uridine kinase activity, demonstrated both *in vivo* and *in vitro*. However, the locus of action of the drug had remained sensitive in the resistant cells, because azauridine-5′-monophosphate inhibited the OMP-decarboxylase from both resistant and sensitive tumors in the same way. Hence, the mechanism of resistance was that given in point (5), listed above.

A development of resistance against 5-FU by neoplasms can be due to a fall in uridine-deoxyuridine phosphorylase activity; the effect can be enhanced by further, though less pronounced, decrease in the activity of uridine kinase (Reichard *et al.*, 1962). Sköld (1963) has shown that the decrease of kinase activity is accompanied by a change in the properties of the enzyme (precipitation at lower concentrations of ammonium sulfate). As long as the uridine kinase activity is maintained, no cross-resistance develops between 5-FU and 5-FUR. Heidelberger and co-workers (1960) obtained results with a 5-FU-resistant tumor that also indicate a decreased affinity of the active inhibitor (5-fluorodeoxyuridine monophosphate; cf. Fig. 152) for thymidylate synthetase. This property of the tumor was not genetically stable, however, and could not be reproduced in later transplantation generations, even after enzyme purification (Hartmann and Heidelberger, 1961). Later, a thymidylate synthetase, unaltered in its properties, was demonstrated by Häggmark (1962) in a series of 5-FU-resistant transplantation tumors. Studies with various 5-FU-resistant human tumors have shown that resistance is not always due to the same reason (Wolberg, 1964).

The resistance to 5-FUDR induced in a tumor in tissue culture was found to be accompanied by decreased utilization of thymidine (Morris and Fischer, 1963), probably because of diminished activity of deoxyuridine-thymidine kinase. In microorganisms, uracil is generally converted directly to the nucleotide by action of a UMP pyrophosphorylase (cf. p. 136); in 5-FU-resistant cells of *E. coli*, this enzyme was found to have disappeared almost completely (Brockman *et al.*, 1960). A 5-FU-resistant strain of *S. faecalis* had lost its uridine phosphorylase and kinase activity (Brockman and Sparks, 1960). The loss of the ability to utilize thymine for DNA synthesis has also been observed in 5-FU-resistant cells (Bloch and Hutchison, 1962). Usually there is cross-resistance between 5-FU and 5-FUR, but generally not with 5-FUDR (Brockman *et al.*, 1960).

Resistance to 8-azaguanine was first attributed to an increased guanine deaminase activity (Hirschberg *et al.*, 1952), but this could not be confirmed in subsequent studies (Mandel, 1959), so that drug resistance here cannot be due to an increased rate of degradation of the drug. On the other hand, resistant cells were found to incorporate much less labeled 8-azaguanine into nucleic acids than normal cells (Bennett *et al.*, 1953). The reason for this is primarily the decreased turnover of nucleotides caused by a reduction of IMP and GMP pyrophosphorylase activity (Brockman *et al.*, 1959a, b); however, some further unknown factors also seem to be involved in the development of resistance. The actual situation seems to be even more complicated, since IMP and GMP pyrophosphorylase activities can be lost either separately or together (Kalle and Gots, 1961). Decreased permeability to 8-azaguanine as a cause of resistance has been excluded in the case of a mouse leukemia L1210 (Davidson, 1958).

Resistance to 6-MP is generally accompanied by a decrease or loss of the ability to convert the drug to the nucleotide form; in microorganisms and also in mammalian cells, a decrease in IMP pyrophosphorylase activity frequently occurs, often coupled with a fall in GMP pyrophosphorylase activity. Although in one 6-MP resistant Ehrlich ascites tumor the enzyme activity was found to be unchanged (Paterson, 1960), effective inhibition of the conversion of IMP to AMP was observed in cell-free systems from two mouse tumors, one of which was resistant to 6-MP (Baker and Bennett, 1964). An obvious explanation would be to attribute the defect in utilizing 6-MP to a change in permeability; however, such a change could not be found in resistant L1210 cells (Davidson, 1958). The inhibition of guanine and hypoxanthine nucleotide pyrophosphorylase reactions (cf. p. 116) was found to be unaltered in Ehrlich ascites tumor cells resistant to 6-MP (Atkinson and Murphy, 1965b).

There are various causes for resistance to 6-TG. Often, as is the case with the other purine analogs, a decrease or even absence of conversion to the nucleotide form, resulting from loss of the necessary enzyme, is responsible (Ellis and LePage, 1963). In addition to a fall in GMP pyrophosphorylase activity, increased breakdown into thiouric acid has also been observed in resistant cells (Sartorelli *et al.*, 1958). Both processes result in a decrease of 6-TG nucleotide formation and subsequent incorporation into nucleic acids. Resistance to 6-TG may be overcome by use of 2′-deoxythioguanosine, if the cause is a decreased GMP pyrophosphorylase activity or an insufficient conversion to the deoxyribose form (LePage *et al.*, 1964). Cross-resistance often exists between 6-MP and 6-TG, as well as 8-azaguanine. This is always the case if the resistance originates in the loss of activity of the purine nucleotide pyrophosphorylase required for nucleotide formation.

The reasons for the development of resistance to azaserine and DON are not known as yet. A surprising phenomenon is the lack of complete cross-resistance between these two glutamine analogs (Hutchison, 1963). Enhanced degradation of the two drugs or decreased uptake in resistant cells could not be demonstrated (Anderson and Jacquez, 1962). At this time, the only possible explanation is the as yet unproved hypothesis of Sartorelli and LePage (1958b) that the blockage of *de novo* purine synthesis by azaserine or DON is counteracted by increased utilization of exogeneous purines (preformed pathway).

Resistance to a particular anticancer drug can give rise to increased sensitivity toward another drug. This phenomenon has been termed "collateral sensitivity" by Szybalski and Bryson (1952) and may, under certain circumstances, permit successful therapy in an otherwise hopeless situation.

Szybalski and Szybalska (1961) have discussed the theoretical possibilities of chemotherapy based on the principle of collateral sensitivity. So far, successful attempts at combination therapy have resulted from empirical combinations. The goal of modern development should be the ability to choose the drug (or drug combination) for optimal treatment on the basis of the known mechanisms of action as well as from knowledge of the metabolic behavior of a tumor or infectious microorganism. In practice, it cannot be predicted whether a given tumor will respond to treatment with a particular drug. Appropriate *in vitro* systems using tumor biopsies may help in deciding whether or not further treatment with a given drug is indicated.

Brockman (1963) has written an excellent and thorough review of the development of resistance to a great variety of anticancer drugs. This is supplemented by Hutchison's (1963) compilation of cross-resistance.

CHAPTER XV
Action of Radiation on Nucleic Acids

The various types of radiation whose effects are being investigated in the field of radiobiology may be divided into three classes; visible light (whose biological effects are not discussed here), ultraviolet light and ionizing radiation. Ultraviolet light has very limited penetrating power; its biological effects have been studied mainly with microorganisms. Ionizing radiation includes radiation resulting from the disintegration of natural and artificial radioisotopes, cosmic radiation, radiation from reactors and particle accelerators, and X-rays. It is now widely assumed that the most profound biological effects of radiation result from its action on the nucleic acids and, to a lesser degree, on other cell constituents like the membrane systems. Only the former will be dealt with here.

ULTRAVIOLET LIGHT

The relationship of nucleic acids to ultraviolet damage first became apparent from the correlation between mutation rates and the wavelengths of the incident light (Knapp *et al.*, 1939; Hollaender and Emmons, 1941). More recently, chemical changes caused by ultraviolet light have been reported. Beukers and Berends (1960) observed the formation of thymine dimers, apparently preceded by radical formation (Eisinger and Schulman, 1963), after irradiation of thymine in frozen solutions; the radical has an unpaired electron on carbon 5 and an additional hydrogen atom on carbon 4. The same type of dimerization was observed following ultraviolet treatment of complete DNA (Wacker *et al.*, 1961a). Figure 157 shows the structure of the thymine dimer as suggested on the basis of the arrangement of thymine in the DNA (Beukers and Berends, 1961; Wacker *et al.*, 1961a). In DNA, dimerization affects only those thymine bases that are immediately adjacent to each other; thymine dimer formation may therefore be helpful in the analysis of base sequences.

While the formation of thymine dimers was found at first only in *in vitro* systems, it was later observed in the DNA of bacterial cells, and a direct relationship between death rate and extent of dimerization was dem-

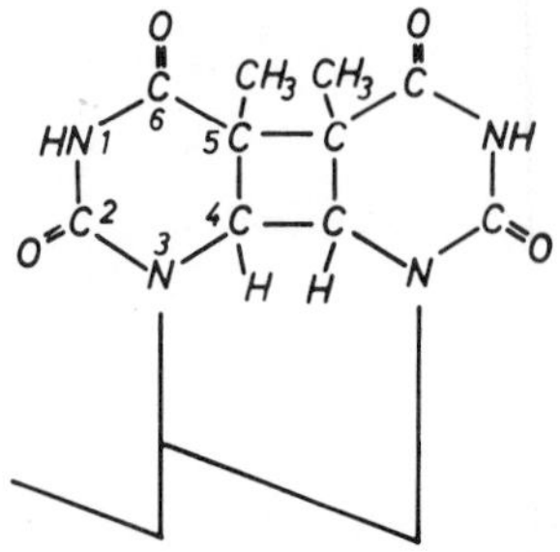

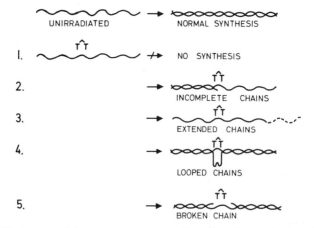

Fig. 157. Dimer formation by two adjacent thymine bases in a DNA strand after exposure to ultraviolet light.

onstrated (Wacker *et al.*, 1962). The increased sensitivity to ultraviolet light during growth is accompanied by an increase in the amount of thymine dimers. However, certain bacteria have such efficient repair mechanisms for the removal of thymine dimers (cf. p. 343) that thymine dimer formation gives very little indication of the inactivation by ultraviolet light (Setlow *et al.*, 1964). Bollum and Setlow (1963) found that ultraviolet-irradiated DNA lost a great deal of its priming efficiency for DNA polymerase. Only about 10% of the initial primer activity was retained. The DNA synthesized at such irradiated templates was found to have a higher G/A ratio, apparently because some of the thymine bases were unavailable due to dimerization. Various models of the effects of thymine dimerization on the process of DNA replication have been presented by Bollum and Setlow (Fig. 158). Models 4 and 5 agree most closely with the experimental data, an analysis of which shows that the

UNIRRADIATED → NORMAL SYNTHESIS

1. → NO SYNTHESIS

2. → INCOMPLETE CHAINS

3. → EXTENDED CHAINS

4. → LOOPED CHAINS

5. → BROKEN CHAIN

Fig. 158. Various models attempting to explain the effect of ultra-violet light on the primer efficiency for DNA polymerase (from Bollum and Setlow, 1963). F. J. Bollum and R. B. Setlow, Ultraviolet inactivation of DNA primer activity. I. Effects of different wavelengths and doses, *Biochim. Biophys. Acta*, 68 (1963) 599–607, Figs. 1–5. Reproduced with permission.

average rate of enzymatic DNA synthesis is dependent upon the number of TT dimers in the DNA template. In a similar manner, ultraviolet light reduces the priming efficiency of isolated DNA for RNA polymerase (Hagen *et al.*, 1965).

The formation of thymine dimers in bacterial DNA may be reversed in various ways. One possibility is the so-called photoreactivation. In this case, an enzyme which needs visible light as a source of energy cleaves the pyrimidine dimers, thus repairing the altered structure of DNA (e.g., Marmur and Grossman, 1961; Wulff and Rupert, 1962; G. K. Setlow, 1966). However, visible light alone can cause a partial reversion, at least in model experiments. Studies by Sztumpf and Shugar (1962) showed that an approximate equilibrium is set up between dimer formation and dissociation. The point of equilibrium was found to be wavelength dependent; thus it can be shifted either in the direction of dimer formation or back toward dissociation. Finally, enzymatic repair of ultraviolet-irradiated DNA may take place without light (dark reactivation), a process of general importance which will be discussed later (cf. p. 343).

The extractability of the DNA of ultraviolet-irradiated bacteria was found to be decreased in a dose-dependent manner; this phenomenon was observed at ultraviolet doses lower than those required for thymine dimerization (Smith, 1962). Donellan and Setlow (1965) detected in the spores of *B. megaterium* after ultraviolet irradiation thymine photoproducts of unknown structure which were not thymine dimers. These observations indicate that in addition to thymine dimerization other effects may contribute (or in certain cases be essential) to the damage of DNA by ultraviolet light (Setlow and Boling, 1965). Ultraviolet-inactivated transforming DNA does not seem to be able to be incorporated into the genome of the receptor cell (Stuy, 1962).

Ultraviolet sensitivity is also exhibited by nucleic acids which contain no thymine, e.g., viral RNA. In model experiments with uracil and uridine, dimer formation has been observed (Wacker *et al.*, 1961c); these dimers, however, are much less stable than that of thymine. In addition to dimer formation, ultraviolet irradiation of uracil also causes the addition of water at the double bond between carbon 4 and carbon 5. If frozen solutions of the pyrimidines are irradiated, there is a shift toward a preferential dimer formation. Irradiation of a mixture of thymine and uracil yields dimers containing both types of pyrimidines. Pseudouridine (5-ribosyluracil; cf. p. 15) is more sensitive to ultraviolet light than uridine (Lis and Allen, 1961); as yet the mechanism of this effect remains unexplained.

The changes in ultraviolet-irradiated polyuridylic acid have been in-

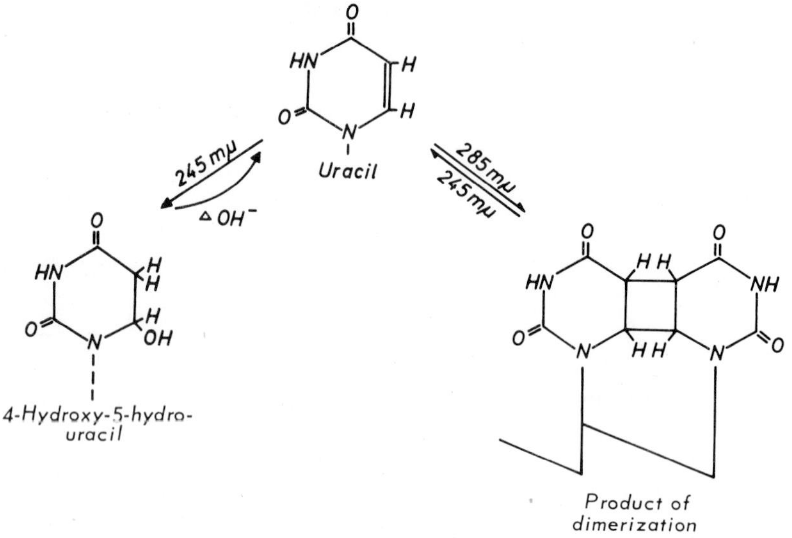

Fig. 159. Diagram showing how ultraviolet light can induce the uracil in polyU either to form a dimer or to add water, depending on the wavelength of the incident irradiation (from Grossman, 1963).

vestigated by Grossman (1963); the two reactions proposed for uracil were confirmed in these studies (Fig. 159). Uracil dimerization is partly reversible by ultraviolet light of shorter wavelengths. In contrast, the addition of water to uracil is enhanced by ultraviolet irradiation of shorter wavelengths; the reaction can be reversed by alkali (Fig. 159). PolyU as a messenger codes the synthesis of polyphenylalanine (see p. 232). As a consequence of ultraviolet-induced H_2O addition in the uracil of the messenger, some of the phenylalanine is replaced by serine, indicating a change in the coding properties. Uracil dimerization, on the other hand, leads to the loss of messenger activity. The addition of water to uracil which occurs under the influence of ultraviolet irradiation can be determined by means of H^3-labeled water (Wacker et al., 1961c). Neither the water addition products of uracil and uridine nor their dimers were found to be utilized for nucleic acid synthesis by a uracil-dependent mutant of E. coli.

Cytosine nucleosides exposed to ultraviolet light add water, just like uridine (Shugar, 1960), and subsequently undergo deamination (Wacker, 1963). Thus, as in nitrous acid treatment, ultraviolet radiation can lead to C $\longrightarrow$ U transitions (cf. p. 241).

Replacement of thymine in the DNA by 5-halogen derivatives of uracil causes a considerable increase in ultraviolet sensitivity. The extent of the radiation sensitization depends on both the extent of the replacement and the type of halogen (Br > I > Cl) (Djordjevic and Szybalski, 1960). It has been shown that in contrast to thymine and uracil, 5-bromouracil does not dimerize under the influence of ultraviolet light (Wacker et al., 1961b). This seems to be a paradox, since the replacement of thymine by 5-BU reduces the number of thymine pairs available for dimerization. Ultraviolet irradiation however, also causes a dose-dependent dehalogenation of DNA, thereby giving rise to uracil in the polynucleotide chain of DNA. Thus uracil can undergo both dimerization and water addition when subjected to ultraviolet radiation (Wacker, 1963). The radiosensitizing effect of the thymine analogs 5-BU or 5-IU is also found in mammalian cells when the analogs are administered as deoxyribosides (Djordjevic and Szybalski, 1960), in transforming DNA (Opara-Kubinska et al., 1961) and in DNA-viruses.

Incorporation of 4-azathymine into bacterial DNA decreases the ultraviolet sensitivity (Günther and Prusoff, 1962; Wacker and Jacherts, 1962), since azathymine can neither dimerize nor add water. 5-Fluorouracil in low concentrations causes a decrease in ultraviolet sensitivity of E. coli, but higher 5-FU concentrations are sensitizing (Ben-Ishal et al., 1962), apparently as a consequence of the inhibition of DNA synthesis by 5-FU.

The sensitivity of bacteria toward ultraviolet light (and other agents that interact with DNA) seems to be dependent on the effectiveness of the cellular repair mechanisms whose actions essentially determine radiosensitivity (cf. Setlow et al., 1966).

The earlier studies on the biological effects of ultraviolet light have been reviewed in detail by Shugar (1960); more recent work has been discussed by Wacker (1963).

IONIZING RADIATION

It is a common property of the various kinds of ionizing radiation that during the process of absorption, energy is transferred to molecules of the penetrated material in such quantities that the molecules can be ionized. The resulting formation of ion pairs is considered the basis of biological action. The extent of radiation damage is dependent on the dose (defined as the amount of absorbed energy per unit mass; 1 rad = 100 ergs/g; formerly 1 R $\cong$ 83 ergs/g), the dose rate (rads per minute or rads per hour), the quality of radiation (e.g., soft or hard X-rays, β-radition) and the sensitivity of the particular biological system being irradiated. The type of radiation can be further characterized by the linear energy transfer (LET) and the relative biological activity (RBA; cf. Rajewsky, 1956). Very little is known of the processes immediately following the absorption of radiation (cf. Zimmer, 1958); they will not be discussed here. The following sections will be limited to the molecular changes observed in nucleic acids following their exposure to ionizing radiation and the impairment of nucleic acid metabolism.

Structural Alterations of Nucleic Acids

Ionizing radiations may cause a variety of biological effects; among the most important and generally observed are those on the genetic apparatus of the cells (mutations, chromosomal alterations). As a consequence of an intensive exposure to X-rays, DNA becomes depolymerized (Sparrow and Rosenfeld, 1946; Taylor et al., 1947). Since the reports of these early findings appeared, radiation effects on nucleic acids (mainly DNA) have been investigated very extensively. Although the great number of experimental results does not as yet permit a unifying concept, there is no doubt about the essential role of DNA alteration in the biological effects of ionizing radiation. Thus, Novelli et al. (1961) observed that X-irradiation destroyed the ability of an E. coli system to induce β-galactosidase; however, by addition of nonirradiated DNA which contained the gene for the β-galactosidase, the capacity for enzyme formation was restored. These findings confirm the long-held hypothesis that the decrease of enzyme activity often observed some time after irradiation is mainly a secondary phenomenon and does not represent direct enzyme damage. They also explain why most radiation effects become apparent after a more or less long lag period. Structural alterations at the DNA template lead to an altered mRNA which is produced only when required to direct protein synthesis, thus accounting for the delay in the expression of radiation damage (cf. Pollard and Vogler, 1961).

Very large doses of radiation are required to damage DNA sufficiently for changes to be observed by chemical means. The effects of lower doses are detectable only by much more sensitive biological tests. Radiation effects are due partly to direct hits and, indirectly, to radical formation in the cell water. The relative proportions of these two effects vary, depending on the tissue studied. When experiments are conducted with very dilute solutions of nucleic acids, the indirect effects predominate; conversely, when dried material is irradiated, the observed effects can be ascribed to direct hits. X-ray treatment of pure nucleic acid bases causes a degradation of purines as well as pyrimidines. Adenine is converted to hypoxanthine by deamination (Ponnamperuma et al., 1961). Under similar conditions, cytosine proved to be less sensitive to radiation; part of the cytosine was deaminated to uracil

(Ponnamperuma *et al.*, 1962). The radiation-induced deamination observed in these experiments may—like the action of nitrous acid (cf. p. 241)—lead to transitions. X-radiation of thymine, unlike the action of ultraviolet light, does not cause dimer formation. Large doses of ionizing radiation destroy thymine, uracil and cytosine in aqueous solution, with urea occurring among the degradation products. In addition to deamination, X-irradiation to a lesser degree converts adenine in solution to adenine-N-oxide (Lochmann *et al.*, 1964). This effect is an indirect one, like the majority of the radiation effects described so far, because it can also be obtained with H_2O_2. Irradiation of nucleosides or nucleotides showed that the bases constitute the most sensitive portion of the molecules (Scholes *et al.*, 1960); in the case of purine nucleosides, the imidazole ring is opened and 4-amino-5-formamidopyrimidines are formed, the glycosidic bond remaining partly intact (Hems, 1960). The extent of these radiation-induced changes is dose dependent and is enhanced by the presence of O_2.

Irradiation of high molecular weight DNA leads to the same type of reactions, accompanied by depolymerization and disruption of the hydrogen bonds (e.g., Cox *et al.*, 1958; Scholes *et al.*, 1960; Hagen, 1962); the most important criteria for these changes are decrease in viscosity and hyperchromicity, and irregular titration curves. The action of fast electrons on dry (or nearly dry) DNA appears to result also in the formation of cross-links (Lett *et al.*, 1961). As expected, the biological activity of transforming DNA is reduced and finally lost, depending on the applied radiation dose (Drew, 1955; Hutchison and Arena, 1960). The effects under *in vivo* conditions are less pronounced but are likely to be the same as those *in vitro*, even though the criteria are less precise (e.g., Cole and Ellis, 1956; Harbers and Backmann, 1956; Ord and Stocken, 1960). From radiation experiments in which the homology with nonirradiated DNA was examined with the technique of DNA-DNA hybridization (cf. p. 276), it has been concluded that in microorganisms DNA is degraded randomly by ionizing radiation, and no part is preferentially selected (Swez and Pollard, 1966).

To evaluate the role of the protein moieties of nucleohistones in the radiosensitivity of the genetic apparatus, comparative studies have been conducted with pure DNA and isolated nucleohistones. Using viscosity measurements as a criterion, isolated nucleohistones were found to be more radiosensitive than DNA alone (Bernstein, 1954). On the other hand, the proteins seem to have a protective effect, because even after fairly high doses very little alteration of DNA bases were detected (Emmerson *et al.*, 1960). This protection was observed only with native nucleohistones; it was considerably decreased when a mixture of DNA and histones had been irradiated. The priming activity of DNA for the *E. coli* RNA polymerase is lowered after exposure to X-rays (Harrington, 1964; Zimmermann *et al.*, 1964). The same effects could be observed with nucleohistones; however, higher doses were required (Vogt and Harbers, 1967). In euchromatin and heterochromatin fractions isolated from intensely irradiated nuclei, no significant differences were observed although several analytical methods had been used; only the template efficiency for the RNA polymerase (increased in euchromatin and decreased in heterochromatin) and the viscosity of heterochromatin (somewhat increased) were found to be changed. After irradiation of whole Ehrlich ascites tumor cells there was less binding of actinomycin to the DNA of euchromatin, while the ratio of the amount of euchromatin to heterochromatin was found to be unchanged (Vogt and Harbers, 1967).

The *in vivo* sensitivity of DNA to ionizing radiation varies widely. Double-stranded DNA is damaged less than single-stranded DNA by the same dosage (Hotz and Müller, 1961; Weisberger and Okada, 1961). This may contribute to the relatively high radiosensitivity of proliferating cells; in growing tissues, a larger proportion of cells is approaching or has entered the S phase during which there is a temporary unwinding of the DNA. In agreement with this, the number of chromosome breaks is highest when radiation occurs during the S phase. In various tissues, a degradation of DNA after irradiation appears to be enhanced by the action of cellular deoxyribonucleases whose activities might be increased (Kurnick, 1952; Stuy, 1961a). The importance of deoxyribonucleases in DNA radiation damage is indicated by the observation that after X-irradiation of radiation-sensitive tissues (spleen, bone marrow, thymus), a marked rise in acid deoxyribonuclease activity per cell occurs, while little or no effect is seen in relatively radiation-resistant tissues (liver, kidney).

Radiation sensitivity is also a function of oxygen concentration (cf. Howard-Flanders, 1958); a threefold increase in sensitivity can be achieved by increasing the oxygen concentration. The higher resistance associated with deoxygenation is accompanied by a lessening of the radiation damage of the DNA (Hutchison and Arena, 1960; Stuy 1961b). The mechanism of this oxygen effect is not yet clearly understood. It seems to involve cell constituents which contain —SH groups whose protective effect is reduced by the presence of O_2 (Hutchison, 1961; Howard-Flanders *et al.*, 1963; Ormedor and Alexander, 1963). The natural level of free —SH groups in the cell helps to determine radiation resistance, as —SH blocking agents increase sensitivity (Harbers, 1961; Bridges, 1962; Howard-Flanders *et al.*, 1963). Cysteine, both *in vitro* (Drew, 1955) and *in vivo*, as well as in experiments with whole cells (e.g., Hutchison and Arena, 1960; Harbers, 1961), exerts a protective effect, provided it is present in sufficient concentration at the time of irradiation.

The radiation sensitivity of microorganisms is affected by the base composition of their DNA (Kaplan and Zavarine, 1962b) which differs considerably in various bacteria (cf. Table 12); the sensitivity increases with higher GC/AT ratio. These results may be explained by the fact that the energy required for the formation of a free radical by X-rays is not uniform for the various nucleic acid bases (A. Müller et al., 1963). Adenine and thymine require the highest energy (1200 and 1400 eV, respectively), whereas the corresponding values for guanine and cytosine are much lower (150 and 350 eV, respectively). In high molecular weight nucleic acids, the energy levels required are lower than those for the free bases (A. Müller, 1962, 1963). The very high resistance of many microorganisms to ionizing radiation is a phenomenon which might be explained by the ability of these cells to repair a considerable part of the DNA alterations produced by irradiation (cf. p. 343). Thus, potential damages—which might even be lethal—seem to be prevented from becoming effective by a continuously active repair mechanism. It is not known whether mammalian cells also possess such a repair system; if they do (cf. Phillips and Tolmach, 1966), the repair mechanism is much less effective than that in microorganisms.

In summary, radiation-induced changes of DNA are due to both direct and indirect effects, the extent of which can be affected by a variety of endogenous (e.g., effectivity of a repair mechanism, amount of —SH compounds) and exogenous factors (e.g., presence or lack of O_2). In addition, the amount of DNA per cell and the degree of ploidy appear to play a role in radiation sensitivity. However, polyploidy does not necessarily cause an increase in radiation resistance, as originally thought (cf. Révész and Norman, 1960; Révész et al., 1963).

To facilitate understanding of the relationship between radiation sensitivity and ploidy, one should study cell populations differing only in degree of ploidy but homogenous in all other respects (Révész et al., 1963). Considerable variation in chromosome number or the amount of DNA per nucleus is often found in neoplastic cells. One might expect to find an increase in the average degree of ploidy during the development of radiation resistance. However, investigations by Richards and Atkin (1959) on human tumors have shown that the development of radiation resistance is accompanied by a selection of a triploid cell type that may well have been resistant initially. Before radiation treatment, tetra- and octaploid cells were also present.

Any consideration of the effects of radiation on the genetic information must also include the possible disturbance of control mechanisms. The occasionally observed increase in enzyme activities or RNA synthesis after irradiation (see below) may perhaps be regarded as an expression of such a disturbance. Investigations of the effects of radiation on nucleic acids have concentrated on DNA; RNA seemed comparatively unimportant and relatively insensitive to radiation. Radiation-induced alterations have been detected, however, in some of the nuclear RNA in ascites tumor cells. RNA (in general single stranded) seems to be more sensitive to radiation than double-stranded DNA. However, radiation damage to RNA can be repaired, provided its DNA template has not also been altered.

The molecular mechanisms of radiation effects have been reviewed by Wacker (1963) and Weiss (1964), and the formation of radicals in nucleic acids due to the action of ionizing radiation has been discussed by A. Müller (1967).

Growth-inhibiting Effects

One of the most important observations of early radiobiological investigations was that ionizing radiation inhibits mitosis (Alberti and Politzer, 1923; Strangeways and Oakley, 1923). In 1942, von Euler and von Hevesy found that DNA synthesis was also inhibited by X-rays. Subsequently, the two effects were often regarded as the expression of a single phenomenon. However, autoradiographic studies by Howard and Pelc (1953) on *Vicia faba* seedlings gave the first indication that the two effects need not be related.

When *Vicia faba* cells were irradiated with 140 rad just before DNA replication, there was no inhibition of DNA synthesis, but a delay in the subsequent cell division. Hence the cells pass through two particularly radiation-sensitive stages. The first precedes DNA replication; the second is during G_2 before mitosis (cf. Fig. 87). The observed independence of these two types of radiation effects was later confirmed by several authors (e.g., Kelly *et al.*, 1957; Casperson *et al.*, 1958; Painter, 1960; Sherman and Quastler, 1960). Frequently cell division shows a higher radiation sensitivity than the process of DNA replication; then within a certain dose range, mitosis may be blocked, while DNA synthesis continues, thus leading to the formation of polyploid nuclei. Conversely, following very high doses to *Habrobracon* eggs, von Borstel (1955) observed continued nuclear divisions where the (DNA specific) Feulgen reaction became progressively weaker and finally negative; in this instance, DNA synthesis had been blocked by radiation, while the division processes were able to continue.

At present, the mechanism of the inhibition of mitosis by ionizing radiation cannot really be explained. For normal cell division, especially for spindle formation, an increase of free sulfhydryl groups is required (Brachet, 1957; Stern, 1958); the mitosis-inhibiting radiation effect may therefore be due to a decrease in the number of available sulfhydryl groups which are extremely radiosensitive.

Many mechanisms have been proposed to explain radiation-induced inhibition of DNA synthesis. Since radiation causes structural alterations of DNA, it seemed logical to assume that the priming efficiency for the DNA polymerase might be lowered. However, in *in vitro* systems the priming activity of DNA could be reduced only by the application of extremely high doses (Stacey, 1961; Wheeler and Okada, 1961). Lower doses of radiation (7500 roentgen) were found to produce a slight increase in enzymatic DNA synthesis. The results of these experiments are inconclusive because the irradiation of isolated cell nuclei or nucleohistones subsequently used in enzymatic systems did lead to a distinct inhibition of enzymatic DNA synthesis (Lehnert and Okada, 1963); nonbasic proteins (cf. p. 90) seem to be primarily responsible for this *in vitro* effect, possibly by way of cross-linking.

Radiation damage of the enzyme DNA polymerase was excluded by a direct approach (Walwick and Main, 1962) after it had been shown (Howard and Pelc, 1953; Beltz *et al.*, 1957) that radiation is effective only if administered in the G_1 phase (before the beginning of DNA replication); these observations argued against specific damage of the DNA polymerase and/or primer DNA. Moreover, if radiation is administered during DNA synthesis, there is practically no inhibition at all. From these results, it follows that once DNA synthesis has begun, it is not significantly influenced by irradiation; thus the inhibition must be attributed to events preceding DNA replication.

A lack of the requisite precursors was considered another possible cause of the inhibition of DNA replication. From early investigations, it seemed that in the case of inhibited DNA synthesis the amount of DNA precursors actually increased because they were less efficiently utilized (Bishop and Davidson, 1957; Ord and Stocken, 1958). Later experimental results showed that this increase in DNA precursors evidently only involves the deoxyribonucleosides, while deoxyribonucleotide concentration falls after irradiation (Jaffe *et al.*, 1959). This agrees with the observation that X-irradiation in the G_1 phase decreases the activity of TMP kinase (Beltz, 1962; Myers, 1962) and deoxycytidine deaminase (Myers, 1962; Stevens and Stocken, 1962), while, under these conditions, TMP synthetase is not radiosensitive. The most important factor in the radiation-induced inhibition of DNA replication seems to be the lowering of the concentration of DNA polymerase, the necessary synthesis of which is prevented or—at lower doses—delayed (Beltz and Applegate, 1959; Bollum *et al.*, 1960). In more general terms, inhibition of DNA synthesis by ionizing radiation can thus be considered a consequence of the blocking of enzyme synthesis, particularly of DNA polymerase and specific deoxyribonucleotide kinases. As small doses of radiation usually cause only a delay in DNA replication, it seems that this is due to a reparable damage of the control mechanism or of the mRNA responsible for enzyme synthesis.

The regeneration of rat liver after partial hepatectomy is delayed or blocked if the animals have been irradiated before the operation is performed (Williams *et al.*, 1952). In the normal liver, little new DNA is synthesized; initiation of DNA replication and subsequent cell division occur only as a result of partial hepatectomy. The irradiation, therefore, must cause alterations in the resting cells which prevent the initiation of regeneration. While a radiation dose of 800 rad in the regenerating rat liver causes a considerable inhibition of DNA synthesis, histone formation apparently remains unaffected, leading to an increase in the histone:DNA ratio (Lehnert and Okada, 1964).

Tissues generally consist of a cell population of which only a certain proportion is in the radiation-sensitive G_1 phase before DNA replication. As a result, the inhibition of DNA synthesis observed after irradiation is less pronounced than with synchronized cells irradiated in the G_1 phage. The incorporation of labeled precursors into DNA decreases as the dose is increased, but reaches a certain limit and then cannot be further reduced by increasing doses (Fig. 160). This phenomenon was at first surprising because it could be observed under conditions when both cell division and DNA synthesis were completely blocked due to the effects of large doses of radiation. Apparently the radiation-insensitive incorporation of labeled precursors into DNA represents a terminal exchange process (cf. p. 167) in which the amount of DNA is not changed. Supporting this interpretation is the observation that the *in vitro* incorporation of labeled precursors into the DNA of ascites tumor cells in a system that allows no cell growth is unchanged after X-irradiation (Harbers and Heidelberger, 1959a); only under conditions that permit cell multiplication—and thus also net synthesis of DNA—can the incorporation of precursors into DNA be inhibited by irradiation. The radiosensitive incorporation of labeled precursors into DNA may thus be a criterion for determining whether mammalian cells can synthesize DNA under certain conditions. The radiation-insensitive incorporation of labeled DNA precursors amounts to 40–50% of that of the unirradiated controls (von Hevesy, 1948; Vermund *et al.*, 1953). This large ($\approx 1:1$) ratio of exchange to synthesis may be ascribed to the fact

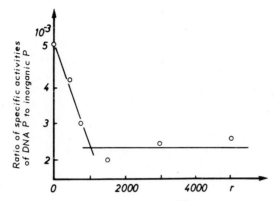

Fig. 160. Incorporation of   P^{32}-orthophosphate
into the DNA of mouse sarcoma 180 after local
treatment with increasing doses of X-rays. Ratio
of specific activities of DNA to that of free inorganic phosphate in the tumor tissue. The labeled phosphate was injected intraperitoneally
immediately after irradiation; the animals were
sacrificed 2 hours later (from Harbers, 1960).

that only a small fraction of a normal cell population reaches the S phase
during the course of an experiment and it is the synthesis in this part of
the cell population which is being compared to the exchange processes
taking place in the entire population. In the approximately synchronized
cells of regenerating rat liver, a sufficient dosage of radiation in the G_1
phase will eliminate almost all incorporation of precursors into DNA
(Beltz et al., 1957); this indicates that the ratio of exchange to synthesis in this system is fairly low.

The delay in cell division caused by lower doses of radiation produces a certain synchronization of the cells (Nygaard and Potter, 1960;
Anderson, 1961). This often causes a temporary increase in the mitotic
rate. Under certain conditions, a true increase in the rate of cell division (James and Müller, 1961) and of DNA synthesis may also occur
(Das and Alfert, 1961). An organism with very unusual behavior in its
cell cycle is the slime mold *Physarum polycephalum*, in which DNA replication takes place shortly after mitosis without any G_1 (cf. p. 185). If
irradiated with 9000 rad at any time during the very long G_2 phase, subsequent cell division and DNA synthesis are delayed (Nygaard and
Güttes, 1962). Radiation (2300 rad) during mitosis prevents further division without, however, inhibiting DNA synthesis. Radiation experiments
done in an atmosphere of either pure O_2 or pure N_2 have shown that
oxygen enhances the radiation-induced block of DNA replication.

Chang and Looney (1967) were the first to study radiation effects on mitochondrial DNA. The authors found that a local X-ray of 1500 R caused a decrease in the incorporation of labeled thymidine into mitochondrial DNA of normal as well as regenerating rat liver.

Studies on the effect of radiation on nucleic acid metabolism have been primarily concerned with DNA; the effect on RNA metabolism has been studied only incidentally. The results reported in the literature are contradictory; in general, only minor effects have been observed.

In most cases, effects on total RNA (i.e., unfractionated) have been investigated. Both a decrease (e.g., von Euler and von Hevesy, 1944; Holmes, 1947; Abrams, 1951) and an increase (Conzelman *et al.*, 1954) in the incorporation of labeled precursors have been reported; frequently, no significant effects have been found (e.g., Lavik and Buckaloo, 1954; Passonneau and Totter, 1955). After fractionation of the tissue (isolated irradiated mouse breast carcinoma), Vermund *et al.* (1953) observed a slight inhibition of P^{32} incorporation into the RNA of cell nuclei and microsomes, but an increased incorporation in the RNA of the soluble cytoplasm. Thomson *et al.* (1954) observed, after whole body irradiation of rats, an increase in the P^{32} uptake by RNA of all cytoplasm fractions in the regenerating liver. Beltz *et al.* (1957), also working with the regenerating rat liver, found that as a result of radiation the incorporation of labeled orotic acid into the RNA pyrimidines in the cytoplasm was inhibited only slightly, with no effect on nuclear RNA. A significant inhibition of the incorporation of C^{14}-uracil into nuclear RNA of Ehrlich ascites tumor cells has been observed, accompanied by an increased incorporation into the cytoplasmic RNA (Harbers

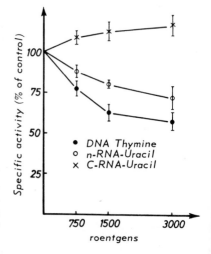

Fig. 161. Incorporation of $2\text{-}C^{14}$-uracil into DNA, nuclear RNA and cytoplasmic RNA of Ehrlich ascites tumor cells after X-irradiation *in vitro*. Immediately after irradiation, labeled uracil was added and incubated at $37^{\circ}C$ for 1 hour (from Harbers and Heidelberger, 1959a).

and Heidelberger, 1959a; Fig. 161). This result seems to indicate that radiation may lead to reduced as well as increased incorporation into the different kinds of RNA in the cell. Fractionation of the cell nuclei showed that the inhibitory effect as seen in Fig. 161 applies only to a certain part of the nuclear RNA.

REPAIR OF RADIATION-DAMAGED DNA

Until recently, alterations of DNA in a living cell were considered irreversible; according to the general view, they should either kill the cell or lead to mutant daughter cells. In 1964, a repair mechanism for DNA was detected in bacteria; this mechanism leads to the disappearance of the thymine dimers which had been produced by ultraviolet light (Boyce and Howard-Flanders, 1964; Setlow and Carrier, 1964); this type of repair should be distinguished from photoreactivation (cf. p. 331). The ability of bacterial cells to repair certain defects in their DNA is a biological phenomenon of general importance and is not limited to radiation effects; it has also been observed after the administration of alkylating agents (e.g., Kohn et al., 1965; Reiter and Strauss, 1965; Strauss et al., 1966). The problem of DNA repair is discussed in this section because radiation, i.e., mainly ultraviolet light, is still the best tool for studying the mechanism of error-correcting systems.

Early observations by Setlow and Carrier (1964) and by Boyce and Howard-Flanders (1964) indicated that during the repair process, an excision of thymine dimers from the altered DNA took place. In comparative experiments, cultures of ultraviolet-sensitive and ultraviolet-resistant bacteria were grown in the presence of labeled thymine which then became incorporated into newly synthesized DNA. After these cells had been exposed to ultraviolet light, it was found that in the sensitive cells the DNA contained labeled thymine dimers, while in the resistant cells the thymine dimers were found in DNA fragments consisting of three nucleotides. Thus the thymine dimers had not been split (as in photoreactivation) but instead had been removed from the DNA chain. These findings led to a theory which postulates DNA repair in four steps (Howard-Flanders and Boyce, 1966; Fig. 162):

(1) Single strands in a DNA double helix may be interrupted in various manners: (a) by radiation-induced single-strand breaks, (b) by an enzymic excision of altered DNA bases, or (c) by a recombination enzyme.

(2) Through the action of an enzyme, nucleotides are released.

(3) A DNA polymerase reconstructs the interrupted DNA strand by adding complementary nucleotides into the gap using the intact opposite strand as a template.

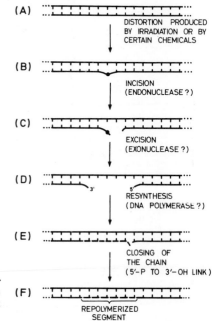

Fig. 162. Suggested multienzyme mechanism to explain the removal of thymine dimers (or other defects) and subsequent repair of DNA.

(4) To complete the repair process, a phosphodiester bond must be restored. As demonstrated in Fig. 162, this model assumes the sequential action of various enzymes. Results of genetic studies revealed that radiation-resistant strains of *E. coli* have at least three genes respon-

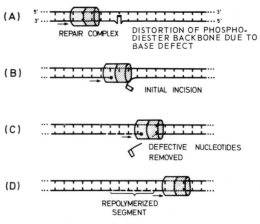

Fig. 163. "Patch-and-cut" repair mechanism by a postulated nuclease-polymerase repair complex, as proposed by Haynes (1966). (Reproduced with permission of Academic Press, Inc.)

sible for the property of radiation resistance, while radiation-sensitive strains lack one or more of those—as yet unidentified—enzymes which are apparently required for the repair of radiation-damaged DNA (Howard-Flanders and Boyce, 1966).

As an alternative principle, Haynes (1966) suggested a "patch-and-cut" repair system, in contrast to the "cut-and-patch" model of Fig. 162. Haynes postulates a single repair enzyme complex, moving in the 3′-5′ direction. According to this model, repair would begin with an incision that cuts the DNA strand near the altered bases. This is followed immediately by a repair replication catalyzed by the same enzyme complex; during this process the defective DNA segment is peeled back and degraded (Fig. 163). The experimental data agree with both models. The latter is more attractive because the whole repair process could then be carried out by a single repair unit moving in only one direction.

DNA fragments can be joined to restore completely the original double-stranded structure in its full length. This process should be possible only if the DNA fragments contain 5′-terminal with overlapping

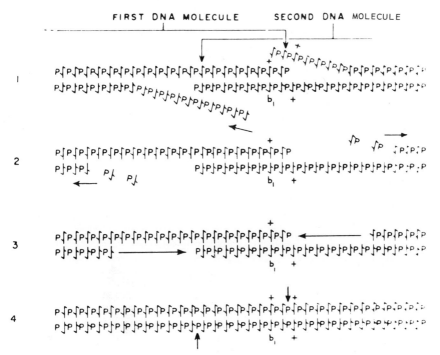

Fig. 164. Suggested mechanism for joining two cut DNA molecules with zones of homologous base sequences that permit base pairing between overlapping single strands (from Howard-Flanders and Boyce, 1966). (Reproduced with permission of National Academy of Sciences)

base sequences which—as a first step—permit base pairing and thus joining the two molecules (Fig. 164, 1). It is postulated that subsequent steps for recombining the DNA molecules are carried out by the same mechanism used in the repair of radiation damage (Fig. 164, 2–3). Thus it appears that a common system of enzymes (or the repair enzyme complex as suggested by Haynes) is involved in genetic recombination as well as in DNA repair (Meselson, 1964; Howard-Flanders and Boyce, 1966). The problem of enzymatic repair of DNA in microorganisms is discussed in detail in various papers presented at a recent meeting (Haynes *et al.*, 1966).

INTERACTION WITH DRUGS

The study of the interaction of drugs with radiation effects is concerned with the protective action of substances such as cysteine (cf. p. 336) as well as with compounds that increase radiation sensitivity. Only compounds of the latter group will be discussed here; these are of particular interest in view of the possibility of an effective therapy combined with radiation. With respect to the mode of action, the following mechanisms must be considered:

(1) Incorporation of base analogs into DNA can increase the radiation sensitivity; this represents a true sensitization.

(2) Inhibition of DNA synthesis by base analogs which are not incorporated into DNA may, under certain conditions, likewise result in an increase of radiation sensitivity. This cannot readily be distinguished from (3).

(3) The additive effects of drugs and radiation.

(4) The possibility of affecting the action of a drug by prolonging its effect in the irradiated cells.

Radiosensitizing effects were first observed after the incorporation of halogenated thymine and thymidine analogs into DNA (Djordjevic and Szybalski, 1960; Kaplan *et al.*, 1961). 5-Chlorodeoxyuridine (5-CUDR), 5-bromodeoxyuridine (5-BUDR) and 5-iododeoxyuridine (5-IUDR; cf. p. 310) increase radiation sensitivity in this order (Cl < Br < I); if the irradiation is done in 10% glycerol, a protective effect is observed (Erikson and Szybalski, 1961). *E. coli*, under O_2 or N_2, may be radiosensitized by 5-bromo- as well as 5-iodouracil (Kaplan *et al.*, 1962c); this has also been observed with the deoxyribonucleoside analog 5-BUDR in mammalian cells in tissue culture (Humphrey *et al.*, 1963). Cysteine was found to reverse the radiosensitizing effect of 5-BUDR on T1 phages (Hotz, 1963). Radiation sensitivity is enhanced when part of the thymine in one DNA strand is replaced by a 5-halogenouracil (Kaplan *et al.*, 1962a).   5-BUDR also increases the biological effects of alkylating agents (extent of chromosomal abberations; Fucik *et al.*, 1963).

Szybalski (1962) suggested that the higher radiosensitivity of DNA in which part of the normal thymine is replaced by 5-bromouracil is due to an instability of the phosphate ester bonds between the 5-BUDR moiety and the adjoining nucleotide on the same strand.   The basis for this instability is considered to be the repulsion between the negatively charged halogen and the nearest phosphate (Fig. 165); the increased

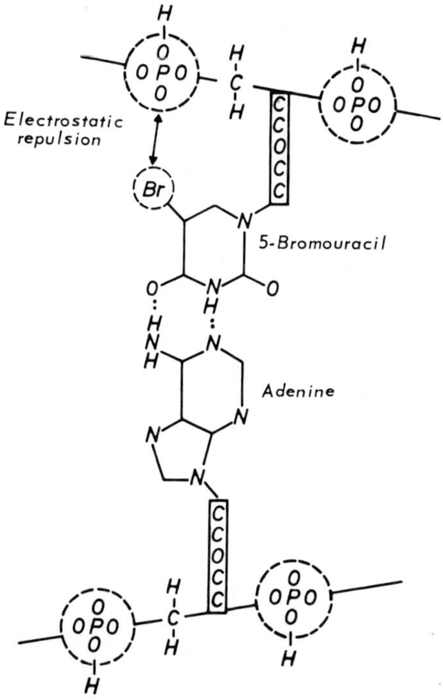

Fig. 165. Steric relationships between halogen and phosphate groups in a DNA molecule where adenine pairs with 5-bromouracil (from Szybalski, 1962).

dehalogenation, observed under irradiation, may also be explained by an increase in this repulsion.   In this connection, it may be important to note that the energy required for radical formation by ionizing radiation is smaller for 5-BU than for normal thymine (Müller *et al.*, 1963; cf. p. 337).

An essential prerequisite for a successful clinical application of these three thymidine analogs in the radiation treatment of neoplasms is a sufficient replacement of thymine in the tumor DNA by halogenated uracil.   This requirement is hard to fulfill because the thymidine analogs

are incorporated into DNA only during cell multiplication with concomitant DNA synthesis. The tumor, therefore, must be allowed to pro-
liferate until a sufficient amount of the analog has been incorporated
(Kaplan *et al.*, 1962a). Since these thymidine analogs are rapidly dehalogenated in the mammalian organism (mainly in the liver), an accumulation in the tumor DNA is still more difficult to obtain (Kriss and
Révész, 1962).

Dehalogenation is less severe with 5-bromodeoxycytidine (5-BCDR),
some of which is converted to the thymidine analog by deamination
(Kriss and Révész, 1962). In this way, a more intensive accumulation
in the DNA is possible. The incorporation can be further increased by
the simultaneous administration of 5-FU (Kriss and Révész, 1961). Because 5-FU inhibits thymine formation (cf. p. 314), the competitive
equilibrium between natural thymidine and its analog is displaced in
favor of the latter.

The more recently developed thymidine analog, 5-trifluoromethyl-2′-
deoxyuridine (F_3 TDR; cf. p. 311) also causes radiosensitization after
incorporation into DNA at levels lower than those required for an equivalent effect with 5-BUDR (Szybalski *et al.*, 1963). Clinical application
of this compound in combined therapy with ionizing radiation appears to
be hopeful since, in contrast to 5-BUDR, it is not dehalogenated and
furthermore it inhibits tumor growth (Heidelberger, 1966).

Nothing is yet known of the manner in which purine analogs such as
6-mercaptopurine, 2-aminopurine and thioguanine increase radiation
sensitivity. It is particularly surprising that this occurs only in the
presence of O_2 and not of N_2 (Kaplan *et al.*, 1962c).

5-FU enhances the *in vivo* effect of X-ray treatment of transplanted
tumors (Heidelberger *et al.*, 1958; Vermund *et al.*, 1961). Clinically,
5-FU has been successfully used in combined radiation therapy of inoperable bronchial carcinomas (e.g., Willett *et al.*, 1961; Gollin *et al.*,
1962; Heidelberger and Ansfield, 1963). The mechanism by which this
drug enhances the effect of X-rays is still not really understood. Several explanations have been proposed. 5-Fluoropyrimidines can cause a
certain synchronization of the cell cycle and thereby bring larger portions of cells simultaneously into a more radiosensitive phase. In microorganisms, the effects of 5-FU correspond closely to the symptoms of
thymineless death of thymine-dependent mutants (cf. p. 315), where
single-strand breaks occur in the DNA due to the lack of sufficient
thymine nucleotides (Mennigmann and Szybalski, 1962; Fig. 166). Because of these single-strand breaks, the DNA becomes more labile toward the action of ionizing radiation which then can produce complete
DNA breaks at a higher rate (Harbers, 1964).

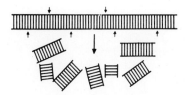

Fig. 166. Postulated structural changes in DNA during thymineless state. The arrows indicate single-strand breaks which cause partial collapse of the molecules during subsequent isolation procedures (from Mennigmann and Szybalski, 1962).

The administration of actinomycin also enhances the effect of ionizing radiation (Chan and Liebner, 1961; Schoeniger *et al.*, 1961; Vermund *et al.*, 1961; Griem and Ranniger, 1962). This can be attributed at least partly to the fact that as a result of the inhibition of DNA synthesis by irradiation, the elimination of the antibiotic is decreased (Harbers *et al.*, 1963). X-rays thus cause actinomycin to be retained longer in the irradiated tissue.

These early results may provide an approach to cancer therapy which will prove to be extremely important in the future (cf., Bagshaw, 1961; Kallmann, 1961).

References

Abbo, F. E. and A. B. Pardee (1960), *Biochim. Biophys. Acta* **39**, 478.

Abbott, M. T., R. J. Kadner, and R. M. Fink (1964), *J. Biol. Chem.* **239**, 156.

Abel, P. and T. A. Trautner (1964), Z. Vererbungslehre **95**, 66.

Abrams, R. (1951), *Arch. Biochem.* **30**, 90.

Abrams, R. (1965), *J. Biol. Chem.* **240**, PC 3697.

Abrams, R. and M. Bentley (1955), *Arch. Biochem.* **58**, 109.

Abrams, R. and M. Bentley (1959), *Arch. Biochem.* **79**, 91.

Acs, G., E. Reich, and S. Valanju (1963), *Biochim. Biophys. Acta* **76**, 68.

Adams, J. M. and M. R. Capecchi (1966), *Proc. Natl. Acad. Sci. U.S.* **55**, 147.

Adler, J., I. R. Lehman, M. J. Bessman, E. S. Simms, and A. Kornberg (1958), *Proc. Natl. Acad. Sci. U.S.* **44**, 641.

Adler, M. and A. B. Gutman (1959), *Science* **130**, 862.

Adler, M., B. Weissman, and A. B. Gutman (1958), *J. Biol. Chem.* **230**, 717.

Agranoff, B. W., M. Colodzin, and R. O. Brady (1954), *Federation Proc.* **13**, 172.

Ahmad, F. A. and A. G. Moat (1963), *Federation Proc.* **23**, 535.

Aksenova, N. N., V. M. Bresler, V. I. Vorobyev, and J. M. Olenov (1962), *Nature* **196**, 443.

Alberti, W. and G. Politzer (1923), *Arch. Mikr. Anat.* **100**, 83.

Alexander, H. E., G. Koch, I. M. Mountain, K. Sprunt, and O. van Damme (1958), *Virology* **5**, 172.

Allan, P. W., H. P. Schnebli, and L. L. Bennett (1966), *Biochim. Biophys. Acta* **114**, 647.

Allen, D. W. and P. C. Zamecnik (1962), *Biochim. Biophys. Acta* **55**, 865.

Allfrey, V. G., V. C. Littau, and A. E. Mirsky (1963), *Proc. Natl. Acad. Sci. U.S.* **49**, 414.

Allfrey, V. G. and A. E. Mirsky (1962), *Proc. Natl. Acad. Sci. U.S.* **48**, 1590.

Allfrey, V. G., B. G. T. Pogo, A. O. Pogo, L. J. Kleinsmith, and A. E. Mirsky (1966), in "Histones, Their Role in the Transfer of Genetic Information," pp. 42–67, CIBA Foundation Study Group No. 24, London, J. & A. Churchill, Ltd.

Alloway, J. L. (1932), *J. Exptl. Med.* **55**, 91.

Altmann, R. (1889), *Arch. Anat. Physiol.*, 524.

Alver, E. and S. Furberg (1959), *Acta Chem. Scand.* **13**, 910.

Ames, B. N. and B. Garry (1959), *Proc. Natl. Acad. Sci. U.S.* **45**, 1453.

Ammann, J., H. Delius, and P. H. Hofschneider (1964), *J. Mol. Biol.* **10**, 557.

Amos, H. (1961), *Biochem. Biophys. Res. Commun.* **5**, 1.

Amos, H. and M. Korn (1958), *Biochim. Biophys. Acta* **29**, 444.

Anderson, E. P. and J. A. Jacquez (1962), *Cancer Res.* **22**, 27.

Anderson, P. A. (1961), *Biochim. Biophys. Acta* **49**, 231.

Anfinsen, C. B. and F. H. White (1961), in "The Enzymes" (P. D. Boyer, H. Lardy, and K. Myrbäck, editors), Vol. 5, pp. 95–122, New York, Academic Press.

Anthony, D. D., J. L. Star, D. S. Kerr, and D. A. Goldthwait (1963), *J. Biol. Chem.* **238**, 690.

Aposhian, H. V. and A. Kornberg (1961), *Federation Proc.* **20**, 361.

Aroidson, H., N. A. Eliasson, E. Hammarsten, P. Reichard, H. v. Ubisch, and S. Bergström (1949), *J. Biol. Chem.* **199**, 169.

Ashman, D. F., R. Lipton, M. M. Melicow, and T. D. Price (1963), *Biochem. Biophys. Res. Commun.* **11**, 330.

Atkinson, D. E. (1965), *Science* **150**, 851.

Atkinson, M. R., J. F. Jackson, and R. K. Morton (1961), *Nature* **192**, 946.

Atkinson, M. R. and A. W. Murray (1965a), *Biochem. J.* **94**, 64.

Atkinson, M. R. and A. W. Murray (1965b), *Biochem. J.* **94**, 71.

Auerbach, C. and J. M. Robson (1946), *Nature* **157**, 302.

Avery, O. T., C. M. Macleoad, and M. McCarthy (1944), *J. Exptl. Med.* **79**, 137.

Axelrod, J. and J. Daly (1962), *Biochim. Biophys. Acta* **61**, 855.

Babich, F. R., A. L. Jacobson, and 3. Dubaoh (1965), *Proc. Natl. Acad. Sci. U.S.* **54**, 1299.

Bach, M. K. and H. G. Johnson (1966), *Nature* **209**, 893.

Bachrach, U. (1957), *J. Gen. Microbiol.* **17**, 1.

Backmann, R. (1963), *Zbl. Allgem. Pathol.* **104**, 581.

Baddiley, J., N. L. Blumson, and A. di Girolamo (1961), *Biochim. Biophys. Acta* **50**, 391.

Baddiley, J., J. G. Buchanan, B. Carss, A. P. Mathias, and A. R. Sanderson (1956), *Biochem. J.* **64**, 599.

Baddiley, J., J. G. Buchanan, and G. R. Greenberg (1957), *Biochem. J.* **66**, 51P.

Baer, D. (1965), *Genetics* **52**, 275.

Bagatell, F. K., E. M. Wright, and H. Z. Sable (1959), *J. Biol. Chem.* **234**, 1369.

Bagshaw, M. A. (1961), *Am. J. Roentgenol. Radium Therapy Nucl. Med.* **85**, 822.

Bailly, O. and J. Gaume (1935), *Bull. Soc. Chim. France* **2**, 354.

Baker, H. T. and L. L. Bennett, Jr. (1964), *Biochim. Biophys. Acta* **80**, 497.

Balis, M. E., G. B. Brown, G. B. Elion, H. H. Hitchings, and H. van der Werff (1951), *J. Biol. Chem.* **188**, 217.

Balis, M. E., D. H. Levin, G. B. Brown, G. B. Elion, H. van der Werff, and G. H. Hitchings (1952), *J. Biol. Chem.* **196**, 729.

Baltimore, D. (1964), *Proc. Natl. Acad. Sci. U.S.* **51**, 450.

Barber, G. A. (1962), *Biochem. Biophys. Res. Commun.* **8**, 204.

Bardos, T. J., J. L. Ambrus, Z. F. Chmielewicz, A. G. Penny, and C. M. Ambrus (1965), *Cancer Res.* **25**, 1238.

Barnes, F. W. and R. Schoenheimer (1943), *J. Biol. Chem.* **151**, 123.

Barnett, W. E. and H. E. Brockman (1962), *Biochem. Biophys. Res. Commun.* **7**, 199.

Barnett, W. E. and D. H. Brown (1967), *Proc. Natl. Acad. Sci. U.S.* **57**, 452.

Barnum, C. P. and R. A. Huseby (1950), *Arch. Biochem.* **29**, 7.

Barnum, C. P., R. A. Huseby, and H. Vermund (1953), *Cancer Res.* **13**, 880.

Baron, S. and C. E. Buckler (1963), *Science* **141**, 1061.

Barr, M. L. and E. G. Bertram (1949), *Nature* **163**, 676.

Barry, R. D., D. R. Ives, and J. G. Cruickshank (1962), *Nature* **194**, 1139.

Basilio, C., A. J. Wahba, P. Lengyel, J. F. Speyer, and S. Ochoa (1962), *Proc. Natl. Acad. Sci. U.S.* **48**, 613.

Bayley, P. M., B. N. Preston, and A. R. Peacocke (1962), *Biochim. Biophys. Acta* **55**, 943.

Bayreuther, K. (1960), *Nature* **186**, 6.

Beck, W. S., M. Goulian, A. Larsson, and P. Reichard (1966), *J. Biol. Chem.* **241**, 2177.

Beck, W. S. and M. Levin (1963), *J. Biol. Chem.* **238**, 702.

Beckwith, J. R., A. B. Pardee, R. Austrian, and F. Jacob (1962), *J. Mol. Biol.* **5**, 618.

Beermann, W. (1952), *Z. Naturforsch.* **7b**, 237.

Beermann, W. (1956), *Cold Spring Harbor Symp. Quant. Biol.* **21**, 217.

Beermann, W. (1962), "Riesenchromosomen," in "Protoplasmatologia, Handbuch der Protoplasmaforschung," Vol. VI D, Vienna, Springer-Verlag.

Begg, D. J. and H. N. Munro (1965), *Nature* **207**, 483.

Behr, W. and G. Hartmann (1965), *Biochem. Z.* **343**, 519.

Behrens, M. (1956), in "Biochemisches Taschenbuch" (H. M. Rauen, editor), pp. 910–918, Berlin-Göttingen-Heidelberg, Springer-Verlag.

Bello, L. J. and M. J. Bessman (1963), *J. Biol. Chem.* **238**, 1777.

Bello, L. J., M. J. V. Ribber, and M. J. Bessman (1961a), *J. Biol. Chem.* **236**, 1467.

Bello, L. J., M. J. V. Bibber, and M. J. Bessman (1961b), *Biochim. Biophys. Acta* **53**, 194.

Belousova, A. K. and M. T. Grigor'eva (1960), *Biochemistry* **25**, 38.

Belser, W. L. (1961), *Biochem. Biophys. Res. Commun.* **4**, 56.

Beltz, R. E. (1962), *Biochem. Biophys. Res. Commun.* **9**, 78.

Beltz, R. E. and R. L. Applegate (1959), *Biochem. Biophys. Res. Commun.* **1**, 298.

Beltz, R. E., J. van Lancker, and V. R. Potter (1957), *Cancer Res.* **17**, 688.

Bendich, A., G. B. Brown, F. S. Philips, and J. B. Thiersch (1950), *J. Biol. Chem.* **183**, 267.

Bendich, A., G. di Mayorca, M. Rosoff, and H. S. Rosenkranz (1959), *Trans. Faraday Soc.* **55**, 491.

Bendich, A., S. S. Furst, and G. B. Brown (1950), *J. Biol. Chem.* **185**, 423.

Bendich, A., H. B. Pahl, and G. B. Brown (1957), in "The Chemical Basis of Heredity" (W. D. McElroy and B. Glass, editors), pp. 378–386, Baltimore, Johns Hopkins Press.

Ben-Ishal, R., H. Goldin, and B. Oppenheim (1962), *Biochim. Biophys. Acta* **55**, 748.

Benjamin, T. L. (1966), *J. Mol. Biol.* **16**, 359.

Bennett, L. L., Jr. and H. E. Skipper (1957), *Cancer Res.* **17**, 370.

Bennett, L. L., Jr., H. E. Skipper, and L. W. Law (1953), *Federation Proc.* **12**, 300.

Bennett, L. L., Jr., H. E. Skipper, C. C. Stock, and C. P. Rhoads (1955), *Cancer Res.* **15**, 485.

Bennett, L. L., Jr., D. Smithers, C. Teague, H. T. Baker, and P. Stutts (1962), *Biochem. Pharmacol.* **11**, 81.

Benoit, J., P. Leroy, C. Vendrely, and R. Vendrely (1958), *Compt. Rend.*, **247**, 1049.

Bentley, R. and A. Neuberger (1952), *Biochem. J.* **52**, 694.

Benzer, S. and S. P. Champe (1962), *Proc. Natl. Acad. Sci. U.S.* **48**, 1114.

Benzer, S. and E. Freese (1958), *Proc. Natl. Acad. Sci. U.S.* **44**, 112.

Benzer, S. and B. Weisblum (1961), *Proc. Natl. Acad. Sci. U.S.* **47**, 1149.

Berenbom, M. (1962), *Cancer Res.* **22**, 1343.

Berg, P. (1956a), *J. Biol. Chem.* **222**, 1015.

Berg, P. (1956b), *J. Biol. Chem.* **222**, 1025.

Berg, P., F. H. Bergmann, E. J. Ofengand, and M. Dieckmann (1961), *J. Biol. Chem.* **236**, 1726.

Berg, P., H. Fancher, and M. Chamberlin (1963), in "Informational Macromolecules" (H. J. Vogel, V. Bryson, and J. O. Lampen, editors), pp. 467–484, New York, Academic Press.

Berg, P., R. D. Kornberg, H. Fancher, and M. Dieckmann (1965), *Biochem. Biophys. Res. Commun.* **18**, 932.

Bergel, F., R. C. Bray, A. Haddow, and I. Lewin (1957), *Ciba Found. Symp. Chem. Biol. Purines* (G. E. W. Wolstenholme and C. M. O'Connor, editors).

Bergkvist, R. (1958), *Acta Chem. Scand.* **12**, 364.

Bergman, W. and D. C. Burke (1955), *J. Org. Chem.* **20**, 1501.

Bergmann, E. D., R. Ben-Ishai, and B. E. Volcani (1952), *J. Biol. Chem.* **194**, 531.

Bergquist, P. L. and R. E. P. Matthews (1962), *Biochem. J.* **85**, 305.

Bernheimer, A. W. (1953), *Biochem. J.* **53**, 53.

Bernheimer, A. W. and J. M. Steele (1955), *Proc. Soc. Exptl. Biol. Med.* **89**, 123.

Bernstein, I. A., D. Fossitt, and D. Sweet (1958), *J. Biol. Chem.* **233**, 1199.

Bernstein, M. N. (1954), *Nature* **174**, 4427.

Bertani, E. L., A. Häggmark, and P. Reichard (1963), *J. Biol. Chem.* **238**, 3407.

Bertino, J. R., D. R. Donohue, B. W. Gabrio, R. Silber, A. Alenty, M. Meyer, and F. M. Huennekens (1962), *Nature* **193**, 140.

Bessman, M. J. and L. J. Bello (1961), *J. Biol. Chem.* **236**, PC 72.

Bessman, M. J. and W. H. Fleming (1966), *Federation Proc.* **25**, 279.

Bessmann, M. J., S. T. Herriott, and M. J. van Bibber Orr (1965), *J. Biol. Chem.* **240**, 439.

Bessman, M. J., I. R. Lehman, J. Adler, S. B. Zimmerman, E. S. Simms, and A. Kornberg (1958), *Proc. Natl. Acad. Sci. U.S.* **44**, 633.

Bethell, M. R. and M. E. Jones (1966), *Federation Proc.* **25**, 279.

Beukers, R. and W. Berends (1960), *Biochim. Biophys. Acta* **41**, 550.

Beukers, R. and W. Berends (1961), *Biochim. Biophys. Acta* **49**, 181.

Bhuyand, B. K. and C. G. Smith (1965), *Proc. Natl. Acad. Sci. U.S.* **54**, 566.

Bianchi, P. A., J. A. V. Butler, A. R. Crathorn, K. V. Shooter (1961), *Biochim. Biophys. Acta* **48**, 213.

Bianchi, P. A., A. R. Crathorn, and K. V. Shooter (1962), *Biochim. Biophys. Acta* **61**, 728.

Bianchi, P. A., M. V. Farina and E. Polli (1964), *Biochim. Biophys. Acta* **91**, 323.

Bielka, H. and A. Graffi (1959), *Acta Biol. Med. Germ.* **3**, 515.

Biemann, K., S. Tsunakawa, J. Sonnenbichler, H. Feldmann, D. Dütting, and H. G. Zachau (1966), *Angew. Chem.* **78**, 600.

Bierling, R. (1960), *Z. Krebsforsch.* **63**, 519.

Biggs, H. and B. Doumas (1963), *J. Biol. Chem.* **238**, 2470.

Birnie, G. D. and S. M. Fox (1965), *Biochem. J.* **95**, 41P.

Biscaro, G. and E. Belloni (1905), *Annuario Soc. Chimica di Milano* **11**, 1.

Bishop, C. W. and J. N. Davidson (1957), *Brit. J. Radiol.* **30**, 367.

Bishop, S. H. and S. Grisolia (1966), *Biochim. Biophys. Acta* **118**, 211.

Biswas, B. B., M. Edmonds, and R. Abrams (1961), *Biochem. Biophys. Res. Commun.* **6**, 146.

Biswas, S. and B. B. Biswas (1965), *Biochim. Biophys. Acta* **108**, 710.

Black, L. M. and R. Markham (1963), *Neth. Plant Pathol.* **69**, 215.

Blair, D. G. R. and V. R. Potter (1961), *J. Biol. Chem.* **236**, 2503.

Blair, D. G. R., J. E. Stone, and V. R. Potter (1960), *J. Biol. Chem.* **235**, 2379.

Blakley, R. L. (1957), *Biochim. Biophys. Acta* **24**, 224.

Blakley, R. L. (1963), *J. Biol. Chem.* **238**, 2113.

Blakley, R. L. and B. M. McDougall (1961), *J. Biol. Chem.* **236**, 1163.

Blakley, R. L., B. V. Ramasastri, and B. M. McDougall (1963), *J. Biol. Chem.* **238**, 3075.

Bloch, A. and J. D. Hutchison (1962), *Federation Proc.* **21**, 371.

Bloch, A. and C. A. Nichol (1964), *Biochem. Biophys. Res. Commun.* **16**, 400.

Bloch, D. P. (1966), in "Protoplasmatologia, Handbuch der Protoplasmaforschung," Vol. V, 3, d, pp. 1–56, Vienna-New York, Springer-Verlag.

Bloemendal, H. and L. Bosch (1962), *Biochim. Biophys. Acta* **55**, 261.

Bock, R. M. (1960), in "The Enzymes" (P. D. Boyer, H. A. Lardy, and K. Myrbäck, editors), Vol. II, New York, Academic Press.

Bockstahler, L. E. and P. Kaesberg (1962), *Biophys. J.* **2**, 1.

Bollum, F. J. (1959), *Ann. N.Y. Acad. Sci.* **81**, 792.

Bollum, F. J. (1962), *J. Biol. Chem.* **237**, 1945.

Bollum, F. J., J. W. Anderegg, A. B. McElya, and V. R. Potter (1960), *Cancer Res.* **20**, 138.

Bollum, F. J. and R. B. Setlow (1963), *Biochim. Biophys. Acta* **68**, 599.

Bonner, J. and R. C. Huang (1963), *J. Mol. Biol.* **6**, 169.

Booth, B. A. and A. C. Sartorelli (1961), *J. Biol. Chem.* **236**, 203.

Booth, B. A. and A. C. Sartorelli (1962), *Science* **138**, 518.

Borek, E. and P. R. Srinivasan (1966), *Ann. Rev. Biochem.* **35**, 275.

Bosch, L., E. Harbers, and C. Heidelberger (1958), *Cancer Res.* **18**, 335.

Boveri, T. (1914), "Zur Frage der Entstehung maligner Tumoren," Jena, Fischer-Verlag.

Boxer, G. E. and C. E. Shonk (1958), *J. Biol. Chem.* **233**, 535.

Boyce, R. P. and P. Howard-Flanders (1964), *Proc. Natl. Acad. Sci. U.S.* **51**, 293.

Boyland, E. and B. Green (1960), *Rep. Brit. Emp. Cancer Camp.* **38**, 49.

Boyland, E. and B. Green (1962), *Brit. J. Cancer* **16**, 507.

Brachet, J. (1940), *Compt. Rend. Soc. Biol.* **133**, 88.

Brachet, J. (1942), *Arch. Biol. (Liège)* **53**, 207.

Brachet, J. (1957), "Biochemical Cytology," New York, Academic Press.

Brachet, J. (1959a), *Exptl. Cell Res., Suppl.* **6**, 78.

Brachet, J. (1959b), in "Handbuch der Histochemie," III/2, pp. 1–42, Stuttgart, Fischer-Verlag.

Brachet, J. (1960), "The Biological Role of Ribonucleic Acids," Amsterdam-London-New York, Elsevier.

Brandenberger, H. (1954a), *Biochim. Biophys. Acta* **15**, 108.

Brandenberger, H. (1954b), *Helv. Chim. Acta* **37**, 641.

Brandt, E. L. and A. C. Griffin (1951), *Cancer* **4**, 1030.

Bransame, E. D., Jr. and E. Chargaff (1964), *Biochim. Biophys. Acta* **91**, 180.

Braun, R., C. Mittermayer, and H. P. Rusch (1965), *Proc. Natl. Acad. Sci. U.S.* **53**, 924.

Braunitzer, G. and V. Rudloff (1962), *Deutsche Med. Wochschr.* **87**, 959.

Brazier, M. A. B., editor (1964), Brain Function, Vol. II, RNA and Brain Function, Memory and Learning," Berkeley and Los Angeles, University of California Press.

Bremer, H. and M. W. Konrad (1964), *Proc. Natl. Acad. Sci. U.S.* **51**, 801.

Bremer, H., M. W. Konrad, K. Gaines, and G. S. Stent (1965), *J. Mol. Biol.* **13**, 540.

Bremer, H., C. Yegian, and M. Konrad (1965), *J. Mol. Biol.* **16**, 94.

Brenner, M., H. R. Müller, and R. W. Pfister (1950), *Helv. Chim. Acta* **33**, 568.

Brenner, S. (1957), *Proc. Natl. Acad. Sci. U.S.* **43**, 687.

Brenner, S., L. Barnett, F. H. C. Crick, and A. Orgel (1961), *J. Mol. Biol.* **3**, 121.

Brenner, S., L. Barnett, E. R. Katz, and F. H. C. Crick (1967), *Nature* **213**, 449.

Brenner, S., G. Streisinger, R. W. Horne, S. P. Champe, L. Barnett, S. Benzer, and M. W. Rees (1959), *J. Mol. Biol.* **1**, 291.

Brenner, S., A. O. W. Stretton, and S. Kaplan (1965), *Nature* **206**, 994.

Bresnick, E. (1962), *Biochim. Biophys. Acta* **61**, 598.

Bresnick, E. and K. Blatchford (1964a), *Arch. Biochem.* **104**, 381.

Bresnick, E. and K. Blatchford (1964b), *Biochim. Biophys. Acta* **81**, 150.

Bresnick, E. and H. G. Hitchings (1961), *Cancer Res.* **21**, 105.

Brewer, E. N. and H. P. Rusch (1965), *Biochem. Biophys. Res. Commun.* **21**, 235.

Bridges, B. A. (1962), *Radiation Res.* **16**, 232.

Brimacombe, R., J. Trupin, M. Nirenberg, P. Leder, M. Bernfield, and J. Jaouni (1965), *Proc. Natl. Acad. Sci. U.S.* **54**, 954.

Brinton, C. C., Jr., P. Gemski, Jr., and J. Carnahan (1964), *Proc. Natl. Acad. Sci. U.S.* **52**, 776.

Britten, R. J. and B. J. McCarthy (1962), *Biophys. J.* **2**, 49.

Britten, R. J., B. J. McCarthy, and R. B. Roberts (1962), *Biophys. J.* **2**, 83.

Brockman, R. W. (1960), *Cancer Res.* **20**, 643.

Brockman, R. W. (1963), *Advan. Cancer Res.* **7**, 129.

Brockman, R. W., L. L. Bennett, Jr., M. S. Simpson, A. R. Wilson, J. R. Thomson, and H. E. Skipper (1959a), *Cancer Res.* **19**, 856.

Brockman, R. W., J. M. Davis, and P. Stutts (1960), *Biochim. Biophys. Acta* **40**, 22.

Brockman, R. W. and M. C. Sparks (1960), *Bact. Proc.*, 71.

Brockman, R. W., C. Sparks, D. J. Hutchison, and H. E. Skipper (1959b), *Cancer Res.* **19**, 177.

Brockmann, H. (1960), *Fortschr. Chem. Organ. Naturstoffe* **18**, 1.

Brooke, M. S., D. Ushiba, and B. Magasanik (1954), *J. Bact.* **68**, 534.

Brookes, P. and C. Heidelberger (1966), in "Molekulare Biologie des Malignen Wachstums," p. 122 (report by C. Heidelberger in the discussion), Berlin-Heidelberg-New York, Springer-Verlag.

Brookes, P. and P. D. Lawley (1964a), *J. Cellular Comp. Physiol.* **64**, Suppl. 1, 111.

Brookes, P. and P. D. Lawley (1964b), *Nature* **202**, 781.

Brown, D. M. and B. Lythgoe (1950), *J. Chem. Soc.*, 1990.

Brown, D. M., D. I. Magrath, and A. R. Todd (1952), *J. Chem. Soc.*, 2708.

Brown, F. and S. J. Martin (1965), *Nature* **208**, 861.

Brown, F. B., J. C. Cain, D. E. Gant, L. F. J. Parker, and E. L. Smith (1955), *Biochem. J.* **59**, 82.

Brown, G. M., R. A. Weisman, and D. A. Molnar (1961), *J. Biol. Chem.* **236**, 2534.

Brown, R. A. and F. T. Ashton (1962), *Arch. Biochem.* **99**, 390.

Brown, S. W. (1966), *Science* **151**, 417.

Brues, A. M., M. M. Tracy, and W. E. Cohn (1944), *J. Biol. Chem.* **155**, 619.

Brunel, A. (1937), *Bull. Soc. Chim. Biol.* **19**, 1027.

Brunschwig, D. (1963), *Surgery* **53**, 423.

Buchanan, J. M., J. C. Sonne, and A. M. Delluva (1948), *J. Biol. Chem.* **173**, 81.

Bucher, N. and D. Mazia (1960), *J. Biophys. Biochem. Cytol.* **7**, 651.

Buckler, C. E., S. Baron, and H. B. Levy (1966), *Science* **152**, 80.

Burchenal, J. H. and G. M. Babcock (1951), *Proc. Soc. Exptl. Biol. Med.* **76**, 382.

Burdette, W. J. (1961), *Science* **133**, 40.

Burdon, R. H., M. A. Billeter, C. Weissmann, R. C. Warner, S. Ochoa, and C. A. Knight (1964), *Proc. Natl. Acad. Sci. U.S.* **52**, 768.

Burdon, R. H. and R. M. S. Smellie (1961), *Biochim. Biophys. Acta* **51**, 153.

Burke, D. C. and J. Ross (1965), *Nature* **208**, 1297.

Burton, A. and R. L. Sinsheimer (1963), *Science* **142**, 962.

Burton, A. and R. L. Sinsheimer (1965), *J. Mol. Biol.* **14**, 327.

Busch, H., P. Byvoet, and K. Smetana (1963), *Cancer Res.* **23**, 313.

Busch, S., P. Chambon, P. Mandel, and J. D. Weill (1962), *Biochem. Biophys. Res. Commun.* **7**, 255.

Bush, M. E. and K. O. Donaldson (1965), *Biochem. Biophys. Res. Commun.* **20**, 635.

Butcher, R. W. and E. W. Sutherland (1962), *J. Biol. Chem.* **237**, 1244.

Byrne, W. L., D. Samuel, E. L. Bennett, M. R. Rosenzweig, E. Wasserman, A. R. Wagner, F. Gardner, R. Galambos, B. D. Berger, D. L. Margules, J. A. Corson, H. E. Enesco, S. L. Chorover, C. E. Holt, III, P. H. Schiller, L. Chiappetta, M. E. Jarvik, R. C. Leaf, J. D. Dutcher, Z. P. Horovitz, and P. L. Carlson (1966), *Science* **153**, 658.

Byvoet, P. (1966), *J. Mol. Biol.* **17**, 311.

Cabib, E. and L. F. Leloir (1954), *J. Biol. Chem.* **206**, 779.

Cabib, E., L. F. Leloir, and C. E. Cardini (1953), *J. Biol. Chem.* **203**, 1055.

Cairns, J. (1962), *J. Mol. Biol.* **4**, 407.

Cairns, J. (1963a), *J. Mol. Biol.* **6**, 208.

Cairns, J. (1963b), *Endeavour* **22**, 141.

Calva, E., J. M. Lowenstein, and P. P. Cohen (1959), *Cancer Res.* **19**, 101.

Cameron, I. L. (1966), *Nature* **209**, 630.

Cameron, I. L. and G. E. Stone (1964), *Exptl. Cell Res.* **36**, 510.

Campbell, A. (1964), in "The Bacteria, V. Heredity" (I. C. Gunsalus and R. Y. Stanier, editors), p. 49, New York, Academic Press.

Campbell, L. L., Jr. (1957), *J. Biol. Chem.* **227**, 693.

Campbell, L. L., Jr. (1958), *J. Biol. Chem.* **233**, 1236.

Campbell, L. L., Jr. (1960), *J. Biol. Chem.* **235**, 2375.

Canellakis, E. S. (1956), *J. Biol. Chem.* **221**, 315.

Canellakis, E. S. (1957a), *Biochim. Biophys. Acta* **25**, 217.

Canellakis, E. S. (1957b), *J. Biol. Chem.* **227**, 329.

Canellakis, E. S. (1957c), *J. Biol. Chem.* **227**, 701.

Canellakis, E. S. and P. P. Cohen (1955), *J. Biol. Chem.* **213**, 385.

Canellakis, E. S. and E. Herbert (1961), *Biochim. Biophys. Acta* **47**, 78.

Canellakis, E. S., J. J. Jaffe, R. Montsavinos, and J. S. Krakow (1959), *J. Biol. Chem.* **234**, 2096.

Canellakis, E. S., H. O. Kammen, and D. R. Morales (1965), *Proc. Natl. Acad. Sci. U.S.* **53**, 184.

Cantarow, A., T. L. Williams, and J. W. Goddard (1966), *Cancer Res.* **26**, 652.

Cantoni, G. L. (1953), *J. Biol. Chem.* **204**, 403.

Cantor, K. P. and J. E. Hearst (1966), *Proc. Natl. Acad. Sci. U.S.* **55**, 642.

Caputto, R., L. F. Leloir, C. E. Cardini, and A. C. Paladini (1950), *J. Biol. Chem.* **184**, 333.

Carbon, J. A. (1962), *Biochem. Biophys. Res. Commun.* **7**, 366.

Carbon, J. A., L. Hung, and D. S. Jones (1965), *Proc. Natl. Acad. Sci. U.S.* **53**, 979.

Cardini, C. E., L. F. Leloir, and J. Chiriboga (1955), *J. Biol. Chem.* **214**, 149.

Cardoso, S. S. and J. J. Jaffe (1961), *Biochem. Pharmacol.* **8**, 252.

Carlo, P. E. and H. G. Mandel (1954), *Cancer Res.* **14**, 459.

Carminatti, H. and E. Carib (1961), *Biochim. Biophys. Acta* **53**, 417.

Carr, D. O. and S. Grisolia (1964), *J. Biol. Chem.* **239**, 160.

Carter, C. E. (1951), *J. Am. Chem. Soc.* **73**, 1537.

Carter, C. E. and W. E. Cohn (1949), *Federation Proc.* **8**, 190.

Carter, W. A. and H. B. Levy (1967), *Science* **155**, 1254.

Caspar, D. L. D. (1966), in "Principles of Biomolecular Organization" (Ciba Foundation Symposium, G. E. W. Wolstenholme, editor), p. 7, London, Churchill, Ltd.

Caspar, D. L. D. and A. Klug (1962), *Cold Spring Harbor Symp. Quant. Biol.* **27**, 1.

Caspersson, T. (1941), *Naturwissenschaften* **29**, 33.

Caspersson, T., E. Klein, and N. R. Ringertz (1958), *Cancer Res.* **18**, 857.

Catlin, B. W. and L. S. Cunningham (1958), *J. Gen. Microbiol.* **19**, 522.

Chakraborsky, K. B. and R. B. Hurlbert (1961), *Biochim. Biophys. Acta* **47**, 607.

Chamberlin, M. and P. Berg (1962), *Proc. Natl. Acad. Sci. U.S.* **48**, 81.

Chambon, P. J., J. D. Weill, and P. Mandel (1963), *Biochem. Biophys. Res. Commun.* **11**, 39.

Champe, S. P. and S. Benzer (1962), *Proc. Natl. Acad. Sci. U.S.* **48**, 532.

Chan, M. and D. J. McCorquodale (1965), *J. Biol. Chem.* **240**, 3116.

Chan, P. Y. M. and E. J. Liebner (1961), *Radiology* **76**, 273.

Chang, L. O. and W. B. Looney (1967), *Intern. J. Radiation Biol.* **12**, 187.

Chantrenne, H. and S. Devreux (1958), *Exptl. Cell Res., Suppl.* **6**, 152.

Chapeville, F., F. Lipmann, G. von Ehrenstein, B. Weisblum, W. T. Ray, and S. Benzer (1962), *Proc. Natl. Acad. Sci. U.S.* **48**, 1086.

Chargaff, E. (1963), "Essays on Nucleic Acids," Amsterdam-London-New York, Elsevier.

Chargaff, E. and J. N. Davidson, editors (1955 and 1960), "The Nucleic Acids," Vols. I–III, New York, Academic Press.

Chaudhuri, N. K., B. J. Montag, and C. Heidelberger (1958), *Cancer Res.* **18**, 318.

Chaudhuri, N. K., K. L. Mukherjee, and C. Heidelberger (1959), *Biochem. Pharmacol.* **1**, 328.

Cheng, P. (1958), *Nature* **181**, 1800.

Chipchase, M. I. H. and M. L. Birnstiel (1963), *Proc. Natl. Acad. Sci. U.S.* **49**, 692.

Cihak, A. and F. Sorm (1964), *Biochim. Biophys. Acta* **80**, 672.

Cihak, A., J. Veselý, and F. Sorm (1965), *Biochim. Biophys. Acta* **108**, 516.

Clark, B. F. C. and T. M. Jaouni (1965), *J. Biol. Chem.* **240**, 3379.

Clark, B. F. C. and K. A. Marcker (1966), *J. Mol. Biol.* **17**, 394.

Clark, J. M., A. Y. Chang, S. Spiegelman, and M. E. Reichmann (1965), *Proc. Natl. Acad. Sci. U.S.* **54**, 1193.

Clarke, D. A., F. S. Philips, S. S. Sternberg, G. H. Hitchings, C. C. Stock, and G. B. Elion (1953), *Cancer Res.* **13**, 593.

Clever, U. (1961), *Chromosoma* **12**, 607.

Cline, M. J., R. Eason, and R. M. S. Smellie (1963), *J. Biol. Chem.* **238**, 1788.

Cocito, C. and A. D. Hershey (1960), *Biochim. Biophys. Acta* **37**, 543.

Cocito, C., A. Prinzie, and P. de Somer (1962), *Experientia* **18**, 218.

Cohen, S. S. (1966), *Progr. Nucleic Acid Res. Mol. Biol.* **5**, 1.

Cohen, S. S. and H. D. Barner (1956), *J. Bacteriol.* **71**, 588.

Cohen, S. S. and H. D. Barner (1957), *J. Biol. Chem.* **226**, 631.

Cohen, S. S. and H. D. Barner (1962), *J. Biol. Chem.* **237**, 1376.

Cohen, S. S., H. D. Barner, and J. Lichtenstein (1961), *J. Biol. Chem.* **236**, 1448.

Cohn, W. E. (1960), *J. Biol. Chem.* **235**, 1488.

Cole, L. J. and M. E. Ellis (1956), *Radiation Res.* **5**, 252.

Colter, J. S., H. H. Bird, and R. A. Brown (1957), *Nature* **179**, 859.

Comb, D. G. (1964), *J. Biol. Chem.* **239**, PC 3597.

Comb, D. G., W. Chin, and S. Roseman (1961), *Biochim. Biophys. Acta* **46**, 394.

Comb, D. G. and S. Roseman (1958), *Biochim. Biophys. Acta* **29**, 653.

Comb, D. G., F. Shimizu, and S. Roseman (1959), *J. Am. Chem. Soc.* **81**, 5513.

Conzelman, G. M., H. G. Mandel, and P. K. Smith (1954), *Cancer Res.* **14**, 100.

Cook, J. L. and M. Vibert (1966), *J. Biol. Chem.* **241**, 158.

Cook, L., A. B. Davidson, D. J. Davis, J. Green, and E. J. Fellows (1963), *Science* **141**, 268.

Cooper, C., R. Wu, and D. W. Wilson (1955), *J. Biol. Chem.* **216**, 37.

Cooper, E. J., M. L. Trautman, and M. Laskowski (1950), *Proc. Soc. Exptl. Biol. Med.* **73**, 219.

Cooper, P. D. (1964), *Virology* **22**, 186.

Cormier, M. J. (1962), *Federation Proc.* **21**, 377.

Corning, W. C. and E. R. John (1961), *Science* **134**, 1363.

Cory, J. G., R. J. Suhadolnik, B. Resnick, and M. A. Rich (1965), *Biochim. Biophys. Acta* **103**, 646.

Coutsogeorgopoulos, C., B. Hacker, and R. Mantsavinos (1966), *Biochem. Biophys. Res. Commun.* **20**, 129.

Cowie, D. B. and B. J. McCarthy (1963), *Proc. Natl. Acad. Sci. U.S.* **50**, 537.

Cox, R. A., W. G. Overend, A. R. Peacocke, and S. Wilson (1958), *Proc. Roy. Soc. London Ser. B* **149**, 511.

Cramer, F. (1966), *Angew. Chem.* **78**, 186.

Cramer, J. W., W. H. Prusoff, A. C. Sartorelli, I. W. Delamore, P. K. Chang, C. F. von Essen, and A. D. Welch (1962), *Proc. Am. Assoc. Cancer Res.* **3**, 312.

Crawford, I., A. Kornberg, and E. S. Simms (1957), *J. Biol. Chem.* **226**, 1093.

Crawford, L. V. (1959), *Virology* **7**, 359.

Creaser, E. H. and J. H. Spencer (1960), *Biochem. J.* **76**, 171.

Creasey, W. A. (1963), *J. Biol. Chem.* **238**, 1772.

Creasey, W. A., M. E. Fink, R. E. Handschumacher, and P. Calabresi (1963), *Cancer Res.* **23**, 444.

Creasey, W. A. and R. E. Handschumacher (1961), *J. Biol. Chem.* **236**, 2058.

Crick, F. H. C. (1958), *Symp. Soc. Exptl. Biol.* **12**, 138.

Crick, F. H. C. (1966), *J. Mol. Biol.* **19**, 548.

Crick, F. H. C., L. Barnett, S. Brenner, and R. J. Watts-Tobin (1961), *Nature* **192**, 1227.

Crick, F. H. C., J. S. Griffith, and L. E. Orgel (1957), *Proc. Natl. Acad. Sci. U.S.* **43**, 416.

Crothers, D. M. and B. H. Zimm (1965), *J. Mol. Biol.* **12**, 525.

Cruft, H. J., C. M. Mauritzen, and E. Stedman (1954), *Nature* **174**, 580.

Dahmus, M. E. and J. Bonner (1965), *Proc. Natl. Acad. Sci. U.S.* **54**, 1370.

Dalal, F. R. and J. S. Gots (1966), *Biochem. Biophys. Res. Commun.* **22**, 340.

Daniel, V. and U. Z. Littauer (1963), *J. Biol. Chem.* **238**, 2102.

Dannenberg, H. (1966), in "Molekulare Biologie des malignen Wachstums," p. 96, Berlin-Heidelberg-New York, Springer-Verlag.

Das, N. K. and M. Alfert (1961), *Proc. Natl. Acad. Sci. U.S.* **47**, 1.

Davern, C. I. and J. Bonner (1958), *Biochim. Biophys. Acta* **29**, 205.

Davern, C. I. and M. Meselson (1960), *J. Mol. Biol.* **2**, 153.

Davey, C. L. (1961), *Arch. Biochem.* **95**, 296.

David, S. and J. Renault (1954), *Compt. Rend.* **239**, 369.

Davidson, J. D. (1958), *Proc. Am. Assoc. Cancer Res.* **2**, 290.

Davidson, J. N. (1965), "The Biochemistry of the Nucleic Acids," 5th Edition, London, Methuen & Co.

Davidson, J. N. and W. E. Cohn, editors "Progress in Nucleic Acid Research and Molecular Biology," New York, Academic Press.

Davidson, J. N., I. A. Leslie, and J. C. White (1951), *Lancet* **II**, 1287.

Davie, E. W., V. V. Koningsberger, and F. Lipmann (1956), *Arch. Biochem.* **65**, 21.

Davies, J., W. Gilbert, and L. Gorini (1964), *Proc. Natl. Acad. Sci. U.S.* **51**, 883.

Davies, J., L. Gorini, and B. D. Davis (1965), *Mol. Pharmacol.* **1**, 93.

Davies, J. W. and G. Harris (1960), *Biochim. Biophys. Acta* **45**, 28.

Davis, F. F. and F. W. Allen (1957), *J. Biol. Chem.* **227**, 907.

Davis, J. E., J. H. Strauss, Jr., and R. L. Sinsheimer (1961), *Science* **134**, 1427.

Davis, J. R. and H. Busch (1959), *Cancer Res.* **19**, 1157.

Davis, J. R. and H. Busch (1960), *Cancer Res.* **20**, 1208.

Davison, P. F., D. Freifelder, R. Hede, and C. Levinthal (1961), *Proc. Natl. Acad. Sci. U.S.* **47**, 1123.

Dawson, M. H. and R. H. P. Sia (1931), *J. Exptl. Med.* **54**, 681.

De Caputto, D. P., W. H. Mosley, J. L. Poyer, and R. Caputto (1961), *J. Biol. Chem.* **236**, 2727.

De Carvalho, S. and H. J. Rand (1961), *Nature* **189**, 815.

De Carvalho, S., H. J. Rand, and D. P. Meyer (1960), *J. Lab. Clin. Med.* **55**, 706.

De Lamirande, G., C. Allard, and A. Cantero (1958), *Cancer Res.* **18**, 952.

De Maeyer, E. and J. de Maeyer-Guignard (1963), Ind. Hyg. (Genève) **21**, 852.

De Recondo, A.-M. (1966), *Biochim. Biophys. Acta* **114**, 338.

De Renzo, E. C. (1956), *Advanc. Enzy. Mol.* **17**, 293.

De Somer, P., A. Prinzie, P. Denys, and E. Schonne (1962), *Virology* **16**, 63.

De Terra, N. (1967), *Proc. Natl. Acad. Sci. U.S.* **57**, 607.

De Verdier, C. H. and V. R. Potter (1960), *J. Natl. Cancer Inst.* **24**, 13.

De Waard, A. (1964), *Biochim. Biophys. Acta.* **87**, 169.

Di Mayorca, G. A., B. E. Eddy, S. E. Stewart, W. S. Hünter, C. Friend, and A. Bendich (1959), *Proc. Natl. Acad. Sci. U.S.* **45**, 1805.

Delbrück, M. and G. S. Stent (1957), in "The Chemical Basis of Heredity" (W. D. McElroy and B. Glass, editors), p. 699, Baltimore, The Johns Hopkins Press.

Denhardt, D. T. and R. L. Sinsheimer (1965a), *J. Mol. Biol.* **12**, 647.

Denhardt, D. T. and R. L. Sinsheimer (1965b), *J. Mol. Biol.* **12**, 674.

Devi, A., P. Lindsay, P. L. Raina, and N. K. Sarkar (1966), *Nature* 212, 474.

Dingman, C. W. and M. B. Sporn (1961), *J. Psychiatr. Res.* 1, 1.

Dingman, C. W. and M. B. Sporn (1965), *Science* 149, 1251.

Dinning, J. S., B. K. Allen, R. S. Young, and P. L. Day (1958), *J. Biol. Chem.* 233, 674.

Dische, Z. (1938), *Naturwissenschaften* 26, 252.

Djordjevic, B. and W. Szybalski (1960), *J. Exptl. Med.* 112, 509.

Dlugajczyk, A. and J. J. Eiler (1963), *Federation Proc.* 23, 470.

Doery, H. M. (1956), *Nature* 177, 381.

Donaldson, K. O. and J. C. Keresztesy (1962), *J. Biol. Chem.* 237, 1298.

Donaldson, K. O., V. F. Scott, and W. Scott (1965), *J. Biol. Chem.* 240, 4444.

Donnellan, J. E., Jr. and R. B. Setlow (1965), *Science* 149, 308.

Doty, P., H. Boedtker, J. Fresco, R. Haselkorn, and M. Litt (1959), *Proc. Natl. Acad. Sci. U.S.* 45, 482.

Doty, P., J. Marmur, J. Eigner, and C. Schildkraut (1960), *Proc. Natl. Acad. Sci. U.S.* 46, 461.

Dounce, A. L. (1955), in "The Nucleic Acids" (E. Chargaff and J. N. Davidson, editors), Vol. II, p. 93, New York, Academic Press.

Dounce, A. L., B. B. Love, J. de Simone, and M. S. Mackay (1966), in "The Cell Nucleus—Metabolism and Radiosensitivity," p. 147, London, Taylor & Francis, Ltd.

Downing, M., A. Adams, and L. Hellenga (1965), *Biochim. Biophys. Acta* 108, 233.

Downing, M. and B. S. Schweigert (1956), *J. Biol. Chem.* 220, 521.

Drew, R. M. (1955), *Radiation Res.* 3, 116.

Druckrey, H., R. Preussmann, S. Ivankovic, and D. Schmähl (1967), *Z. Krebsforsch.* 69, 103.

Drummond, G. I., N. T. Iyer, and J. Keith (1962), *J. Biol. Chem.* 237, 3535.

Dubin, D. T. and B. D. Davis (1962), *Biochim. Biophys. Acta* 55, 793.

Duggan, D. E. and E. Titus (1960), *J. Pharmacol. Exptl. Therap.* 130, 375.

Dukes, P. P. and C. E. Sekeris (1965), *Z. Physiol. Chem.* 341, 149.

Dulbecco, R. (1960), *Cancer Res.* 20, 751.

Dulbecco, R. (1967), Lecture at the occasion of receiving the Paul Ehrlich-Anschlusspreis, Frankfurt/Main, March 14.

Dulbecco, R., L. H. Hartwell, and M. Vogt (1965), *Proc. Natl. Acad. Sci. U.S.* 53, 403.

Dulbecco, R. and M. Vogt (1963), *Proc. Natl. Acad. Sci. U.S.* 50, 236.

Dunn, D. B. (1959), *Biochim. Biophys. Acta* 34, 286.

Dunn, D. B. (1961), *Biochim. Biophys. Acta* 46, 198.

Dunn, D. B. and J. D. Smith (1954a), *Biochem. J.* 58, X.

Dunn, D. B. and J. D. Smith (1954b), *Nature* 174, 305.

Dunn, D. B. and J. D. Smith (1957), *Biochem. J.* 67, 494.

Dunn, D. B. and J. D. Smith (1958), *Biochem. J.* 68, 627.

Dunn, D. B., J. D. Smith, and P. F. Spahr (1960), *J. Mol. Biol.* 2, 113.

Edgar, R. S. and W. B. Wood (1966), *Proc. Natl. Acad. Sci. U.S.* 55, 498.

Edmonds, M. and R. Abrams (1960), *J. Biol. Chem.* 235, 1142.

Edmonds, M. and R. Abrams (1962), *J. Biol. Chem.* 237, 2636.

Edson, N. L., H. A. Krebs, and A. Model (1936), *Biochem. J.* 30, 1380.

Eisinger, J. R. and G. Schulman (1963), *Proc. Natl. Acad. Sci. U.S.* 50, 694.

Eker, P. (1965), *J. Biol. Chem.* **240**, 2607.

Elion, G. B., S. Bieber, and G. H. Hitchings (1954), *Ann. N. Y. Acad. Sci.* **60**, 297.

Elion, G. B., E. Burgi, and G. H. Hitchings (1952), *J. Am. Chem. Soc.* **74**, 411.

Elion, G. B., S. Callahan, H. Nathan, S. Bieber, R. W. Rundles, and G. H. Hitchings (1963), *Biochem. Pharmacol.* **12**, 85.

Elion, G. B. and G. H. Hitchings (1950), *J. Biol. Chem.* **187**, 511.

Elion, G. B. and G. H. Hitchings (1955), *J. Am. Chem. Soc.* **77**, 1676.

Elion, G. B., A. Kovensky, and G. H. Hitchings (1966), *Biochem. Pharmacol.* **15**, 863.

Elion, G. B., S. Singer, and G. H. Hitchings (1954), *J. Biol. Chem.* **208**, 477.

Ellis, D. B. and G. A. LePage (1963), *Cancer Res.* **23**, 403.

Ellis, D. B. and G. A. LePage (1965), *Mol. Pharmacol.* **1**, 231.

Elson, D. (1965), *Ann. Rev. Biochem.* **34**, 449.

Elzina, N. V. and V. A. Engelhardt (1958), *Biokhimiya* **23**, 453.

Emanuel, C. F. and I. L. Chaikoff (1957), *Biochem. Biophys. Acta* **24**, 254.

Embden, G. and M. Zimmermann (1927), *Z. Physiol. Chem.* **167**, 137.

Emmelot, P. and O. Mühlbock, editors (1964), "Cellular Control Mechanisms and Cancer," Amsterdam-London-New York, Elsevier.

Emmerson, P., G. Scholes, D. H. Thomson, J. F. Ward, and J. J. Weiss (1960), *Nature* **187**, 319.

Engelhardt, D. L., R. E. Werster, R. C. Wilhelm, and N. D. Zinder (1965), *Proc. Natl. Acad. Sci. U. S.* **54**, 1791.

Erikson, R. L. and W. Szybalski (1961), *Biochem. Biophys. Res. Commun.* **4**, 258.

Falaschi, A., J. Adler, and H. G. Khorana (1963), *J. Biol. Chem.* **238**, 3080.

Falaschi, A. and A. Kornberg (1964), *Federation Proc.* **23**, 940.

Farber, S., L. K. Diamond, R. D. Mercer, R. F. Sylvester, Jr., and J. A. Wolff (1948), *New Engl. J. Med.* **238**, 787.

Feigelson, P., M. Feigelson, and C. Fancher (1959), *Biochim. Biophys. Acta* **32**, 133.

Feingold, D. S., E. F. Neufeld, and W. Z. Hassid (1960), *J. Biol. Chem.* **235**, 910.

Feix, G., H. Slor, and C. Weissmann (1967), *Proc. Natl. Acad. Sci. U. S.* **57**, 1401.

Feldmann, H. and H. G. Zachau (1964), *Biochem. Biophys. Res. Commun.* **15**, 13.

Fenwick, M. L., R. L. Erikson, and R. M. Franklin (1964), *Science* **146**, 527.

Fernandes, J. F., G. A. LePage, and A. Lindner (1956), *Cancer Res.* **16**, 154.

Feulgen, R. (1923), "Chemie und Physiologie der Nukleinstoffe nebst Einführung in die Chemie der Purinkörper," Berlin, Verlag Borntraeger.

Feulgen, R. and H. Rossenbeck (1924), *Z. Physiol. Chem.* **135**, 203.

Fiala, S. and A. Fiala (1961), *Biochim. Biophys. Acta* **49**, 228.

Ficq, A., C. Pavan, and J. Brachet (1958), *Exptl. Cell. Res., Suppl.* **6**, 105.

Fiel, R. J., T. J. Bardos, Z. F. Chmielewicz, and J. L. Ambrus (1965), *Cancer Res.* **25**, 1244.

Field, W. S. and W. Abbott, editors (1963), "Information Storage and Neural Control," Springfield, Ill., Charles Thomas, Publisher.

Fiers, W. and H. G. Khorana (1963a), *J. Biol. Chem.* **238**, 2780.

Fiers, W. and H. G. Khorana (1963b), *J. Biol. Chem.* **238**, 2789.

Fiers, W. and R. L. Sinsheimer (1962), *J. Mol. Biol.* **5**, 424.

Finamore, F. J. and A. H. Warner (1963), *J. Biol. Chem.* **238**, 344.

Fink, K., R. B. Henderson, and R. M. Fink (1952), *J. Biol. Chem.* **197**, 442.

Fink, R. M. (1963), *J. Biol. Chem.* **238**, 1764.

Fink, R. M. and K. Fink (1962a), *Federation Proc.* **21**, 377.

Fink, R. M. and K. Fink (1962b), *J. Biol. Chem.* **237**, 2289.

Fischer, E. (1897), *Ber. Deut. Chem. Ges.* **30**, 2226.

Fischer, G. A. (1961), *Biochem. Pharmacol.* **7**, 75.

Flaks, J. G. and S. S. Cohen (1959), *J. Biol. Chem.* **234**, 1501.

Flaks, J. G., M. J. Erwin, and J. M. Buchanan (1957a), *J. Biol. Chem.* **228**, 201.

Flaks, J. G., M. J. Erwin, and J. M. Buchanan (1957b), *J. Biol. Chem.* **229**, 603.

Flamm, W. G. and M. L. Birnstiel (1964), *Exptl. Cell Res.* **33**, 621.

Flavin, M. and S. Graff (1951), *J. Biol. Chem.* **191**, 55.

Fleissner, E. and E. Borek (1962), *Proc. Natl. Acad. Sci. U. S.* **48**, 1199.

Fleming, W. H. and M. J. Bessman (1965), *J. Biol. Chem.* **240**, PC 4108.

Flexner, J. B., L. B. Flexner, and E. Stellar (1963), *Science* **141**, 57.

Florkin, M. and G. Duchateau (1942), *Arch. Intern. Physiol.* **52**, 261.

Foster, D. W. and V. Ginsberg (1961), *Biochim. Biophys. Acta* **54**, 376.

Foster, T. S. and H. Stern (1958), *Science* **128**, 653.

Fox, C. F., W. S. Robinson, R. Haselkorn, and S. B. Weiss (1964), *J. Biol. Chem.* **239**, 186.

Fox, C. F. and S. B. Weiss (1964), *J. Biol. Chem.* **239**, 175.

Fox, J. J., K. A. Watanabe, and A. Bloch (1966), *Progr. Nucleic Acid Res. Mol. Biol.* **5**, 251.

Fox, M. S. (1957), *Biochim. Biophys. Acta* **26**, 83.

Fox, M. S. (1962), *Proc. Natl. Acad. Sci. U.S.* **48**, 1043.

Fraenkel-Conrat, H. (1961), *Biochim. Biophys. Acta* **49**, 169.

Fraenkel-Conrat, H. and R. C. Williams (1955), *Proc. Natl. Acad. Sci. U. S.* **41**, 690.

Francke, B. and P. H. Hofschneider (1966), *Proc. Natl. Acad. Sci. U. S.* **56**, 1883.

Francki, R. J. B. (1960), *Virology* **10**, 374.

Francki, R. J. B. and R. E. F. Matthews (1962), *Virology* **17**, 22.

Franklin, R. E., A. Klug, and K. C. Holmes (1957), in "The Nature of Viruses" (G. E. W. Wolstenholme and E. C. P. Miller, editors) p. 39, London, Churchill.

Fraser, D., H. R. Mahler, A. L. Shug, and C. A. Thomas, Jr. (1957), *Proc. Natl. Acad. Sci. U. S.* **43**, 939.

Fredericq, E., A. Oth, and F. Fontaine (1961), *J. Mol. Biol.* **3**, 11.

Freese, E. (1959a), *J. Mol. Biol.* **1**, 87.

Freese, E. (1959b), *Proc. Natl. Acad. Sci. U. S.* **45**, 622.

Freese, E. (1963), in "Molecular Genetics," Part I (J. H. Taylor, editor), p. 207, New York-London, Academic Press.

Freese, E. and E. Bautz Freese (1966); *Radiation Res., Suppl.* **6**, 97.

Freese, E. B. and E. Freese (1967), *Proc. Natl. Acad. Sci. U. S.* **57**, 650.

Frenster, J. H. (1965), *Nature* **206**, 680.

Frenster, J. H., V. G. Allfrey, and A. E. Mirsky (1963), *Proc. Natl. Acad. Sci. U. S.* **50**, 1026.

Fresco, J. and B. M. Alberts (1960), *Proc. Natl. Acad. Sci. U. S.* **46**, 311.

Fresco, J. R. and D. B. Straus (1962), *Am. Scientist* **50**, 158.

Friedkin, M. (1950), *J. Biol. Chem.* **184**, 449.

Friedkin, M. (1963), *Ann. Rev. Biochem.* **32**, 185.

Friedkin, M., E. Crawford, S. R. Humphreys, and A. Goldin (1962), *Cancer Res.* **22**, 600.

Friedkin, M. and D. Roberts (1954a), *J. Biol. Chem.* **207**, 245.

Friedkin, M. and D. Roberts (1954b), *J. Biol. Chem.* **207**, 257.

Friedkin, M. and D. Roberts (1956), *J. Biol. Chem.* **220**, 653.

Friedland, M. D. and W. Visser (1961), *Biochim. Biophys. Acta* **51**, 148.

Friedmann, H. C. and B. Vennesland (1960), *J. Biol. Chem.* **235**, 1526.

Fritzson, P. and A. Pihl (1957), *J. Biol. Chem.* **226**, 229.

Frydman, R. B.,and C. E. Cardini (1964), *Biochem. Biophys. Res. Commun.* **14**, 353.

Fuchs, E., R. Millette, W. Zillig, and G. Walter (1967), *Europ. J. Biochem.* **3**, 183.

Fuchs, E., W. Zillig, P. H. Hofschneider, and A. Preuss (1964), *J. Mol. Biol.* **10**, 546.

Fucik, V., A. Michaelis, and R. Rieger (1963), *Biochem. Biophys. Res. Commun.* **13**, 366.

Fuller, W., M. H. Wilkins, H. R. Wilson, and L. D. Hamilton (1965), *J. Mol. Biol.* **12**, 60.

Furberg S. (1950); *Acta Chem. Scand.* **4**, 751.

Furth, J. J., J. Hurwitz, and M. Goldmann (1961a), *Biochem. Biophys. Res. Commun.* **4**, 362.

Furth, J. J., J. Hurwitz, and M. Goldmann (1961b), *Biochem. Biophys. Res. Commun.* **4**, 431.

Futterman, S. (1957), *J. Biol. Chem.* **228**, 1031.

Gabriel, O. and G. Ashwell (1961), *Biochem. Biophys. Res. Commun.* **6**, 89.

Gaetani, S. and M. A. Spadoni (1961), *Nature* **191**, 1296.

Gaito, J., editor (1966), "Macromolecules and Behavior," Amsterdam, North-Holland Publishing Company.

Gamble, W. and L. D. Wright (1965), *Biochem. J.* **97**, 340.

Gander, J. E., W. E. Petersen, and P. D. Boyer (1956), *Arch. Biochem.* **60**, 259.

Ganesan, A. T. and J. Lederberg (1965), *Biochem. Biophys. Res. Commun.* **18**, 824.

Garen, A. and O. Siddiqi (1962), *Proc. Natl. Acad. Sci. U. S.* **48**, 173.

Gartler, S. M. (1959), *Nature* **184**, 1505.

Gaulden, M. E. and R. P. Perry (1958), *Proc. Natl. Acad. Sci. U. S.* **44**, 553.

Geiger, A. and S. Yamasaki (1957), *J. Neurochem.* **1**, 93.

Gellert, M, M. N. Lipsett, and D. R. Davies (1962), *Proc. Natl. Acad. Sci. U.S.* **48**, 2013.

Geren, W., A. Bendich, O. Bodansky, and G. B. Brown (1950), *J. Biol. Chem.* **183**, 21.

Gerhart, J. C. and H. Holoubek (1967), *J. Biol. Chem.* **242**, 2886.

Gerhart, J. C. and A. B. Pardee (1962), *J. Biol. Chem.* **237**, 891.

Gerhart, J. C. and H. K. Schachman (1965), *Biochemistry* **4**, 1054.

Gibor, A. and S. Granick (1964), *Science* **145**, 890.

Gibson, D. M., P. Ayengar, and D. R. Sanadi (1955), *Biochim. Biophys. Acta* **16**, 536.

Gibson, D. M., P. Ayengar, and D. R. Sanadi (1956), *Biochim. Biophys. Acta* **21**, 86.

Giebler, P. (1958), *Z. Naturforsch.* **13b**, 238.

Gierer, A. (1963), *J. Mol. Biol.* **6**, 148.

Gierer, A. and G. Schramm (1956a), *Z. Naturforsch.* **11b**, 138.

Gierer, A. and G. Schramm (1956b), *Nature* **177**, 702.

Gifford, G. E. and E. Heller (1963), *Nature* **200**, 50.

Gilbert, W. (1963), *J. Mol. Biol.* **6**, 389.

Gilbert, W. and B. Müller-Hill (1966), *Proc. Natl. Acad. Sci. U. S.* **56**, 1891.

Gillespie, D. and S. Spiegelman (1965), *J. Mol. Biol.* **12**, 829.

Ginoza, W. (1958), *Nature* **181**, 958.

Ginoza, W. and B. H. Zimm (1961), *Proc. Natl. Acad. Sci. U. S.* **47**, 639.

Ginsburg, V. and H. N. Korkman (1958), *J. Am. Chem. Soc.* **80**, 3481.

Ginsburg, V., P. J. O'Brien, and C. W. Hall (1962a), *J. Biol. Chem.* **237**, 497.

Ginsburg, V., P. J. O'Brien, and C. W. Hall (1962b), *Biochem. Biophys. Res. Commun.* **7**, 1.

Ginsburg, V., P. K. Stumpf, and W. Z. Hassid (1956), *J. Biol. Chem.* **223**, 977.

Giovanella, B. C., L. E. McKinney, and C. Heidelberger (1964), *J. Mol. Biol.* **8**, 20.

Giovanella, B. C., R. J. Palzer, and C. Heidelberger (1966), *Proc. Am. Assoc. Cancer Res.* **7**, 24.

Glaser, L. and S. Kornfeld (1961), *J. Biol. Chem.* **236**, 1795.

Glasky, A. J. and L. N. Simon (1966), *Science* **151**, 702.

Gold, M. and J. Hurwitz (1964a), *J. Biol. Chem.* **239**, 3858.

Gold, M. and J. Hurwitz (1964b), *J. Biol. Chem.* **239**, 3866.

Gold, M., J. Hurwitz, and M. Anders (1963), *Proc. Natl. Acad. Sci. U. S.* **50**, 164.

Goldberg, I. H. and M. Rabinowitz (1961), *Biochim. Biophys. Acta* **54**, 202.

Goldberg, I. H. and M. Rabinowitz (1962), *Science* **136**, 315.

Goldberg, I. H. and M. Rabinowitz (1963a), *Biochim. Biophys. Acta* **72**, 116.

Goldberg, I. H. and M. Rabinowitz (1963b), *J. Biol. Chem.* **238**, 1793.

Goldberg, I. H., M. Rabinowitz, and E. Reich (1962), *Proc. Natl. Acad. Sci. U. S.* **48**, 2094.

Goldé, A. and P. Vigier (1961), *Compt. Rend.* **252**, 1693.

Goldfeder, A. and L. A. Miller (1963), *Intern. J. Radiation Biol.* **6**, 575.

Goldstein, A., J. B. Kirschbaum, and A. Roman (1965), *Proc. Natl. Acad. Sci. U. S.* **54**, 1669.

Goldstein, L. (1958), *Exptl. Cell Res.* **15**, 635.

Goldstein, L. and J. Micou (1959), *J. Biophys. Biochem. Cytol.* **6**, 1.

Goldstein, L. and W. Plaut (1955), *Proc. Natl. Acad. Sci. U. S.* **41**, 874.

Goldthwait, D. A. (1956), *J. Biol. Chem.* **222**, 1051.

Goldthwait, D. A., R. A. Peabody, and G. R. Greenberg (1956), *J. Biol. Chem.* **221**, 555.

Gollin, F. F., F. J. Ansfield, A. R. Curreri, C. Heidelberger, and H. Vermund (1962), *Cancer* **15**, 1209.

Gomatos, P. J., R. M. Krug, and I. Tamm (1964), *J. Mol. Biol.* **9**, 193.

Gomatos, P. J., R. M. Krug, and I. Tamm (1965), *J. Mol. Biol.* **13**, 802.

Gomatos, P. J. and I. Tamm (1963), *Proc. Natl. Acad. Sci. U. S.* **49**, 707.

Gomatos, P. J., I. Tamm, S. Dales, and R. M. Franklin (1962), *Virology* **17**, 441.

Gonzales, N. S. and H. G. Pontis (1963), *Biochim. Biophys. Acta* **69**, 179.

Gordon, J. and H. G. Boman (1964), *J. Mol. Biol.* **9**, 638.

Gordon, M., J. M. Ravel, R. E. Eakin, and W. Shive (1948), *J. Am. Chem. Soc.* **70**, 878.

Gordon, M. P. and C. Smith (1960), *J. Biol. Chem.* **235**, PC 28.

Gordon, M. P. and M. Staehelin (1959), *Biochim. Biophys. Acta* **36**, 351.

Gots, J. S. (1950), *Federation Proc.* 9, 178.

Gottschling, H. and C. Heidelberger (1963), *J. Mol. Biol.* 7, 541.

Goulian, M. and W. S. Beck (1966a), *Federation Proc.* 25, 280.

Goulian, M. and W. S. Beck (1966b), *J. Biol. Chem.* 241, 4233.

Graham, A. F. and L. Siminovitch (1957), *Biochim. Biophys. Acta* 26, 427.

Granick, S. and A. Gibor (1967), *Progr. Nucleic Acid Res. Mol. Biol.* 6, 143.

Gray, E. D., S. M. Weissman, J. Richards, D. Bell, H. M. Keir, R. M. S. Smellie, and J. N. Davidson (1960), *Biochim. Biophys. Acta* 45, 111.

Green, B. R. and M. P. Gordon (1966), *Science* 152, 1071.

Green, M. and M. Pina (1963), *Proc. Natl. Acad. Sci. U. S.* 50, 44.

Greenberg, G. R. (1951), *J. Biol. Chem.* 190, 611.

Greenberg, G. R. (1956), *J. Biol. Chem.* 219, 423.

Greenberg, G. R., R. L. Somerville, and S. de Wolf (1962), *Proc. Natl. Acad. Sci. U. S.* 48, 242.

Greenlee, L. and P. Handler (1964), *J. Biol. Chem.* 239, 1090.

Greenlees, J. and G. A. LePage (1956), *Cancer Res.* 16, 808.

Greenman, D. L., W. D. Wicks, and F. T. Kenney (1965), *J. Biol. Chem.* 240 4420.

Greenspan, E. M., A Goldin, and E. B. Schoenbach (1950), *Cancer* 3, 856.

Greenstein, J. P. (1956), *Cancer Res.* 16, 641.

Griem, M. L. and K. Ranninger (1962), *Radiation Res.* 17, 92.

Griffith, F. (1928), *J. Hyg.* 27, 113.

Grisolia, S. and S. S. Cardoso (1957), *Biochim. Biophys. Acta* 25, 430.

Gross, J. D. and L. G. Caro (1966), *J. Mol. Biol.* 16, 269.

Grossman, L. (1963), *Proc. Natl. Acad. Sci. U. S.* 50, 657.

Groth, D. P. and N. Jiang (1966), *Biochem. Biophys. Res. Commun.* 22, 62.

Grunberg-Manago, M. and A. M. Michelson (1964), *Biochim. Biophys. Acta* 80, 431.

Grunberg-Manago, M., P. J. Ortiz, and S. Ochoa (1955), *Science* 122, 907.

Guarino, A. J. and H. Z. Sable (1955), *J. Biol. Chem.* 215, 515.

Guild, W. R. (1961), *Proc. Natl. Acad. Sci. U. S.* 47, 1560.

Gulland, J. M., D. O. Jordan, and H. F. W. Taylor (1947), *J. Chem. Soc.* 1131.

Günther, H. L. and W. H. Prusoff (1962), *Biochim. Biophys. Acta* 55, 778.

Guthrie, G. D. and R. L. Sinsheimer (1960), *J. Mol. Biol.* 2, 297.

Güttes, E. and S. Güttes (1961), *Intern. Congr. Biochem., Moscow, Abstr.*, 68.

Güttes, E., S. Güttes, and H. P. Rusch (1961), *Develop. Biol.* 3, 588.

Habel, K. (1961), *Proc. Soc. Exptl. Biol. Med.* 106, 722.

Habel, K. (1966), *Cancer Res.* 26, 2018.

Hackmann, Ch. (1953), *Strahlentherapie* 90, 296.

Haddow, A., G. de Lamirande, F. Bergel, R. C. Bray, and D. A. Gilbert (1958), *Nature* 182, 1144.

Haddow, A. and S. Weinhouse, editors, "Advances in Cancer Research," New York, Academic Press.

Hagen, U. (1962), *Nature* 194, 182.

Hagen, U., K. Keck, H. Kröger, F. Zimmermann, and T. Lücking (1965), *Biochim. Biophys. Acta* 95, 418.

Häggmark, A. (1962), *Cancer Res.* 22, 568.

Hakala, M. T., S. F. Zakrzewski, and C. A. Nichol (1961), *J. Biol. Chem.* 236, 952.

Hall, B. D. and S. Spiegelman (1961), *Proc. Natl. Acad. Sci. U. S.* 47, 137.

Hall, D. H. and I. Tessman (1966), *Federation Proc.* **25**, 776.

Hall, J. B. and F. W. Allen (1960), *Biochim. Biophys. Acta* **45**, 163.

Hall, L. M., R. L. Metzenberg, and P. P. Cohen (1958), *J. Biol. Chem.* **230**, 1013.

Hall, R. H. (1963a), *Biochem. Biophys. Res. Commun.* **12**, 361.

Hall, R. H. (1963b), *Biochem. Biophys. Res. Commun.* **12**, 429.

Hall, R. H. (1963c), *Biochem. Biophys. Res. Commun.* **13**, 394.

Hamers, R. and C. Hamers-Casterman (1961), *J. Mol. Biol.* **3**, 166.

Hamilton, L. and G. B. Elion (1954), *Ann. N. Y. Acad. Sci.* **60**, 304.

Hamilton, L. D., W. Fuller, and E. Reich (1963), *Nature* **198**, 538.

Hamilton, T. H. (1963), *Proc. Natl. Acad. Sci. U. S.* **49**, 373.

Hammarsten, E., P. Reichard, and E. Saluste (1950), *J. Biol. Chem.* **183**, 105.

Hampton, A. (1963), *J. Biol. Chem.* **238**, 3068.

Hanawalt, P. and R. Haynes (1965), *Biochem. Biophys. Res. Commun.* **19**, 462.

Handschumacher, R. E. (1957), *Biochim. Biophys. Acta* **23**, 428.

Handschumacher, R. E. (1958), *Federation Proc.* **17**, 237.

Handschumacher, R. E. (1963), *Cancer Res.* **23**, 634.

Handschumacher, R. E., P. Calabresi, A. D. Welch, V. Bono, H. Fallon, and E. Frei (1962), *Cancer Chemotherapy Rept.* **21**, 1.

Hannig, K. (1964), *Z. Physiol. Chem.* **338**, 211.

Hannig, K., W. Klofat, and H. Endres (1964), *Z. Naturforsch.* **19b**, 1072.

Hansen, H. J. and S. B. Nadler (1961), *Proc. Soc. Exptl. Biol. Med.* **107**, 324.

Hansen, R. G. and E. Hagemann (1956), *Arch. Biochem.* **62**, 511.

Harbers, E. (1958), in "Handbuch der Histochemie" (W. Graumann and K. Neumann, editors), Vol. I/1, p. 400, Stuttgart, Fischer Verlag.

Harbers, E. (1960), *Strahlentherapie* **112**, 333.

Harbers, E. (1961), *Radiation Res.* **14**, 789.

Harbers, E. (1964), *Sonderbd. Strahlenther.* **57**, 182.

Harbers, E. (1966), in "Molekulare Biologie des malignen Wachstums," p. 61, Berlin-Heidelberg-New York, Springer-Verlag.

Harbers, E. and R. Backmann (1956), *Exptl. Cell Res.* **10**, 125.

Harbers, E., H. Bujard, and W. Müller (1964), "Proceedings of the Third International Congress of Chemotherapy," p. 995, Stuttgart, Georg Thieme Verlag.

Harbers, E., N. K. Chaudhuri, and C. Heidelberger (1959), *J. Biol. Chem.* **234**, 1255.

Harbers, E. and C. Heidelberger (1959a), *J. Biol. Chem.* **234**, 1249.

Harbers, E. and C. Heidelberger (1959b), *Biochim. Biophys. Acta* **35**, 381.

Harbers, E., B. Lederer, W. Sandritter, and U. Spaar (1968), in press.

Harbers, E. and W. Müller (1962), *Biochem. Biophys. Res. Commun.* **7**, 107.

Harbers, E., W. Müller, and R. Backmann (1963), *Biochem. Z.* **337**, 224.

Harbers, E. and W. Sandritter (1968), *Deut. Med. Wochschr.* **93**, 269.

Harbers, E. and U. Spaar (1968), in press.

Harbers, E. and M. Vogt (1966), in "The Cell Nucleus—Metabolism and Radiosensitivity," p. 165, London, Taylor & Francis, Ltd.

Harrington, H. (1964), *Proc. Natl. Acad. Sci. U. S.* **51**, 59.

Hartman, S. C. (1963), *J. Biol. Chem.* **238**, 3036.

Hartman, S. C. and J. M. Buchanan (1958a), *J. Biol. Chem.* **233**, 451.

Hartman, S. C. and J. M., Buchanan (1958b), *J. Biol. Chem.* **233**, 456.

Hartman, S. C. and J. M. Buchanan (1959), *J. Biol. Chem.* **234**, 1812.

Hartmann, G. and U. Coy (1961a), *Z. Physiol. Chem.* **323**, 285.

Hartmann, G. and U. Coy (1961b), *Biochim. Biophys. Acta* **51**, 205.

Hartmann, G. and U. Coy (1962), *Angew. Chem.* **74**, 501.

Hartmann, K. -U. and C. Heidelberger (1961), *J. Biol. Chem.* **236**, 3006.

Haruna, I. and S. Spiegelman (1965), *Proc. Natl. Acad. Sci. U. S.* **54**, 1189.

Haruna, I. and S. Spiegelman (1966), *Proc. Natl. Acad. Sci. U. S.* **55**, 1256.

Hashimoto, T. and H. Yoshikawa (1961), *Biochem. Biophys. Res. Commun.* **5**, 71.

Hassid, W. Z., E. F. Neufeld, and D. S. Feingold (1959), *Proc. Natl. Acad. Sci. U. S.* **45**, 905.

Hatfield, D., R. A. Greenland, H. L. Stewart, and J. B. Wyngaarden (1964), *Biochim. Biophys. Acta* **91**, 163.

Hatfield, D. and J. B. Wyngaarden (1964a), *J. Biol. Chem.* **239**, 2580.

Hatfield, D. and J. B. Wyngaarden (1964b), *J. Biol. Chem.* **239**, 2587.

Hauschka, T. S. (1963), *Exptl. Cell Res., Suppl.* **9**, 86.

Hayaishi, O. and A. Kornberg (1952), *J. Biol. Chem.* **197**, 717.

Hayashi, M., M. N. Hayashi, and S. Spiegelman (1963), *Proc. Natl. Acad. Sci. U. S.* **50**, 664.

Hayes, D. H., A. M. Michelson, and A. R. Todd (1955), *J. Chem. Soc.* 808.

Hayes, W. (1964), "The Genetics of Bacteria and their Viruses," Oxford, Blackwell Scientific Publications.

Hayes, W. (1967), *Endeavour* **26**, 33.

Haynes, R. H. (1966), *Radiation Res., Suppl.* **6**, 231.

Haynes, R. H., S. Wolff, and J. Till, editors, (1966), *Radiation Res., Suppl.* **6**.

Haywood, A. M. and R. L. Sinsheimer (1963), *J. Mol. Biol.* **6**, 247.

Hearst, J. E. and W. H. Stockmayer (1962), *J. Chem. Phys.* **37**, 1425.

Heath, E. C. (1960): *Biochim. Biophys. Acta* **39**, 377.

Hecht, L. I., M. L. Stephenson, and P. C. Zamecnik (1958), *Biochim. Biophys. Acta* **29**, 460.

Hecht, L. I., M. L. Stephenson, and P. C. Zamecnik (1959), *Proc. Natl. Acad. Sci. U. S.* **45**, 505.

Heidelberger, C. (1964), *J. Cellular Comp. Physiol.* **64**, Suppl. 1, 129.

Heidelberger, C. (1965), *Progr. Nucleic Acid Res. Mol. Biol.* **4**, 1.

Heidelberger, C. (1966), in "Molekulare Biologie des malignen Wachstums," p. 156, Berlin-Heidelberg-New York, Springer-Verlag.

Heidelberger, C. and S. W. Anderson (1964), *Cancer Res.* **24**, 1979.

Heidelberger, C. and F. J. Ansfield (1963), *Cancer Res.* **23**, 1226.

Heidelberger, C. and G. R. Davenport (1961), *Acta Unio Intern. Contra Cancrum* **27**, 55.

Heidelberger, C., A. Ghobar, R. K. Baker, and K. L. Mukherjee (1960), *Cancer Res.* **20**, 897.

Heidelberger, C., H. Gottschling, and G. D. Birnie (1963), *Federation Proc.* **22**, 532.

Heidelberger, C., L. Griesbach, B. J. Montag, D. Mooren, and O. Cruz (1958), *Cancer Res.* **18**, 305.

Heidelberger, C., E. Harbers, K. C. Leibman, Y. Takagi, and V. R. Potter (1956), *Biochim. Biophys. Acta* **20**, 445.

Heidelberger, C., G. Kaldor, K. L. Mukherjee, and P. B. Danneberg (1960), *Cancer Res.* **20**, 903.

Heidelberger, C., K. C. Leibman, E. Harbers, and P. M. Bhargava (1957), *Cancer Res.* **17**, 399.

Heidelberger, C. and M. G. Moldenhauer (1956), *Cancer Res.* **16**, 442.

Heidelberger, C., D. Parsons, and D. C. Remy (1962), *J. Am. Chem. Soc.* 84, 3597.

Heinle, R. W. and A. D. Welch (1948), *J. Clin. Invest.* 27, 539.

Heinrich, M. R. and D. W. Wilson (1950), *J. Biol. Chem.* 186, 447.

Heinrikson, R. L. and D. Goldwasser (1963), *J. Biol. Chem.* 238, PC 485.

Hems, G. (1960), *Radiation Res.* 13, 777.

Henderson, J. F. (1962), *J. Biol. Chem.* 237, 2631.

Henderson, J. F. (1963), *Biochem. Pharmacol.* 12, 551.

Henderson, J. F. and I. G. Junga (1960), *Biochem. Pharmacol.* 5, 167.

Henderson, J. F. and I. G. Junga (1961), *Cancer Res.* 21, 118.

Henderson, J. F. and M. K. Y. Khoo (1965), *J. Biol. Chem.* 240, 2349.

Henderson, J. F. and G. A. LePage (1958), *Chem. Rev.* 58, 645.

Henderson, J. F., A. R. P. Paterson, J. C. Caldwell, and M. Hori (1967), *Cancer Res.* 27, 715.

Heppel, L. A. (1963), *J. Biol. Chem.* 238, 357.

Heppel, L. A. (1966), in "Procedures in Nucleic Acid Research" (G. L. Cantoni and D. R. Davies, editors), p. 31, New York, Harper and Row.

Heppel, L. A. and R. J. Hilmoe (1951), *J. Biol. Chem.* 188, 665.

Heppel, L. A. and R. J. Hilmoe (1953), *J. Biol. Chem.* 202, 217.

Heppel, L. A., P. J. Ortiz, and S. Ochoa (1957a), *J. Biol. Chem.* 229, 679.

Heppel, L. A., P. J. Ortiz, and S. Ochoa (1957b), *J. Biol. Chem.* 229, 695.

Herriott, R. M. (1951), *J. Bacteriol* 61, 752.

Herrmann, R. L. and J. L. Farrley (1957), *J. Biol. Chem.* 227, 1109.

Hershey, A. D. (1954), *J. Gen. Physiol.* 37, 1.

Hershey, A. D. and M. Chase (1952), *J. Gen. Physiol.* 36, 39.

Hertz, R., D. M. Bergenstal, M. B. Lipsett, E. B. Price, and T. F. Hilbisch (1958), *J. Am. Med. Assoc.* 168, 845.

Hiatt, H. H. (1958), *J. Clin. Invest.* 37, 1453.

Hiatt, H. H. and T. B. Bojarski (1960), *Biochem. Biophys. Res. Commun.* 2, 35.

Hiatt, H. H., E. C. Henshaw, C. A. Hirsch, M. Revel, and R. Finkel (1965), *Israel J. Med. Sci.* 1, 1323.

Hiatt, H. H. and J. Lareau (1958), *J. Biol. Chem.* 233, 1023.

Hiatt, H. H. and J. Lareau (1960), *J. Biol. Chem.* 235, 1241.

Higashi, T. and F. Hasegawa (1960), *J. Biochem. (Tokyo)* 48, 71.

Hignett, R. C. (1966), *Biochim. Biophys. Acta* 114, 559.

Hill, M., A. Miller-Faurès, and M. Errera (1964), *Biochim. Biophys. Acta* 80, 39.

Hillman, N. W. and M. C. Niu (1963), *Proc. Natl. Acad. Sci. U. S.* 50, 486.

Hilmoe, R. J. (1960), *J. Biol. Chem.* 235, 2117.

Himes, R. H. and T. Wilder (1965), *Biochim. Biophys. Acta* 99, 464.

Hindley, J. (1963), *Biochem. Biophys. Res. Commun.* 12, 175.

Hirono, I. (1960), *Nature* 186, 1059.

Hirono, I., H. Kachi, and A. Ohashi (1962), *Gann* 53, 73.

Hirs, C. W. H., S. Moore, and W. H. Stein (1953), *J. Biol. Chem.* 200, 493.

Hirschberg, E., J. Kream, and A. Gellhorn (1952), *Cancer Res.* 12, 524.

Hitchings, G. H., E. A. Falco, and M. B. Sherwood (1945), *Science* 102, 251.

Hoagland, M. B. (1955), *Biochim. Biophys. Acta* 16, 288.

Hoagland, M. B. (1959), *Brookhaven Symp. Biol.* 12, 40.

Hoagland, M. B. (1960), in "The Nucleic Acids," Vol. III, p. 349, New York, Academic Press.

Hoagland, M. B. and L. T. Comly (1960), *Proc. Natl. Acad. Sci. U. S.* 46, 1554.

Hoagland, M. D., E. B. Keller, and P. C. Zamecnik (1956), *J. Biol. Chem.* **218**, 345.

Hodes, M. E., C. G. Palmer, and J. D. Hubbard (1962), *Proc. Am. Assoc. Cancer Res.* **3**, No. 4, 329.

Hoffer, M., R. Dushinsky, J. J. Fox, and N. Yung (1959), *J. Am. Chem. Soc.* **81**, 4112.

Hoffmann, H. -D. and W. Müller (1966), *Biochim. Biophys. Acta* **123**, 421.

Hofschneider, P. H. (1960), *Z. Naturforsch.* **15b**, 441.

Hofschneider, P. H. (1963), *Proc. Intern. Congr. Biochem. V, 1961*, **1**, 115.

Hofschneider, P. H. (1963b), *Z. Naturforsch.* **18b**, 203.

Hogeboom, G. H. and W. C. Schneider (1955), in "The Nucleic Acids," Vol. II, p. 199, New York, Academic Press.

Hogenkamp, H. P. C. (1963), *J. Biol. Chem.* **238**, 477.

Hogenkamp, H. P. C., J. N. Ladd, and H. A. Barker (1962), *J. Biol. Chem.* **237**, 1950.

Hollaender, A. and C. W. Emmons (1941), *Cold Spring Harbor Symp. Quant. Biol.* **9**, 179.

Holland, J. J. (1962a), *Biochem. Biophys. Res. Commun.* **9**, 556.

Holland, J. J. (1962b), *Proc. Natl. Acad. Sci. U. S.* **48**, 2044.

Holland, J. J., L. C. McLaren, and J. T. Syverton (1959), *J. Gen. Physiol.* **110**, 65.

Holley, R. W., J. Apgar, G. A. Everett, J. T. Madison, M. Marquisee, S. H. Merrill, J. R. Pennswick, and A. Zamir (1965), *Science* **147**, 1462.

Hollmann, S. (1961), "Nicht-glykolytische Stoffwechselwege der Glucose," Stuttgart, Thieme Verlag.

Hollmann, S. and O. Touster (1957), *J. Biol. Chem.* **225**, 87.

Holmes, B. (1947), *Brit. J. Radiol.* **20**, 450.

Holmes, W. L. (1956), *J. Biol. Chem.* **223**, 677.

Holmgren, A., P. Reichard, and L. Thelander (1965), *Proc. Natl. Acad. Sci. U. S.* **54**, 830.

Holoubek, V. and R. R. Rueckert (1964), *Biochem. Biophys. Res. Commun.* **15**, 166.

Holzer, A. and A. W. Holldorf, editors (1966), "Molekulare Biologie des malignen Wachstums," Berlin-Heidelberg-New York, Springer-Verlag.

Homburger, F., editor, "Progress in Experimental Tumor Research," Basel-New York, S. Karger Verlag.

Honjo, M., Y. Kanai, Y. Furukawa, Y. Mizuno, and Y. Sanno (1964), *Biochim. Biophys. Acta* **87**, 698.

Hopkins, J. W., V. G. Allfrey, and A. E. Mirsky (1961), *Biochim. Biophys. Acta* **47**, 194.

Horecker, B. L. (1962), "Pentose Metabolism in Bacteria," New York, John Wiley & Sons.

Horecker, B. L., G. Domagk, and H. H. Hiatt (1958), *Arch. Biochem.* **78**, 510.

Hori, K., H. Fujiki, and Y. Takagi (1966), *Nature* **210**, 604.

Horowitz, J., J. J. Saukkonen, and E. Chargaff (1958), *Biochim. Biophys. Acta* **29**, 222.

Horowitz, J., J. J. Saukkonen, and E. Chargaff (1960), *J. Biol. Chem.* **235**, 3266.

Hoshino, A., S. Kurita, T. Okita, and K. Kimura (1960), *Gann* **51**, Suppl. 1, 42.

Hotchkiss, R. D. (1951), *Cold Spring Harbor Symp. Quant. Biol.* **16**, 457.

Hotchkiss, R. D. (1957), in "The Chemical Basis of Heredity" (W. C. McElroy and B. Glass, editors), p. 321, Baltimore, Md., The Johns Hopkins Press.

Hotchkiss, R. D. and J. Marmur (1954), *Proc. Natl. Acad. Sci. U. S.* **40**, 55.

Hotz, G. (1963), *Biochem. Biophys. Res. Commun.* **11**, 393.

Hotz, G. and A. Müller (1961), *Z. Naturforsch.* **16b**, 282.

Hough, L. and J. K. N. Jones (1951), *Nature* **167**, 180.

Howard, A. and S. R. Pelc (1953), *Heredity* **6**, Suppl., 261.

Howard, F. L. (1932), *Ann. Botany* **46**, 461.

Howard-Flanders, P. (1958), *Advan. Biol. Med. Phys.* **6**, 553.

Howard-Flanders, P. (1960), *Nature* **186**, 485.

Howard-Flanders, P. and R. P. Boyce (1966), *Radiation Res., Suppl.* **6**, 156.

Howard-Flanders, P., J. Levin, and L. Theriot (1963), *Radiation Res.* **18**, 593.

Hoyer, B. H., B. J. McCarthy, and E. T. Bolton (1964), *Science* **144**, 959.

Huang, R. C. and J. Bonner (1962), *Proc. Natl. Acad. Sci. U. S.* **48**, 1216.

Huang, R. C. and J. Bonner (1965), *Proc. Natl. Acad. Sci. U. S.* **54**, 960.

Huberman, J. A. and A. D. Riggs (1966), *Proc. Natl. Acad. Sci. U. S.* **55**, 599.

Huebner, R. J., W. T. Lane, A. D. Welch, P. Calabresi, R. W. McCollum, and W. H. Prusoff (1963), *Science* **142**, 488.

Huebner, R. J., W. P. Rowe, and W. T. Lane (1962), *Proc. Natl. Acad. Sci. U. S.* **48**, 2051.

Huguley, C. M., Jr., J. A. Bain, S. L. Rivers, and R. B. Scoggins (1959), *Blood* **14**, 615.

Humm, D. G. and J. H. Humm (1966), *Proc. Natl. Acad. Sci. U. S.* **55**, 114.

Humphrey, R. W., W. C. Dewey, and A. Cork (1963), *Nature* **198**, 268.

Huppert, J., F. Lacour, J. Lacour, and J. Harel (1961), *Compt. Rend.* **252**, 1876.

Hurlbert, R. B. (1962), *Federation Proc.* **21**, 383.

Hurlbert, R. B. and H. O. Kammen (1960), *J. Biol. Chem.* **235**, 443.

Hurwitz, J. (1959), *J. Biol. Chem.* **234**, 2351.

Hurwitz, J., M. Anders, and I. Smith (1965), *J. Biol. Chem.* **240**, 1256.

Hurwitz, J., J. J. Furth, M. Anders, and A. Evans (1962), *J. Biol. Chem.* **237**, 3752.

Hurwitz, J., J. J. Furth, M. Malamy, and M. Alexander (1962), *Proc. Natl. Acad. Sci. U. S.* **48**, 1222.

Hurwitz, J., M. Gold, and M. Anders (1964a), *J. Biol. Chem.* **239**, 3462.

Hurwitz, J., M. Gold, and M. Anders (1964b), *J. Biol. Chem.* **239**, 3474.

Hutchison, D. J. (1963), *Advan. Cancer Res.* **7**, 235.

Hutchison, F. (1961), *Radiation Res.* **14**, 721.

Hutchison, F. and J. Arena (1960), *Radiation Res.* **13**, 137.

Hydén, H. (1962), *Endeavour* **21**, 144.

Hydén, H. (1967), *Progr. Nucleic Acid Res. Mol. Biol.* **6**, 187.

Hydén, H. and E. Egyházi (1962), *Proc. Natl. Acad. Sci. U. S.* **48**, 1366.

Hydén, H. and E. Egyházi (1963), *Proc. Natl. Acad. Sci. U. S.* **49**, 618.

Ingram, V. M. and J. A. Sjöquist (1963), *Cold Spring Harbor Symp. Quant. Biol.* **28**, 133.

Inoué, S. and H. Sato (1962), *Science* **136**, 1122.

Irino, S., Z. Ota, T. Sezaki, M. Suzaki, and K. Kiraki (1963), *Gann* **54**, 225.

Ito, Y. and C. A. Evans (1961), *J. Exptl. Med.* **114**, 485.

Ives, D. H., P. A. Morse, Jr., and V. R. Potter (1963), *J. Biol. Chem.* **238**, 1467.

Iyer, V. N. and W. Szybalski (1963), *Proc. Natl. Acad. Sci. U. S.* **50**, 355.

Izawa, M., V. G. Allfrey, and A. E. Mirsky (1963), *Science* **140**, 382.

Jacob, F. (1966), *Science* **152**, 1470.

Jacob, F. and S. Brenner (1963), *Compt. Rend.* **256**, 298.

Jacob, F. and J. Monod (1961a), *J. Mol. Biol.* **3**, 318.

Jacob, F. and J. Monod (1961b), *Cold Spring Harbor Symp. Quant. Biol.* **26**, 193.

Jacob, F. and J. Monod (1963), in "Cytodifferentiation and Macro-molecular Synthesis" (M. Locke, editor), p. 30, New York, Academic Press.

Jacob, F. and J. Monod (1965), *Biochem. Biophys. Res. Commun.* **18**, 693.

Jacob, F. and E. L. Wollman (1961), "Sexuality and the Genetics of Bacteria," New York, Academic Press.

Jacobs, E. M., J. K. Luce, and R. Cailleau (1966), *Cancer* **19**, 869.

Jacobson, A. L., C. Fried, and S. D. Horowitz (1966), *Nature* **209**, 599.

Jacobson, B. and E. A. Davidson (1962), *J. Biol. Chem.* **237**, 638.

Jacquez, J. A. and J. -H. Sherman (1962), *Cancer Res.* **22**, 56.

Jaenicke, L. (1961), *Angew. Chem.* **73**, 449.

Jaenicke, L. and E. Brode (1961), *Biochem. Z.* **334**, 108.

Jaenisch, R., P. H. Hofschneider, and A. Preuss (1966), *J. Mol. Biol.* **21**, 501.

Jaffe, J. J., L. G. Lajtha, J. Lascelles, M. G. Ord, and L. A. Stocken (1959), *Intern. J. Radiation Biol.* **1**, 241.

James, A. P. and I. Müller (1961), *Radiation Res.* **14**, 779.

Jardetzky, C. D. and O. Jardetzky (1960), *J. Am. Chem. Soc.* **82**, 222.

Jenny, E. and D. M. Greenberg (1963), *J. Biol. Chem.* **238**, 3378.

Jervell, K. F. (1963), *Acta Endocrinol. Suppl.* **44**, 88.

Jesaitis, M. A. (1956), *Nature* **178**, 637.

Jiang, N. S. and D. P. Groth (1962), *J. Biol. Chem.* **237**, 3339.

Jones, D. S., S. Nishimura, and H. G. Khorana (1966), *J. Mol. Biol.* **16**, 454.

Jones, M. E., A. D. Anderson, C. Anderson, and S. Hol (1961), *Arch. Biochem.* **95**, 499.

Jones, M. E. and L. Spector (1960), *J. Biol. Chem.* **235**, 2897.

Jones, M. E., L. Spector, and F. Lipmann (1955), *J. Am. Chem. Soc.* **77**, 819.

Jones, O. W. and M. W. Nirenberg (1962), *Proc. Natl. Acad. Sci. U. S.* **48**, 2115.

Jones, W. (1920), "Nucleic Acids—Their Chemical Properties and Physiological Conduct," 2nd Edition, London, Longmans, Green and Co.

Jordan, D. O. (1960), "The Chemistry of Nucleic Acids," London, Butterworths & Co.

Josse, J., A. D. Kaiser, and A. Kornberg (1961), *J. Biol. Chem.* **236**, 864.

Josse, J. and A. Kornberg (1962), *J. Biol. Chem.* **237**, 1968.

Julian, G. R., R. G. Wolfe, and F. J. Reithel (1961), *J. Biol. Chem.* **236**, 754.

Kadenbach, B. (1966), *Biochim. Biophys. Acta* **134**, 430.

Kaerner, H. C. and H. Hoffmann-Berling (1964), *Nature* **202**, 1012.

Kahan, F. M. and J. Hurwitz (1962), *J. Biol. Chem.* **237**, 3778.

Kalckar, H. M. (1947), *J. Biol. Chem.* **167**, 461.

Kalckar, H. M. and E. S. Maxwell (1958), *Physiol. Rev.* **38**, 77.

Kalle, G. P. and J. S. Gots (1961), *Biochim. Biophys. Acta* **51**, 130.

Kallen, R. G., M. Simon, and J. Marmur (1962), *J. Mol. Biol.* **5**, 248.

Kallmann, R. F., editor (1961), "Research in Radiotherapy. Approaches to Chemical Sensitization," National Academy of Sciences—National Research Council, Washington, D. C.

Kalman, S. M., P. H. Duffield, and T. Brzozowski (1966), *J. Biol. Chem.* **241**, 1871.

Kaplan, H. S. and L. A. Heppel (1956), *J. Biol. Chem.* **222**, 907.

Kaplan, H. S., K. C. Smith, and P. Tomlin (1961), *Nature* **190**, 794.

Kaplan, H. S., K. C. Smith, and P. Tomlin (1962), *Radiation Res.* **16**, 98.

Kaplan, H. S. and R. Zavarine (1962), *Biochem. Biophys. Res. Commun.* **8**, 432.

Kaplan, H. S., R. Zavarine, and J. Earle (1962), *Nature* **194**, 662.

Kaplan, J. G., F. Lacroute, M. Duphil, and P. Slonimski (1966), *Federation Proc.* **25**, 279.

Kapoor, M. and E. R. Waygood (1962), *Biochem. Biophys. Res. Commun.* **9**, 7.

Karlson, P. (1961), *Deut. Med. Wochschr.* **86**, 668.

Karlson, P. (1963), *Deut. Med. Wochschr.* **88**, 1029.

Karlson, P. (1966), *Naturwissenschaften* **53**, 445.

Karlson, P., C. E. Sekeris, and R. Maurer (1964), *Z. Physiol. Chem.* **336**, 100.

Katanuma, N. (1958), *Arch. Biochem.* **76**, 547.

Kaufman, H. E. and C. Heidelberger (1964), *Science* **145**, 585.

Kay, L. D., M. J. Osborn, Y. Hatefi, and F. M. Huennekens (1960), *J. Biol. Chem.* **235**, 195.

Keir, H. M., H. Omura, and J. B. Shepherd (1963), *Biochem. J.* **89**, 425.

Kellenberger, E. (1966), in "Principles of Biomolecular Organization" (G. E. W. Wolstenholme, editor), p. 192, London, J. & A. Churchill, Ltd.

Kelly, L. S., D. Hirsch, G. Beach, and N. L. Petrakis (1957), *Proc. Soc. Exptl. Biol. Med.* **94**, 83.

Kennedy, E. P. (1960), in "The Enzymes" (P. D. Boyer, H. Lardy, and K. Myrbäck, editors), p. 63, New York, Academic Press.

Kennedy, E. P. and S. B. Weiss (1956), *J. Biol. Chem.* **222**, 193.

Kenney, F. T. and F. J. Kull (1963), *Proc. Natl. Acad. Sci. U. S.* **50**, 493.

Kerr, I. M., E. A., Pratt, and I. R. Lehman (1965), *Biochem. Biophys. Res. Commun.* **20**, 154.

Kerr, S. E., K. Seraidarian, and G. B. Brown (1951), *J. Biol. Chem.* **188**, 207.

Kersten, H., W. Kersten, G. Leopold, and B. Schnieders (1964), *Biochim. Biophys. Acta* **80**, 521.

Kersten, W. (1961), *Biochim. Biophys. Acta* **47**, 610.

Kersten, W., H. Kersten, and H. M. Rauen (1960), *Nature* **187**, 60.

Khorana, H. G. (1961a), in "The Enzymes" (P. D. Boyer, H. Lardy, and K. Myrbäck, editors), Vol. 5, p. 79, New York, Academic Press.

Khorana, H. G. (1961b), "Some Recent Developments in the Chemistry of Phosphate Esters of Biological Interest," New York, John Wiley & Sons.

Khorana, H. G., J. F. Fernandes, and A. Kornberg (1958), *J. Biol. Chem.* **230**, 941.

Khorana, H. G., A. F. Turner, and J. P. Viszolyi (1961), *J. Am. Chem. Soc.* **83**, 686.

Khorana, H. G. and J. P. Viszsolyi (1961), *J. Am. Chem. Soc.* **83**, 675.

Khym, J. X., D. G. Doherty, E. Volkin, and W. E. Cohn (1953), *J. Am. Chem. Soc.* **75**, 1262.

Kidson, C. (1966), *J. Mol. Biol.* **17**, 1.

Kim, J-H. and A. G. Perez (1965), *Nature* **207**, 974.

Kinoshita, J. H. (1957), *J. Biol. Chem.* **228**, 247.

Kirby, K. S. (1957), *Biochem. J.* **66**, 495.

Kirk, J. M. (1960), *Biochim. Biophys. Acta* **42**, 167.

Kit, S. (1960a), *Arch. Biochem.* **87**, 330.

Kit, S. (1960b), *Biochem. Biophys. Res. Commun.* **3**, 361.

Kit, S., D. R. Dubbs, and P. M. Frearson (1965), *J. Biol. Chem.* **240**, 2565.

Kit, S., D. R. Dubbs, L. J. Piekarski, and T. C. Hsu (1963), *Exptl. Cell Res.* **31**, 297.

Kit, S. and T. C. Hsu (1961), *Biochem. Biophys. Res. Commun.* **5**, 120.

Kizer, D. E., B. Cox, C. A. Lovig, and S. F. de Estrugo (1963), *J. Biol. Chem.* **238**, 3048.

Kizer, D. E., C. A. Lovig, B. A. Howell, and B. Cox (1964), *Cancer Res.* **24**, 1050.

Klamerth, O. (1957), *Biochem. Z.* **328**, 443.

Klein, F. and J. A. Szirmai (1963), *Biochim. Biophys. Acta* **72**, 48.

Klein, G. (1959), *Can. Cancer Conf.* **3**, 215.

Kleinschmidt, A. K., A. Burton, and R. L. Sinsheimer (1963), *Science* **142**, 961.

Kleinschmidt, A. K., D. Lang, D. Jacherts, and R. Zahn (1962), *Biochim. Biophys. Acta* **61**, 857.

Klemperer, H. G. (1963a), *Biochim. Biophys. Acta* **72**, 403.

Klemperer, H. G. (1963b), *Biochim. Biophys. Acta* **72**, 416.

Klenow, H. (1963a), *Biochim. Biophys. Acta* **76**, 347.

Klenow, H. (1963b), *Biochim. Biophys. Acta* **76**, 354.

Klenow, H. and E. Lichtler (1957), *Biochim. Biophys. Acta* **23**, 7.

Kleppe, K., T. Sanner, and A. Pihl (1966), *Biochim. Biophys. Acta* **118**, 210.

Klug, A., J. T. Finch, R. Leberman, and W. Longley (1966), in "Principles of Biomolecular Organization" (G. E. W. Wolstenholme, editor), p. 158, London, J. & A. Churchill, Ltd.

Knapp, E., A. Reuss, O. Risse, and H. Schreiber (1939), Naturwissenschaften **27**, 304.

Knight, C. A. and B. R. Woody (1958), *Arch. Biochem.* **78**, 460.

Kobata, A. (1962), *Biochem. Biophys. Res. Commun.* **7**, 346.

Koch, A. L. and G. Vallee (1959), *J. Biol. Chem.* **234**, 1213.

Koch, G., S. Koenig, and H. E. Alexander (1960), *Virology* **10**, 329.

Koerner, J. F. and R. L. Sinsheimer (1957), *J. Biol. Chem.* **228**, 1049.

Koerner, J. F., M. S. Smith, and J. M. Buchanan (1960), *J. Biol. Chem.* **235**, 2691.

Kohn, K. W., C. L. Spears, and P. Doty (1966), *J. Mol. Biol.* **19**, 266.

Kohn, K. W., N. H. Steigbigel, and C. L. Spears (1965), *Proc. Natl. Acad. Sci. U. S.* **53**, 1154.

Korn, D. and A. Weissbach (1964), *J. Biol. Chem.* **239**, 3849.

Kornberg, A., I. R. Lehman, M. J. Bessman, and E. S. Simms (1956), *Biochim. Biophys. Acta* **21**, 197.

Kornberg, A., I. Lieberman, and E. S. Simms (1955), *J. Biol. Chem.* **215**, 417.

Kornberg, S. R., I. R. Lehman, M. J. Bessman, E. S. Simms, and A. Kornberg (1958), *J. Biol. Chem.* **233**, 159.

Kornberg, S. R., S. B. Zimmerman, and A. Kornberg (1961), *J. Biol. Chem.* **236**, 1487.

Korner, A. (1964), *Biochem. J.* **92**, 449.

Kornfeld, S. and L. Glaser (1961), *J. Biol. Chem.* **236**, 1791.

Kornfeld, S. and L. Glaser (1962), *J. Biol. Chem.* **237**, 3052.

Koschel, K., G. Hartmann, W. Kersten, and H. Kersten (1966), *Biochem. Z.* **344**, 76.

Kossel, A. and A. Neumann (1894), *Ber. Ges. Dtsch. Chem.* **27**, 2215.

Kossel, A. and H. Steudel (1902–03), *Z. Physiol. Chem.* **37**, 177.

Kossel, A. and H. Steudel (1903), *Z. Physiol. Chem.* **38**, 49.

Kössel, H. (1965), *Z. Physiol. Chem.* **340**, 210.

Kowlessar, O. D., S. Okada, J. L. Potter, and K. I. Altman (1957), *Arch. Biochem.* **68**, 231.

Koyama, G., K. Maeda, and H. Umezawa (1966), *Tetrahedron Letters* 6, 597.

Kozinski, A. W. and P. B. Kozinski (1965), *Proc. Natl. Acad. Sci. U. S.* **54**, 634.

Kozloff, L. M. (1960), *Ann. Rev. Biochem.* **29**, 475.

Krakow, G. and B. Vennesland (1961), *J. Biol. Chem.* **236**, 142.

Krakow, J. S., C. Coutsogeorgopoulos, and E. S. Canellakis (1962), *Biochim. Biophys. Acta* 55, 639.

Krakow, J. S., H. O. Kammen, and E. S. Canellakis (1961), *Biochim. Biophys. Acta* 53, 52.

Krakow, J. S. and S. Ochoa (1963), *Proc. Natl. Acad. Sci. U. S.* **49**, 88.

Kramer, G., H. G. Wittmann, and H. Schuster (1964), *Z. Naturforsch.* 19b, 46.

Krebs, H. A. and R. Hems (1955), *Biochem. J.* **61**, 435.

Krenitsky, T. A., M. Barclay, and J. A. Jacquez (1964), *J. Biol. Chem.* **239**, 805.

Krenitsky, T. A., G. B. Elion, R. A. Strelitz, and G. H. Hitchings (1967), *J. Biol. Chem.* **242**, 2675.

Krenitsky, T. A., J. W. Mellors, and R. K. Barclay (1965), *J. Biol. Chem.* **240**, 1281.

Kriss, J. P. and L. Révész (1961), *Cancer Res.* **21**, 1141.

Kriss, J. P. and L. Révész (1962), *Cancer Res.* **22**, 254.

Kriss, J. P., L. Tung, and S. Bond (1962), *Cancer Res.* **22**, 1257.

Kröger, H. (1963), *Kolloqu. Biochem. u. Strahlenmutation, u. Struktur d. DNS, Freiburg*, 8.–9.3.

Kröger, H. and B. Greuer (1965), *Biochem. Z.* **341**, 190.

Krüger, M. and G. Salomon (1898), *Z. Physiol. Chem.* **24**, 364.

Kunitz, M. (1940), *J. Gen. Physiol.* **24**, 15.

Kunitz, M. (1948), *Science* **108**, 19.

Kunitz, M. (1950), *J. Gen. Physiol.* **33**, 363.

Kunkee, R. E. and A. B. Pardee (1956), *Biochim. Biophys. Acta* 19, 236.

Kuno, S. and I. R. Lehman (1962), *J. Biol. Chem.* **237**, 1266.

Kuramitsu, H. K., S. Udaka, and H. S. Moyed (1964), *J. Biol. Chem.* **239**, 3425.

Kurita, S. (1960), *J. Nagoya Med. Assoc.* **81**, 1156.

Kurnick, N. B. (1952), in "The Chemistry and Physiology of the Nucleus," *Exptl. Cell Res., Suppl.* 2, 266.

Lacour, F., J. Lacour, J. Harel, and J. Huppert (1960), *J. Natl. Cancer Inst.* **24**, 301.

Lagerkvist, U. (1953), *Arkiv Kemi* 5, 569.

Lagerkvist, U. (1958a), *J. Biol. Chem.* **233**, 138.

Lagerkvist, U. (1958b), *J. Biol. Chem.* **233**, 143.

Lajtha, L. G. and J. R. Vane (1958), *Nature* **182**, 191.

Lake, J. A. and W. W. Beeman (1967), *Science* **156**, 1371.

Lamfrom, H., C. S. McLaughlin, and A. Sarabhai (1966), *J. Mol. Biol.* **22**, 355.

Lang, K. and G. Siebert (1952), *Biochem. Z.* **322**, 360.

Lang, N. and C. E. Sekeris (1966), cited by C. E. Sekeris (1966).

Langen, P. (1966), in "Molekulare Biologie des malignen Wachstums," p. 179, Berlin-Heidelberg-New York, Springer-Verlag.

Langridge, R. and A. Rich (1963), *Nature* **198**, 725.

Langridge, R., H. R. Wilson, C. W. Hooper, M. H. F. Wilkins, L. D. Hamilton, W. E. Seede, and D. A. Marvin (1960), *J. Mol. Biol.* 2, 38.

Lanning, M. C. and S. S. Cohen (1954), *J. Biol. Chem.* **207**, 193.

Lansford, E. M., R. B. Turner, C. J. Weathersbee, and W. Shive (1964), *J. Biol. Chem.* **239**, 497.

Lark, K. G. (1962), *Proc. Intern. Union Physiol. Sci., XXII Intern. Congress, Leiden, Holland*, 766.

Larrabee, A. R., S. Rosenthal, R. E. Cathow, and J. M. Buchanan (1963), *J. Biol. Chem.* **238**, 1025.

Larsen, A. (1963), *J. Biol. Chem.* **238**, 3414.

Larsson, A. and J. B. Neilands (1966), *Biochem. Biophys. Res. Commun.* **25**, 222.

Larsson, A. and P. Reichard (1966a), *J. Biol. Chem.* **241**, 2533.

Larsson, A. and P. Reichard (1966b), *J. Biol. Chem.* **241**, 2540.

Laskowski, M., Jr. (1966), in "Procedures in Nucleic Acid Research" (G. L. Cantoni and D. R. Davis, editors), p. 85, New York, Harper and Row.

Laskowski, M., Jr. (1967), *Advan. Enzymol.* **29**, 165.

Laster, L. and A. Blair (1963), *J. Biol. Chem.* **238**, 3348.

Latarjet, R., N. Rebeyrotte, and E. Moustacchi (1958), *Compt. Rend.* **264**, 853.

Laurent, T. C., E. C. Moore, and P. Reichard (1964), *J. Biol. Chem.* **239**, 3436.

Lavik, P. S. and G. W. Buckaloo (1954), *Radiation Res.* **1**, 221.

Law, L. W. (1950), *Cancer Res.* **10**, 186.

Law, L. W. (1952), *Texas Rept. Biol. Med.* **10**, 571.

Lawley, P. D. (1957), *Biochim. Biophys. Acta* **26**, 450.

Lawley, P. D. (1962), in "The Molecular Basis of Neoplasia," p. 123, Austin, Texas, University of Texas Press.

Lawley, P. D. (1966), *Progr. Nucleic Acid Res. Mol. Biol.* **5**, 89.

Lawley, P. and P. Brookes (1965), *Nature* **206**, 480.

Lederberg, J. and E. L. Tatum (1946), *Nature* **158**, 558.

Lee, Y. P. (1957), *J. Biol. Chem.* **227**, 987.

Lee-Huang, S. and L. F. Cavalieri (1963), *Proc. Natl. Acad. Sci. U. S.* **50**, 1116.

Le Gal, M. L., Y. le Gal, J. Roche, and J. Hedegaard (1967), *Biochem. Biophys. Res. Commun.* **27**, 618.

Legier, J. F. (1964), *Cancer* **17**, 730.

Lehman, I. R. (1959), *Ann. N. Y. Acad. Sci.* **81**, 745.

Lehman, I. R., M. J. Bessman, E. S. Simms, and A. Kornberg (1958), *J. Biol. Chem.* **233**, 163.

Lehman, I. R. and E. A. Pratt (1960), *J. Biol. Chem.* **235**, 3254.

Lehman, I. R. and Ch. C. Richardson (1964), *J. Biol. Chem.* **239**, 233.

Lehnert, S. M. and S. Okada (1963), *Nature* **199**, 1108.

Lehnert, S. M. and S. Okada (1964), *Intern. J. Radiation Biol.* **8**, 75.

Leloir, L. F. (1951), *Arch. Biochem.* **33**, 186.

Leloir, L. F. and C. E. Cardini (1957), *J. Am. Chem. Soc.* **79**, 6340.

Leloir, L. F. and C. E. Cardini (1960), in "The Enzymes" (P. D. Boyer, H. Lardy, and K. Myrbäck, editors), p. 39, New York, Academic Press.

Leon, S. A. and T. D. Brock (1967), *J. Mol. Biol.* **24**, 391.

LePage, G. A. (1960), *Cancer Res.* **20**, 403.

LePage, G. A. and M. Jones (1961), *Cancer Res.* **21**, 1590.

LePage, G. A., I. Junga, and B. Bowman (1964), *Cancer Res.* **24**, 835.

LePage, G. A. and W. W. Umbreit (1943), *J. Biol. Chem.* **148**, 255.

Lerman, L. S. (1961), *J. Mol. Biol.* **3**, 18.

Lerman, L. S. (1963), *Proc. Natl. Acad. Sci. U. S.* **49**, 94.

Lett, J. T., K. A. Stacey, and P. Alexander (1961), *Radiation Res.* **14**, 349.

Leuchtenberger, C., F. Schrader, D. R. Weir, and D. P. Gentile (1953), *Chromosoma* 6, 61.

Levenberg, B. and J. M. Buchanan (1957), *J. Biol. Chem.* 224, 1019.

Levenberg, B., S. C. Hartman, and J. M. Buchanan (1956), *J. Biol. Chem.* 220, 379.

Levenberg, B., I. Melnick, and J. M. Buchanan (1957), *J. Biol. Chem.* 225, 163.

Levene, P. A. (1938), *J. Biol. Chem.* 126, 63.

Levene, P. A. and L. W. Bass (1931), "Nucleic Acids," New York, The Chemical Catalog Company, Inc., Book Department.

Levene, P. A. and S. A. Harris (1933), *J. Biol. Chem.* 101, 419.

Levene, P. A. and W. A. Jacobs (1909a), *Ber. Ges. Deut. Chem.* 42, 335.

Levene, P. A. and W. A. Jacobs (1909b), *Ber. Ges. Deut. Chem.* 42, 2102.

Levene, P. A., L. A. Mikeska, and T. Mori (1930), *J. Biol. Chem.* 85, 785.

Levey, S., T. Woods, and W. E. Abbott (1963), *Metabolism* 12, 148.

Levinthal, C. and H. R. Crane (1956), *Proc. Natl. Acad. Sci. U. S.* 42, 436.

Levinthal, C., A. Keynan, and A. Higa (1962), *Proc. Natl. Acad. Sci. U. S.* 48, 1631.

Levis, A.G., G. A. Danieli, and E. Piccinni (1965), *Nature* 207, 608.

Lewin, S. (1964), *J. Chem. Soc.*, 792.

Liao, S. (1965), *J. Biol. Chem.* 240, 1236.

Lieberman, J. (1956a), *J. Biol. Chem.* 222, 765.

Lieberman, J. (1956b), *J. Biol. Chem.* 223, 327.

Lieberman, I., R. Abrams, N. Hunt, and P. Ove (1963), *J. Biol. Chem.* 238, 3955.

Lieberman, I., R. Abrams, and P. Ove (1963), *J. Biol. Chem.* 238, 2141.

Lieberman, I. and A. Kornberg (1954), *J. Biol. Chem.* 207, 911.

Lieberman, I., A. Kornberg, and E. S. Simms (1955), *J. Biol. Chem.* 215, 403.

Lin, S. C. and D. M. Greenberg (1954), *J. Gen. Physiol.* 38, 181.

Lindner, A. (1959), *Cancer Res.* 19, 189.

Lipmann, F., W. C. Hülsmann, G. Hartmann, H. G. Boman, and G. Acs (1959), *J. Cellular Comp. Physiol.* 54, Suppl. 1, 75.

Lipsett, M. N. (1965), *J. Biol. Chem.* 240, 3975.

Lipton, A. and A. Weissbach (1966), *Biochem. Biophys. Res. Commun.* 23, 436.

Lis, A. W. and F. W. Allen (1961), *Biochim. Biophys. Acta* 49, 190.

Lis, A. W. and F. W. Allen (1962), *Biochim. Biophys. Acta* 61, 250.

Lis, A. W. and F. W. Allen (1962), *Biochim. Biophys. Acta* 61, 799.

Lis, A. W. and E. W. Lis (1963), *Federation Proc.* 23, 470.

Litman, R. M. and A. B. Pardee (1956), *Nature* 178, 529.

Litman, R. M. and W. Szybalski (1963), *Biochem. Biophys. Res. Commun.* 10, 473.

Litt, M., J. Marmur, H. Ephrussi-Taylor, and P. Doty (1958), *Proc. Natl. Acad. Sci. U. S.* 44, 144.

Littau, V. C., C. J. Burdick, V. G. Allfrey, and A. E. Mirsky (1965), *Proc. Natl. Acad. Sci. U. S.* 54, 1204.

Littlefield, J. W. and D. B. Dunn (1958), *Biochem. J.* 70, 642.

Littlefield, J. W., K. B. Margeson, and A. P. McCovern (1962), *Biochim. Biophys. Acta* 55, 783.

Lloyd, P. H. and A. R. Peacocke (1965), *Biochim. Biophys. Acta* 95, 522.

Lochmann, E. R., D. Weinblum, and A. Wacker (1964), *Biophysik* 1, 396.

Loeb, T. and N. D. Zinder (1961), *Proc. Natl. Acad. Sci. U. S.* 47, 282.

Loftfield, R. B. and E. A. Eigner (1965), *J. Biol. Chem.* 240, PC 1482.

Lohmann, K. (1929), *Naturwissenschaften* 17, 624.

Lohmann, K. (1935), *Biochem. Z.* 282, 109.

Lohrmann, R. and H. G. Khorana (1966), *J. Am. Chem. Soc.* 88, 829.

Loring, H. S. and J. G. Pierce (1944), *J. Biol. Chem.* 153, 61.

Lorkiewicz, Z. and W. Szybalski (1960), *Biochem. Biophys. Res. Commun.* 2, 413.

Love, S. H. and B. Levenberg (1959), *Biochim. Biophys. Acta* 35, 367.

Lowenstein, J. M. and P. P. Cohen (1956), *J. Biol. Chem.* 220, 57.

Lowy, B. A., J. A. Davoll, and G. B. Brown (1952), *J. Biol. Chem.* 197, 591.

Lowy, B. A., M. K. Williams, and I. M. London (1961), *J. Biol. Chem.* 236, 1439.

Luck, D. J. L. and E. Reich (1964), *Proc. Natl. Acad. Sci. U. S.* 52, 931.

Lukens, L. N. and J. M. Buchanan (1959), *J. Biol. Chem.* 234, 1799.

Lukens, L. N. and K. A. Herrington (1957), *Biochim. Biophys. Acta* 24, 432.

Luria, S. E. (1959), *Brookhaven Symp. Biol.* 12, 95.

Luttges, M., T. Johnson, C. Buck, J. Holland, and J. McGaugh (1966), *Science* 151, 834.

Lwoff, A. (1953), *Bacteriol. Rev.* 17, 269.

Lythgoe, B. and A. R. Todd (1944), *J. Chem. Soc.*, 592.

Maassab, H. F. (1959), *Proc. Natl. Acad. Sci. U. S.* 45, 877.

Madison, J. T., G. A. Everett, and H. Kung (1966), *Science* 153, 531.

Madison, J. T. and R. W. Holley (1965), *Biochem. Biophys. Res. Commun.* 18, 153.

Magasanik, B. (1958), in "The Chemical Basis of Development" (W. D. McElroy and B. Glass, editors), p. 485, Baltimore, Md., The Johns Hopkins Press.

Magasanik, B., H. S. Moyed, and L. B. Gehring (1957), *J. Biol. Chem.* 226, 339.

Magee, P. N. (1966), in "Molekulare Biologie des malignen Wachstums," p. 79, Berlin-Heidelberg-New York, Springer-Verlag.

Magee, P. N. and J. M. Barnes (1956), *Brit. J. Cancer* 10, 114.

Magee, P. N. and E. Farber (1962), *Biochem. J.* 83, 114.

Mager, J. and B. Magasanik (1960), *J. Biol. Chem.* 235, 1474.

Maley, F. (1958), *Federation Proc.* 17, 267.

Maley, F. (1960), *Proc. Natl. Acad. Sci. U. S.* 46, 632.

Maley, F. (1962), *Arch. Biochem.* 96, 550.

Maley, F. and G. F. Maley (1959), *Biochim. Biophys. Acta* 31, 577.

Maley, F. and G. F. Maley (1960), *J. Biol. Chem.* 235, 2968.

Maley, F. and G. F. Maley (1961), *Biochim. Biophys. Acta* 47, 181.

Maley, G. F., D. U. Guarino, and F. Maley (1967), *J. Biol. Chem.* 242, 3517.

Maley, G. F. and F. Maley (1960), *J. Biol. Chem.* 235, 2964.

Maley, G. F. and F. Maley (1961), *Biochem. Biophys. Res. Commun.* 5, 439.

Maley, G. F. and F. Maley (1963), *Arch. Biochem.* 101, 342.

Maley, G. F. and F. Maley (1964), *J. Biol. Chem.* 239, 1168.

Maley, G. F. and F. Maley (1966a), *J. Biol. Chem.* 241, 2176.

Maley, G. F. and F. Maley (1966b), *Federation Proc.* 25, 279.

Maller, R. K. and C. Heidelberger (1957), *Cancer Res.* 17, 284.

Mandel, H. G. (1959), *Pharmacol. Rev.* 11, 743.

Mandel, H. G., R. Markham, and R. E. F. Matthews (1957), *Biochim. Biophys. Acta* 24, 205.

Mandel, L. R. and E. Borek (1961), *Biochem. Biophys. Res. Commun.* 6, 138.

Mandel, L. R., P. R. Srinivasa, and E. Borek (1966), *Nature* 209, 586.

Manson, L. A. (1960), *J. Biol. Chem.* 235, 2955.

Mansour, A. M. and M. C. Niu (1965), *Proc. Natl. Acad. Sci. U. S.* **53**, 764.

Marcker, K. and F. Sanger (1964), *J. Mol. Biol.* **8**, 835.

Markham, R. and J. D. Smith (1952a), *Biochem. J.* **52**, 558.

Markham, R. and J. D. Smith (1952b), *Biochem. J.* **52**, 565.

Markowitz, A. (1961), *Biochem. Biophys. Res. Commun.* **6**, 250.

Marmur, J. and L. Grossman (1961), *Proc. Natl. Acad. Sci. U. S.* **47**, 778.

Marmur, J. and D. Lane (1960), *Proc. Natl. Acad. Sci. U. S.* **46**, 453.

Marmur, J., C. L. Schildkraut, and P. Doty (1961), *J. Chim. Phys.* **58**, 945.

Marrian, D. H. (1954), *Biochim. Biophys. Acta* **13**, 278.

Marroquin, R. F. and E. Farber (1962), *Biochim. Biophys. Acta* **55**, 403.

Marshall, J. A. (1966), cited by S. W. Brown, *Science* **151**, 417.

Marshall, M. and P. P. Cohen (1961), *J. Biol. Chem.* **236**, 718.

Marshall, M., R. L. Metzenberg, and P. P. Cohen (1958), *J. Biol. Chem.* **233**, 102.

Marshall, M., R. L. Metzenberg, and P. P. Cohen (1961), *J. Biol. Chem.* **236**, 2229.

Marvin, D. A. and H. Schaller (1966), *J. Mol. Biol.* **15**, 1.

Massie, H. R. and B. H. Zimm (1965), *Proc. Natl. Acad. Sci. U. S.* **54**, 1636.

Mathews, C. K. and S. S. Cohen (1963a), *J. Biol. Chem.* **238**, 367.

Mathews, C. K. and S. S. Cohen (1963b), *J. Biol. Chem.* **238**, PC 853.

Mathews, C. K. and F. M. Huennekens (1963), *J. Biol. Chem.* **238**, 3436.

Mathieson, A. R. and S. Matty (1957), *J. Polymer Sci.* **23**, 747.

McCarthy, B. J. and E. T. Bolton (1963), *Proc. Natl. Acad. Sci. U. S.* **50**, 156.

McCarthy, B. J. and J. J. Holland (1965), *Proc. Natl. Acad. Sci. U. S.* **54**, 880.

McCollister, R. J., W. R. Gilbert, Jr., D. M. Ashton, and J. B. Wyngaarden (1964), *J. Biol. Chem.* **239**, 1560.

McCollister, R. J., W. R. Gilbert, Jr., and J. B. Wyngaarden (1962), *J. Clin. Invest.* **41**, 1383.

McConnell, J. V., A. L. Jacobson, and D. P. Kimble (1959), *J. Comp. Physiol. Psychol.* **51**, 1.

McCorkindale, J. and N. L. Edson (1954), *Biochem. J.* **57**, 518.

McCully, K. S. and G. L. Cantoni (1961), *Biochim. Biophys. Acta* **51**, 190.

Melnick, I. and J. M. Buchanan (1957), *J. Biol. Chem.* **225**, 157.

Mendicino, J. (1962), *J. Biol. Chem.* **237**, 165.

Mennigmann, H. -D. and W. Szybalski (1962), *Biochem. Biophys. Res. Commun.* **9**, 398.

Merker, P. C., C. Reyes, and P. Anido (1962), *Cancer Res.* **22**, 1163.

Meselson, M. (1964), *J. Mol. Biol.* **9**, 734.

Meselson, M. and F. W. Stahl (1958), *Proc. Natl. Acad. Sci. U. S.* **44**, 671.

Metzenberg, R. L., M. Marshall, and P. P. Cohen (1958), *J. Biol. Chem.* **233**, 1560.

Metzenberg, R. L., M. Marshall, W. K. Paik, and P. P. Cohen (1961), *J. Biol. Chem.* **236**, 162.

Metzner, H. (1952), *Biol. Zbl.* **71**, 257.

Michelson, A. M., W. Drell, and H. K. Mitchell (1951), *Proc. Natl. Acad. Sci. U. S.* **37**, 396.

Michelson, A. M. and A. R. Todd (1955), *J. Chem. Soc.*, 816.

Miescher, F. (1871a), Hoppe-Seyler's Medicinisch-chemische Untersuchungen 441.

Miescher, F. (1871b), Hoppe-Seyler's Medicinisch-chemische Untersuchungen 502.

Miescher, F. (1874), *Verhandl. Naturforsch. Ges. Basel* **6**, 138.

Miescher, F. (1897), "Die histochemischen und physiologischen Arbeiten. Gesammelt und herausgegeben von seinen Freunden," Leipzig, F. C. W. Vogel Verlag, 2 vols.

Mil, S. and S. Ochoa (1957), *Biochim. Biophys. Acta* **26**, 445.

Miller, A. and H. Waelsch (1957), *J. Biol. Chem.* **228**, 397.

Miller, E. C. (1951), *Cancer Res.* **11**, 100.

Miller, E. C. and J. A. Miller (1947), *Cancer Res.* **7**, 468.

Miller, E. C. and J. A. Miller (1953), *Advan. Cancer Res.* **1**, 339.

Miller, J. A. and E. C. Miller (1961), *Can. Cancer Conf.* **4**, 57.

Miller, R. W. and J. M. Buchanan (1962a), *J. Biol. Chem.* **237**, 485.

Miller, R. W. and J. M. Buchanan (1962b), *J. Biol. Chem.* **237**, 491.

Miller, R. W., L. N. Lukens, and J. M. Buchanan (1957), *J. Am. Chem. Soc.* **79**, 1513.

Miller, R. W., L. N. Lukens, and J. M. Buchanan (1959), *J. Biol. Chem.* **234**, 1806.

Misra, D. K., S. R. Humphreys, M. Friedkin, A. Goldin, and E. J. Crawford (1961), *Nature* **189**, 39.

Mitchell, H. K. and M. B. Houlahan (1947), *Federation Proc.* **6**, 506.

Mitchell, H. K., M. B. Houlahan, and J. F. Nyc (1948), *J. Biol. Chem.* **172**, 525.

Mitra, S., M. D. Enger, and P. Kaesberg (1963), *Proc. Natl. Acad. Sci. U. S.* **50**, 68.

Mittermayer, C., R. Braun, and H. P. Rusch (1964), *Biochim. Biophys. Acta* **91**, 399.

Mittermayer, C., R. Braun, and H. P. Rusch (1966), *Biochim. Biophys. Acta* **114**, 536.

Miura, K. -I., I. Kimura, and N. Suzuki (1966), *Virology* **28**, 571.

Mohberg, J. and H. P. Rusch (1965), personal communication.

Mokrasch, L. C. and S. Grisolia (1959), *Biochim. Biophys. Acta* **34**, 165.

Monier, R., S. Naono, D. Hayes, F. Hayes, and F. Gros (1962), *J. Mol. Biol.* **5**, 311.

Monod, J. (1962), in "Basic Problems in Neoplastic Disease" (A. Gellhorn and E. Hirschberg, editors), p. 218, New York and London, Columbia University Press.

Monod, J., J. P. Changeux, and F. Jacob (1963), *J. Mol. Biol.* **6**, 306.

Montgomery, J. A., F. M. Schabel, Jr., and H. E. Skipper (1962), *Cancer Res.* **22**, 504.

Montgomery, J. A. and H. J. Thomas (1962), *Advan. Carbohydrate Chem.* **17**, 301.

Moore, E. C. and S. S. Cohen (1967), *J. Biol. Chem.* **242**, 2116.

Moore, E. C. and R. B. Hurlbert (1961), *Cancer Res.* **21**, 257.

Moore, E. C. and R. B. Hurlbert (1966), *J. Biol. Chem.* **241**, 4802.

Moore, E. C. and G. A. LePage (1958), *Cancer Res.* **18**, 1075.

Moore, E. C., P. Reichard, and L. Thelander (1964a), *J. Biol. Chem.* **239**, 3445.

Moore, E. C., P. Reichard, and L. Thelander (1964b), *J. Biol. Chem.* **239**, 3453.

Morishima, A., M. M. Grumbach, and J. H. Taylor (1962), *Proc. Natl. Acad. Sci. U. S.*

Morris, N. R. and G. A. Fischer (1960), *Biochim. Biophys. Acta* **42**, 183.

Morris, N. R. and G. A. Fischer (1963), *Biochim. Biophys. Acta* **68**, 84.

Morris, N. R., P. Reichard, and G. A. Fischer (1963), *Biochim. Biophys. Acta* **68**, 93.

Mosley, W. H. and R. Caputto (1958), *J. Am. Chem. Soc.* **80**, 4746.

Moyed, H. S. and B. Magasanik (1957), *J. Biol. Chem.* **226**, 351.

Mueller, G. C. and E. Stubblefield (1962), *Proc. Am. Assoc. Cancer Res.* 3, No. 4, 347.

Mueller, G. C. (1965), in "Mechanisms of Hormone Action" (P. Karlson, editor), p. 228, Stuttgart, Georg Thieme Verlag.

Mueller, G. C. and K. Kajiwara (1966), *Biochim. Biophys. Acta* **114**, 108.

Mukherjee, K. L. and C. Heidelberger (1962), *Cancer Res.* **22**, 815.

Müller, A. (1962), *Intern. J. Radiation Biol.* **5**, 199.

Müller, A., (1963), *Intern. J. Radiation Biol.* **6**, 137.

Müller, A. (1967), *Progr. Biophys. Mol. Biol.* **17** (in press).

Müller, A., W. Köhnlein, and K. G. Zimmer (1963), *J. Mol. Biol.* **7**, 92.

Müller, W. (1962), *Naturwissenschaften* **49**, 156.

Müller, W., H. Bujard, and E. Harbers (1963), *Ann. Chem.* **668**, 129.

Müller, W. and I. Emme (1965), *Z. Naturforsch.* 20b, 835.

Müller, W. and H. -C. Spatz (1965), *Z. Naturforsch.* 20b, 842.

Munyon, W. and N. P. Salzman (1962), *Virology* **18**, 95.

Murata, T. and T. Akazawa (1964), *Biochem. Biophys. Res. Commun.* **16**, 6.

Murray, K. (1964), *Biochemistry* **3**, 10.

Myers, D. K. (1962), *Can. J. Biochem.* **40**, 619.

Nakada, D. and B. Magasanik (1964), *J. Mol. Biol.* **8**, 105.

Nakamoto, T. and S. B. Weiss (1962), *Proc. Natl. Acad. Sci. U. S.* **48**, 880.

Nass, G. and F. C. Neidhardt (1967), *Biochim. Biophys. Acta* **134**, 347.

Nass, S. and M. M. K. Nass (1964), *J. Natl. Cancer Inst.* **33**, 777.

Nathanson, R. B. (1962), *Science* **135**, 916.

Nazario, M. and J. L. Reissig (1964), *Biochem. Biophys. Res. Commun.* **16**, 42.

Nemeth, A. M., C. S. Russell, and D. Shemin (1957), *J. Biol. Chem.* **229**, 415.

Neubert, D. (1966a), *Arch. Exptl. Pathol. Pharmakol.* **253**, 152.

Neubert, D. (1966b), in "Molekulare Biologie des malignen Wachstums," p. 69, Berlin-Heidelberg-New York, Springer-Verlag.

Neubert, D., H. Helge, and R. Bass (1965), *Arch. Exptl. Pathol. Pharmakol.* **252**, 258.

Neubert, D., H. Helge, and H. -J. Merker (1965), *Biochem. Z.* **343**, 44.

Neufeld, E. F. and S. M. Iloes (1962), *Federation Proc.* **21**, 80.

Neumann, J. and M. E. Jones (1964), *Arch. Biochem.* **104**, 438.

Newton, B. A. (1965), *Ann. Rev. Microbiol.* **19**, 209.

Nichols, J. L. and B. G. Lane (1966), *Biochim. Biophys. Acta* **119**, 649.

Nierlich, D. P. and B. Magasanik (1961), *J. Biol. Chem.* **236**, PC 32.

Nierlich, D. P. and B. Magasanik (1965a), *J. Biol. Chem.* **240**, 358.

Nierlich, D. P. and B. Magasanik (1965b), *J. Biol. Chem.* **240**, 366.

Nierlich, D. P. and E. McFall (1963), *Biochim. Biophys. Acta* **76**, 469.

Nikaido, H. and H. Jokura (1961), *Biochem. Biophys. Res. Commun.* **6**, 304.

Nirenberg, M. W. and P. Leder (1964), *Science* **145**, 1399.

Nirenberg, M., P. Leder, M. Bernfield, R. Brimacombe, J. Trupin, F. Rottman, and C. O'Neal (1965), *Proc. Natl. Acad. Sci. U. S.* **53**, 1161.

Nirenberg, M. W. and J. H. Matthaei (1961), *Proc. Natl. Acad. Sci. U.S.* **47**, 1588.

Nishimura, S., D. S. Jones, E. Ohtsuka, H. Hayatsu, T. M. Jacob, and H. G. Khorana (1965), *J. Mol. Biol.* **13**, 283.

Niu, M. C. (1963), *Federation Proc.* **22**, 354.

Niu, M. C., C. C. Cordova, L. C. Niu, and C. L. Radbill (1962), *Proc. Natl. Acad. Sci. U. S.* **48**, 1964.

Noll, H. (1965), in "Developmental and Metabolic Control Mechanisms and Neoplasia," p. 67, Baltimore, Md., The Williams and Wilkins Company.

Noll, H., T. Staehelin, and F. O. Wettstein (1963), *Nature* **198**, 632.

Nomura, M., B. D. Hall, and S. Spiegelman (1960), *J. Mol. Biol.* **2**, 306.

Norton, S. J., J. M. Ravel, and W. Shive (1962), *Biochim. Biophys. Acta* **55**, 222.

Notani, G. W., D. L. Engelhardt, W. Konigsberg, and N. D. Zinder (1965), *J. Mol. Biol.* **12**, 439.

Novelli, G. D., T. Kameyama, and J. M. Eisenstadt (1961), *J. Cell. Comp. Physiol.* **58**, Suppl. 1, 225.

Nowell, P. C. and D. A. Hungerford (1961), *J. Natl. Cancer Inst.* **27**, 1031.

Nygaard, O. F. and S. Güttes (1962); *Intern. J. Radiation Biol.* **5**, 33.

Nygaard, O. F., S. Güttes, and H. P. Rusch (1960), *Biochim. Biophys. Acta* **38**, 298.

Nygaard, O. F. and R. L. Potter (1960), *Radiation Res.* **12**, 120.

O'Brien, J. S. (1962), *Cancer Res.* **22**, 267.

Ochoa, S. and L. A. Heppel (1957), in "The Chemical Basis of Heredity" (W. D. McElroy and B. Glass, editors), p. 615, Baltimore, Md., The Johns Hopkins Press.

Ohtaka, Y. and S. Spiegelman (1963), *Science* **142**, 493.

Okada, Y., E. Terzaghi, G. Streisinger, J. Emrich, M. Inouye, and A. Tsugita (1966), *Proc. Natl. Acad. Sci. U. S.* **56**, 1692.

Okazaki, R. and A. Kornberg (1964b), *J. Biol. Chem.* **239**, 269.

Okazaki, R. and A. Kornberg (1964c), *J. Biol. Chem.* **239**, 275.

Okazaki, R., T. Okazaki, and K. Sakabe (1966), *Biochem. Biophys. Res. Commun.* **22**, 611.

Okazaki, R., T. Okazaki, J. L. Strominger, and A. M. Michelson (1962), *J. Biol. Chem.* **237**, 3014.

Okazaki, T. and A. Kornberg (1964a), *J. Biol. Chem.* **239**, 259.

Okubo, S., M. Stodolsky, K. Bott, and B. Strauss (1963), *Proc. Natl. Acad. Sci. U. S.* **50**, 679.

O'Neal, M. A., V. Ward. and A. C. Griffin (1962), *Proc. Am. Assoc. Cancer Res.* **3**, No. 4, 349.

Opara-Kubinska, Z. Z. Lorkiewicz, and W. Szybalski (1961), *Biochem. Biophys. Res. Commun.* **4**, 288.

Ord, M. G. and L. A. Stocken (1958), *Biochim. Biophys. Acta* **29**, 201.

Ord, M. G., and L. A. Stocken (1960), *Biochim. Biophys. Acta* **37**, 352.

Ord, M. G., and L. A. Stocken (1966), *Biochem. J.* **98**, 888.

Orloff, J. and J. S. Handler (1961), *Biochem. Biophys. Res. Commun.* **5**, 63.

Ormedor, M. G. and P. Alexander (1963), *Radiation Res.* **18**, 495.

Osborn, M. J. and F. M. Huennekens (1958), *J. Biol. Chem.* **233**, 969.

Otaka, E., S. Osawa, and Y. Oota (1961), *J. Mol. Biol.* **3**, 693.

Oth, A. (1955), *Bull. Soc. Chim. Belg* **64**, 484.

Overgaard-Hansen, K. (1964), *Biochim. Biophys. Acta* **80**, 504.

Overgaard-Hansen, K. and H. Klenow (1961), *Proc. Natl. Acad. Sci. U. S.* **47**, 680.

Paege, L. M. and F. Schlenk (1952), *Arch. Biochem.* **40**, 42.

Paigen, K. (1962), *Cancer Res.* **22**, 1290.

Painter, R. B. (1960), *Radiation Res.* **13**, 726.

Pardee, A. B. (1964), *Natl. Cancer Inst. Monograph* **14**, 7.

Pardee, A. B. and L. S. Prestidge (1961), *Biochim. Biophys. Acta* **49**, 77.

Park, J. T. and J. L. Strominger (1957), *Science* **125**, 99.

Passonneau, J. V. and J. R. Totter (1955), *Radiation Res.* **2**, 304.

Pasternak, C. A., N. A. Fischer, and R. E. Handschumacher (1961), *Cancer Res.* **21**, 110.

Pasternak, C. A. and R. E. Handschumacher (1959), *J. Biol. Chem.* **234**, 2992.

Paterson, A. R. P. (1959), *Can. J. Biochem. Physiol.* **37**, 1011.

Paterson, A. R. P. (1960), *Can. J. Biochem. Physiol.* **38**, 1129.

Paul, J. and R. S. Gilmour (1966), *J. Mol. Biol.* **16**, 242.

Pelling, C. (1959), *Nature* **184**, 655.

Penman, S., K. Scherrer, Y. Becker and J. E. Darnell (1963), *Proc. Natl. Acad. Sci. U. S.* **49**, 654.

Perry, R. P. (1962), *Proc. Natl. Acad. Sci. U. S.* **48**, 2179.

Perry, R. P. (1967), *Progr. Nucleic Acid. Res. Mol. Biol.* **6**, 219.

Perry, R. P., R. R. Srinivasan, and D. Kelley (1964), *Science* **145**, 504.

Perutz, M. F. (1962), "Proteins and Nucleic Acids—Structure and Function," Amsterdam-New York, Elsevier.

Peterkofsky, A., C. Jessensky, A. Bank, and A. Mehler (1964), *J. Biol. Chem.* **239**, 2918.

Petermann, M. L. and M. G. Hamilton (1955), *J. Biophys. Biochem. Cytol.* **1**, 469.

Peters, J. M. and D. M. Greenberg (1958), *J. Am. Chem. Soc.* **80**, 6679.

Pfau, C. J. and J. F. McCrea (1962), *Biochim. Biophys. Acta* **55**, 271.

Pfefferkorn, E. and H. Amos (1958), *Virology* **6**, 299.

Phares, E. F. (1951), *Arch. Biochem.* **33**, 173.

Philips, F. S. (1950), *Pharmacol. Rev.* **2**, 281.

Phillips, D. M. P. (1962), *Progr. Biophys. Biophys. Chem.* **12**, 211.

Phillips, D. M. P. (1963), *Biochem. J.* **87**, 258.

Phillips, R. A. and L. J. Tolmach (1966), *Radiation Res.* **29**, 413.

Pietrzykowska, I. and D. Shugar (1966), *Biochem. Biophys. Res. Commun.* **25**, 567.

Pirie, N. W. (1966), in "Principles of Biomolecular Organization" (G. E. W. Wolstenholme, editor), p. 136, London, J. & A. Churchill, Ltd.

Pitot, H. C. (1966a), in "Molekulare Biologie des malignen Wachstums," p. 43, Berlin-Heidelberg-New York, Springer-Verlag.

Pitot, H. C. (1966b), *Ann. Rev.* **35**, 335.

Pitot, H. C. and C. Heidelberger (1963), *Cancer Res.* **23**, 1694.

Pitot, H. C. and C. Peraino (1964), *Biochem. Clin.* **3**, 139.

Pitot, H. C., C. Peraino, C. Lamar, and A. L. Kenman (1965), *Proc. Natl. Acad. Sci. U. S.* **54**, 845.

Pizer, L. I. and S. S. Cohen (1962), *J. Biol. Chem.* **237**, 1251.

Plaut, W. (1958), *Exptl. Cell Res., Suppl.* **6**, 69.

Plaut, W. and D. Mazia (1956), *J. Biophys. Biochem. Cytol.* **2**, 573.

Plaut, W. and A. Sagan (1958), *J. Biophys. Biochem. Cytol.* **4**, 843.

Plotnikoff, N. (1966), *Science* **151**, 703.

Pogo, A. O., B. G. T. Pogo, V. C. Littau, V. G. Allfrey, A. E. Mirsky, and M. G. Hamilton (1962), *Biochim. Biophys. Acta* **55**, 849.

Pogo, B. G. T., V. G. Allfrey, and A. E. Mirsky (1966), *Proc. Natl. Acad. Sci. U. S.* **55**, 805.

Pollak, J. K. and H. R. V. Arnstein (1962), *Biochim. Biophys. Acta* **55**, 798.

Pollard, E. and C. Vogler (1961), *Radiation Res.* 15, 109.

Pomales, R., S. Bieber, R. Friedman, and G. H. Hitchings (1963), *Biochim. Biophys. Acta* 72, 119.

Ponnamperuma, C., R. M. Lemmon, E. L. Bennett, and M. Calvin (1961), *Science* 134, 113.

Ponnamperuma, C. A., R. M. Lemmon, and M. Calvin (1962), *Science* 137, 605.

Pontis, H. G. (1955), *J. Biol. Chem.* 216, 195.

Pontis, H. G., A. L. James, and J. Baddiley (1960), *Biochem. J.* 75, 428.

Pontremoli, S., A. DeFlora, E. Grazi, G. Mangiarotti, A. Bonsignore, and B. L. Horecker (1961), *J. Biol. Chem.* 236, 2936.

Potter, R. L. and O. F. Nygaard (1963), *J. Biol. Chem.* 238, 2150.

Potter, V. R. (1948), *Acta Unio Intern. Contra Cancrum* 6, 301.

Potter, V. R. (1958), *Federation Proc.* 17, 691.

Potter, V. R. (1960), "Nucleic Acid Outlines," Vol. I: Structure and Metabolism, Minneapolis, Minn., Burgess Publishing Co.

Potter, V. R. (1962), in "The Molecular Basis of Neoplasia," p. 367, Austin, Texas, University of Texas Press.

Potter, V. R. (1964), in "Cellular Control Mechanisms and Cancer" (P. Emmelot and O. Mühlbock, editors), p. 190, Amsterdam, Elsevier.

Potter, V. R., A. F. Brumm, and F. J. Bollum (1958), *Proc. Am. Assoc. Cancer Res.* 2, 336.

Potter, V. R. and C. A. Elvehjem (1936), *J. Biol. Chem.* 114, 495.

Potter, V. R., L. I. Hecht, and E. Herbert (1956), *Biochim. Biophys. Acta* 20, 439.

Potter, V. R., H. C. Pitot, A. B. McElya, and P. A. Morse (1960), *Federation Proc.* 19, 312.

Preiss, J. (1962), *Biochem. Biophys. Res. Commun.* 9, 235.

Prescott, D. M. (1959), *J. Biophys. Biochem. Cytol.* 6, 203.

Prescott, D. M. and M. A. Bender (1962), *Exptl. Cell Res.* 26, 260.

Prescott, D. M. and R. F. Kimball (1961), *Proc. Natl. Acad. Sci. U. S.* 47, 686.

Pricer, W. E., Jr., and B. L. Horecker (1960), *J. Biol. Chem.* 235, 1292.

Prusoff, W. H. (1960), *Biochim. Biophys. Acta* 39, 327.

Prusoff, W. H., W. L. Holmes, and A. D. Welch (1954), *Cancer Res.* 14, 570.

Prusoff, W. H., J. J. Jaffe, and H. Gunther (1960), *Biochem. Pharmacol.* 3, 110.

Ptashne, M. (1967a), *Proc. Natl. Acad. Sci. U. S.* 57, 306.

Ptashne, M. (1967b), *Nature* 214, 232.

Pullman, B. and A. Pullman (1961), *Nature* 189, 725.

Raacke, I. D. and J. Fiala (1964), *Proc. Natl. Acad. Sci. U. S.* 51, 323.

Rabinowitz, J. C. (1956), *J. Biol. Chem.* 218, 175.

Rabinowitz, J. C. and H. A. Barker (1956), *J. Biol. Chem.* 218, 161.

Rabinowitz, J. C. and W. E. Pricer, Jr. (1956), *J. Am. Chem. Soc.* 78, 5702.

Rabinowitz, M., and J. M. Fisher (1962), *J. Biol. Chem.* 237, 477.

Rabinowitz, M. and I. H. Goldberg (1961), *J. Biol. Chem.* 236, PC 79.

Rada, B. and V. Gregusová (1964), *Biochem. Biophys. Res. Commun.* 15, 324.

Rahman, Y. E. (1966), *Biochim. Biophys. Acta* 119, 470.

RajBhandary, U. L., S. H. Chang, A. Stuart, R. D. Faulkner, R. M. Hoskinson, and H. G. Khorana (1967), *Proc. Natl. Acad. Sci. U. S.* 57, 751.

Rajewsky, B. (1956), "Strahlendosis und Strahlenwirkung. Tafeln und Erläuterungen, Unterlagen für den Strahlenschutz," 2. Aufl., Stuttgart, Georg Thieme Verlag.

Ramasatri, B. V. and R. L. Blakley (1962), *J. Biol. Chem.* **237**, 1982.

Rappaport, I. (1959), *Nature* **184**, 1732.

Ratliff, R. L., R. H. Weaver, H. A. Lardy, and S. A. Kuby (1964), *J. Biol. Chem.* **239**, 301.

Rauen, H. M., H. Kersten, and W. Kersten (1960), *Z. Physiol. Chem.* **321**, 139.

Ravin, A. W. (1961), *Advanc. Genet.* **10**, 61.

Razell, W. E. (1961a), *J. Biol. Chem.* **240**, 3028.

Razell, W. E. (1961b), *J. Biol. Chem.* **240**, 3031.

Recondo, E. and L. F. Leloir (1961), *Biochem. Biophys. Res. Commun.* **6**, 85.

Reddi, K. K. (1961), *Science* **133**, 1367.

Reddi, K. L. (1963), *Proc. Natl. Acad. Sci. U. S.* **50**, 75.

Rees, K. R., J. I. Clifford, and J. S. Varcoe (1966), in "The Cell Nucleus—Metabolism and Radiosensitivity," p. 335, London, Taylor & Francis, Ltd.

Reich, E. and R. M. Franklin (1961), *Proc. Natl. Acad. Sci. U. S.* **47**, 1212.

Reich, E., R. M. Franklin, A. J. Shatkin, and E. L. Tatum (1961), *Science* **134**, 556.

Reich, E. and I. H. Goldberg (1964), *Progr. Nucleic Acid. Res. Mol. Biol.* **3**, 183.

Reichard, P. (1960), *Biochim. Biophys. Acta* **41**, 368.

Reichard, P. (1961), *J. Biol. Chem.* **236**, 2511.

Reichard, P. and B. Estborn (1951), *J. Biol. Chem.* **188**, 839.

Reichard, P. and U. Lagerkvist (1953), *Acta Chem. Scand.* **7**, 1207.

Reichard, P. and O. Sköld (1957), *Acta Chem. Scand* **11**, 17.

Reichard, P. and O. Sköld (1958), *Biochim. Biophys. Acta* **28**, 376.

Reichard, P., O. Sköld, and G. Klein (1959), *Nature* **183**, 939.

Reichard, P., O. Sköld, G. Klein, L. Révész, and P. -H. Magnusson (1962), *Cancer Res.* **22**, 235.

Reichman, M. E., S. A. Rice, C. A. Thomas, and P. Doty (1954), *J. Am. Chem. Soc.* **76**, 3047.

Reis, J. (1937), *Enzymologia* **2**, 183.

Reiter, H. and B. Strauss (1965), *J. Mol. Biol.* **14**, 179.

Remy, C. N. (1960), *Federation Proc.* **19**, 313.

Rendi, R. and S. Ochoa (1961), *Science* **133**, 1367.

Rendi, R. and S. Ochoa (1962), *J. Biol. Chem.* **237**, 3711.

Révész, L., U. Glass, and G. Hilding (1963), *Nature* **198**, 260.

Révész, L. and U. Norman (1960), *J. Natl. Cancer Inst.* **25**, 1041.

Reyes, P. and C. Heidelberger (1965), *Mol. Pharmacol.* **1**, 14.

Rhodes, W. C. and W. D. McElroy (1958), *J. Biol. Chem.* **233**, 1528.

Rich, A. (1958), *Biochim. Biophys. Acta* **29**, 502.

Rich, A. (1959), *Brookhaven Symp. Biol.* **12**, 17.

Rich, A., D. R. Davies, F. H. C. Crick, and J. D. Watson (1961), *J. Mol. Biol.* **3**, 71.

Rich, M. A., A. G. Perez, and M. L. Eidinoff (1962), *Cancer Res.* **22**, 3.

Richards, B. M. and N. B. Atkin (1959), *Brit. J. Cancer* **13**, 788.

Richardson, C. C., R. B. Inman, and A. Kornberg (1964), *J. Mol. Biol.* **9**, 46.

Richardson, C. C., C. L. Schildkraut, H. V. Aposhian, and A. Kornberg (1964), *J. Biol. Chem.* **239**, 222.

Risebrough, R. W., A. Tissières, and J. D. Watson (1962), *Proc. Natl. Acad. Sci. U. S.* **48**, 430.

Ritossa, F. M. and S. Spiegelman (1965), *Proc. Natl. Acad. Sci. U. S.* **53**, 737.

Rittenberg, D. and L. Ponticorvo (1962), *J. Biol. Chem.* **237**, PC 2709.

Robbins, P. W. and J. B. Hammond (1962), *J. Biol. Chem.* **237**, PC 1329.
Robbins, P. W. and B. M. Kinsey (1963), *Fed. Proc.* **22**, 229.
Robbins, P. W. and F. Lipmann (1958), *J. Biol. Chem.* **233**, 681.
Roberts, J. J. and G. P. Warwick (1966), *Intern. J. Cancer* **1**, 179.
Roberts, P. A. (1965), *Nature* **205**, 725.
Robinson, W. S., A. Pitkanen, and H. Rubin (1965), *Proc. Natl. Acad. Sci. U. S.* **54**, 137.
Roger, M. and R. D. Hotchkiss (1961), *Proc. Natl. Acad. Sci. U. S.* **47**, 653.
Rogers, H. J. and H. R. Perkins (1960), *Biochem. J.* **77**, 448.
Rogers, St. (1966), *Nature* **212**, 1220.
Rolfe, R. (1962), *J. Mol. Biol.* **4**, 22.
Roll, P. M., H. Weinfeld, and E. Carroll (1956), *J. Biol. Chem.* **220**, 455.
Roscoe, H. G. and W. L. Nelson (1964), *J. Biol. Chem.* **239**, 8.
Rose, S. M. (1958), *Sci. Am.* **199**, No. 6, 36.
Roseman, S. (1962), *Proc. Natl. Acad. Sci. U. S.* **48**, 437.
Rosenberg, F. (1965), *Proc. Natl. Acad. Sci. U. S.* **53**, 836.
Ross, W. C. J. (1958), *Ann. N. Y. Acad. Sci.* **68**, 669.
Rotem, Z., R. A. Cox, and A. Isaacs (1963), *Nature* **197**, 564.
Roth, J. S. (1957), *J. Biol. Chem.* **227**, 591.
Roth, J. S. and G. Buccino (1963), *Arch. Biochem.* **101**, 516.
Rottman, F. and P. P. Cerutti (1966), *Proc. Natl. Acad. Sci. U. S.* **55**, 960.
Rottman, F. and A. J. Guarino (1964), *Biochim. Biophys. Acta* **89**, 465.
Roush, A. H. and R. E. Betz (1958), *J. Biol. Chem.* **233**, 261.
Roush, A. H. and T. R. Shieh (1963), *Federation Proc.* **22**, 292.
Rubin, H. (1964), *J. Cellular Comp. Physiol.* **64**, Suppl. 1, 173.
Rubinstein, I., C. A. Thomas, and A. D. Hershey (1961), *Proc. Natl. Acad. Sci. U.S.* **47**, 1113.
Rudkin, G. T. and P. S. Woods (1959), *Proc. Natl. Acad. Sci. U. S.* **45**, 997.
Rundles, R. W., G. B. Elion, and G. H. Hitchings (1966), *Bull. Rheumatic Diseases* **16**, 400.
Rushizky, G. W. and H. A. Sober (1963), *J. Biol. Chem.* **238**, 371.
Rutman, R. J., A. Cantarow, and K. E. Paschkis (1954), *Cancer Res.* **14**, 119.
Rutman, R. J., W. J. Steele, and C. C. Price (1961), *Biochem. Biophys. Res. Commun.* **4**, 278.

Sable, H. Z. (1952), *Biochim. Biophys. Acta* **8**, 687.
Sachsenmaier, W. (1962), *Oesterr. Chemiker-Z.* **63**, 295.
Sachsenmaier, W. (1966), in "Probleme der biologischen Reduplikation," p. 139, Berlin-Heidelberg-New York, Springer-Verlag.
Sacks, J. (1955), *Biochim. Biophys. Acta* **16**, 436.
Sager, R. and M. R. Ishida (1963), *Proc. Natl. Acad. Sci. U.S.* **50**, 725.
Salas, M., M. A. Smith, W. M. Stanley, Jr., A. J. Wahba, and S. Ochoa (1965), *J. Biol. Chem.* **240**, 3988.
Salser, J. S., J. D. Hutchison, and M. E. Balis (1960), *J. Biol. Chem.* **235**, 429.
Salzman, N. P., H. Eagle, and E. D. Sebring (1958), *J. Biol. Chem.* **230**, 1001.
Sandberg, A. A. (1962), *Blood* **20**, 393.
Sanders, F. K. (1960), *Nature* **185**, 802.
Sandritter, W. (1958), in "Handbuch der Histochemie," Vol. I/1, p. 220, Stuttgart, Fischer Verlag.
Sandritter, W. (1961), *Deut. Med. Wochschr.* **86**, 2177.

Sandritter, W., K. Jobst, L. Rakow, and K. Bosselmann (1965), *Histochemie* 4, 420.

Sandritter, W., G. Kiefer, and G. Schlüter (1968), in press.

Sandritter, W., D. Müller, and O. Gensecke (1960), *Acta Histochem.* 10, 139.

Sarabhai, A. S., A. O. Stretton, S. Brenner, and A. Bollé (1964), *Nature* 201, 13.

Sarin, P. S. and P. C. Zamecnik (1965), *Biochem. Biophys. Res. Commun.* 20, 400.

Sartorelli, A. C. and B. A. Booth (1962), *Biochim. Biophys. Acta* 55, 214.

Sartorelli, A. C. and B. A. Booth (1965), *Cancer Res.* 25, 1393.

Sartorelli, A. C. and G. A. LePage (1958a), *Cancer Res.* 18, 457.

Sartorelli, A. C. and G. A. LePage (1958b), *Cancer Res.* 18, 1329.

Sartorelli, A. C., G. A. LePage, and E. C. Moore (1958), *Cancer Res.* 18, 1232.

Sato, T., Y. Kyogoku, S. Higuchi, Y. Mitsui, Y. Iitaka, M. Tsuboi, and K. Miura (1966), *J. Mol. Biol.* 16, 180.

Scarano, E. (1953), *Nature* 172, 951.

Scarano, E. (1960), *J. Biol. Chem.* 235, 706.

Scarano, E., L. Bonaduce, and B. de Petrocellis (1960), *J. Biol. Chem.* 235, 3556.

Scarano, E., L. Bonaduce, and B. de Petrocellis (1962), *J. Biol. Chem.* 237, 3742.

Scarano, E., G. Geraci, A. Polzella, and E. Campanile (1963), *J. Biol. Chem.* 238, PC 556.

Scarano, E. and M. Talarico (1961), *Biochim. Biophys. Acta* 51, 173.

Schachman, H. K., J. Adler, C. M. Radding, l. R. Lehman, and A. Kornberg (1960), 235, 3242.

Schaefer, P. (1964), in "The Bacteria," Vol. V: "Heredity" (I. C. Gunsalus and R. Y. Stanier, editors), p. 87, New York, Academic Press.

Scharff, M. D., A. J. Shatkin, and L. Levintow (1963), *Proc. Natl. Acad. Sci. U.S.* 50, 686.

Schildkraut, C. L., J. Marmur, and P. Doty (1961), *J. Mol. Biol.* 3, 595.

Schildkraut, C. L., C. C. Richardson, and A. Kornberg (1964), *J. Mol. Biol.* 9, 24.

Schildkraut, C. L., K. L. Wierzchowski, J. Marmur, D. M. Green, and P. Doty (1962), *Virology* 18, 43.

Schimke, R. T. (1962), *J. Biol. Chem.* 237, 459.

Schlegel, D. E. (1960), *Virology* 11, 329.

Schlenk, F. and H. von Euler (1936), *Naturwissenschaften* 24, 794.

Schmidt, C. G. (1963), *Ergebn. Inn. Med. Kinderheilk.* 20, 283.

Schmidt, C. G. (1966), *Arch. Exptl. Pathol. Pharmakol.* 253, 176.

Schmidt, G., R. Cubiles, N. Zöllner, L. Hecht, N. Strickler, K. Seraidarian, M. Seraidarian, and S. J. Thannhauser (1951), *J. Biol. Chem.* 192, 715.

Schmitz, H., V. R. Potter, R. B. Hurlbert, and D. M. White (1954), *Cancer Res.* 14, 66.

Schneider, J. H. and V. R. Potter (1958), *J. Biol. Chem.* 233, 154.

Schneider, W. C. and E. L. Kuff (1965), *Proc. Natl. Acad. Sci. U.S.* 54, 1650.

Schneider, W. C. and J. Rotherham (1961), *J. Biol. Chem.* 236, 2764.

Schoenbach, E. B., E. M. Greenspan, and J. Colsky (1950), *J. Am. Med. Assoc.* 144, 1558.

Schoenheimer, R. (1942), "The Dynamic State of Body Constituents," Cambridge, Mass., Harvard University Press.

Schoeniger, E. L., R. P. Salerne, and H. L. Friedell (1961), *Radiation Res.* 14, 499.

Scholes, G., J. F. Ward, and J. Weiss (1960), *J. Mol. Biol.* 2, 379.

Schrecker, A. W. and M. J. Urshel (1966), *Federation Proc.* 25, 782.

Schulman, M. P. (1961), in "Metabolic Pathways," Vol. II (D. M. Greenberg, editor), p. 389, New York, Academic Press.

Schulman, M. P. and J. M. Buchanan (1952), *J. Biol. Chem.* 196, 513.

Schulman, M. P., J. M. Buchanan, and C. S. Miller (1950), *Federation Proc.* 9, 225.

Schultze, B., W. Oehlert, and W. Maurer (1958), *Beitr. Pathol. Anat. Allgem. Pathol.* 120, 58.

Schuster, H. (1960), in "The Nucleic Acids," Vol. III, p. 245, New York, Academic Press.

Schwartz, H. S. (1962), *Proc. Am. Assoc. Cancer Res.* 3, No. 4, 359.

Schwarz, M. R. and W. O. Rieke (1962), *Science* 136, 152.

Schweiger, H. G. and H. J. Bremer (1961), *Biochim. Biophys. Acta* 51, 50.

Scott, J. F. and E. B. Taft (1958), *Biochim. Biophys. Acta* 28, 45.

Searashi, T. and B. Strauss (1965), *Biochem. Biophys. Res. Commun.* 20, 680.

Šebesta, K., J. Bauerová, and Z. Šormová (1961), *Biochim. Biophys. Acta* 50, 393.

Segal, S. J., O. W. Davidson, and K. Wada (1965), *Proc. Natl. Acad. Sci. U.S.* 54, 782.

Sekeris, C. E. (1966), in "The Cell Nucleus—Metabolism and Radiosensitivity," p. 67, London, Taylor & Francis, Ltd.

Sekeris, C. E., P. P. Dukes, and W. Schmid (1965), *Z. Physiol. Chem.* 341, 152.

Sekeris, C. E. and N. Lang (1964a), *Life Sciences* 3, 169.

Sekeris, C. E. and N. Lang (1964b), *Life Sciences* 3, 625.

Sekeris, C. E., N. Lang, and P. Karlson (1965), *Z. Physiol. Chem.* 341, 36.

Setlow, J. K. (1966), *Radiation Res., Suppl.* 6, 141.

Setlow, J. K. and M. E. Boling (1965), *Biochim. Biophys. Acta* 108, 259.

Setlow, R. B. and W. L. Carrier (1964), *Proc. Natl. Acad. Sci. U.S.* 51, 226.

Shapot, V. and H. C. Pitot (1966), *Biochim. Biophys. Acta* 119, 37.

Shatkin, A. J. (1965), *Proc. Natl. Acad. Sci. U.S.* 54, 1721.

Shatkin, A. J., E. Reich, R. M. Franklin, and E. L. Tatum (1962), *Biochim. Biophys. Acta* 55, 277.

Sheinin, R. (1966), *Virology* 28, 621.

Sheperdson, M. and A. B. Pardee (1960), *J. Biol. Chem.* 235, 3233.

Sherman, F. G. and H. Quastler (1960), *Exptl. Cell Res.* 19, 343.

Shipp, W. and R. Haselkorn (1964), *Proc. Natl. Acad. Sci. U.S.* 52, 401.

Shive, W., W. W. Ackermann, M. Gordon, M. E. Getzendoner, and R. E. Eakin (1947), *J. Am. Chem. Soc.* 69, 725.

Shnider, B. I., E. Frei, III, J. H. Ruohy, J. Gorman, E. J. Freireich, C. O. Brindley, Jr., and J. Clements (1960), *Cancer Res.* 20, 28.

Shooter, K. V. and J. A. V. Butler (1956), *Trans Faraday Soc.* 52, 734.

Shortman, K. (1961), *Biochim. Biophys. Acta* 51, 37.

Shortman, K. and I. R. Lehman (1964), *J. Biol. Chem.* 239, 2964.

Shugar, D. (1960), in "The Nucleic Acids," Vol. III, p. 39, New York, Academic Press.

Shuster, L. (1957), *J. Biol. Chem.* 229, 289.

Shuster, L. and A. Goldin (1958), *J. Biol. Chem.* 230, 883.

Shuster, L. and N. O. Kaplan (1953), *J. Biol. Chem.* **201**, 535.

Siebert, G. (1956), in "Biochemisches Taschenbuch" (H. M. Rauen, editor), p. 904, Berlin-Göttingen-Heidelberg, Springer-Verlag.

Siebert, G. (1963), *Exptl. Cell Res., Suppl.* 9, 389.

Siebert, G. (1966a), *Z. Klin. Chem.* 3, 13.

Siebert, G. (1966b), *Hoppe-Seyler, (Thierfelder) Handbuch der physiologisch- und pathologisch-chemischen Analyse* (K. Lang and E. Lehnartz, editors), **6B**, 1061.

Siebert, G. (1966c), *Hoppe-Seylers Thierfelder, Handbuch der physiologisch- und pathologisch-chemischen Analyse* (K. Lang and E. Lehnartz, editors), **6B**, 721.

Siebert, G. (1966d), *Hoppe-Seyler-Thierfelder, Handbuch der physiologisch- und pathologisch-chemischen Analyse* (K. Lang and E. Lehnartz, editors), **6C**, 439.

Siedler, A. J. and M. T. Holtz (1963), *J. Biol. Chem.* **238**, 697.

Signer, E. R., J. R. Beckwith, and S. Brenner (1965), *J. Mol. Biol.* **14**, 153.

Silagi, S. (1965), *Cancer Res.* **25**, 1446.

Silberman, H. R. and J. B. Wyngaarden (1961), *Biochim. Biophys. Acta* **47**, 178.

Silver, S. D. (1963), *J. Mol. Biol.* **6**, 349.

Siminovitch, C. and A. F. Graham (1956), *H. Histochem. Cytochem.* **4**, 508.

Simon, E. (1961), *Bact. Proc.*, 148.

Simpson, L., L. L. Bennett, Jr., and J. Golden (1962), *Proc. Am. Assoc. Cancer Res.* 3, No. 4, 361.

Sinsheimer, R. L. (1954), *Science* **120**, 551.

Sinsheimer, R. L. (1959), *J. Mol. Biol.* **1**, 43.

Sinsheimer, R. L. (1961), *J. Chim. Phys.* **58**, 986.

Sinsheimer, R. L., B. Starman, C. Nagler, and S. Guthrie (1962), *J. Mol. Biol.* **4**, 142.

Sirlin, J. L. (1962), *Progr. Biophys. Biophys. Chem.* **12**, 25.

Sirlin, J. L., K.-I. Kato, and K. W. Jones (1961), *Biochim. Biophys. Acta* **48**, 421.

Sirotnak, F. M. and D. J. Hutchison (1959), *Biochim. Biophys. Acta* **36**, 246.

Sjögren, H. O., I. Hellström, and G. Klein (1961), *Cancer Res.* **21**, 329.

Skipper, H. E. and F. M. Schabel, Jr. (1962), *Cancer Chemotherapy Rept.* **22**, 1.

Skoda, J., J. Kara, Z. Šormova, and F. Šorm (1959), *Biochim. Biophys. Acta* **33**, 579.

Sköld, O. (1958), *Biochim. Biophys. Acta* **29**, 651.

Sköld, O. (1960a), *J. Biol. Chem.* **235**, 3273.

Sköld, O. (1960b), *Biochim. Biophys. Acta* **44**, 1.

Sköld, O. (1963), *Biochim. Biophys. Acta* **76**, 160.

Sluyser, M. (1966), in "The Cell Nucleus—Metabolism and Radiosensitivity," p. 75, London, Taylor and Francis, Ltd.

Sluyser, M., P. J. Thung, and P. Emmelot (1965), *Biochim. Biophys. Acta* **108**, 249.

Smith, C. G. and I. A. Bernstein (1961), *Biochim. Biophys. Acta* **52**, 184.

Smith, D. A. and D. W. Visser (1965), *J. Biol. Chem.* **240**, 446.

Smith, E. E. B., B. Galloway, and G. T. Mills (1961), *Biochem. Biophys. Res. Commun.* 5, 148.

Smith, E. E. B. and G. T. Mills (1954), *Biochim. Biophys. Acta* **13**, 386.

Smith, E. E. B., G. T. Mills, and E. M. Harper (1957), *Biochim. Biophys. Acta* **23**, 662.

Smith, J. D. and D. B. Dunn (1959a), *Biochem. J.* **72**, 294.

Smith, J. D. and D. B. Dunn (1959b), *Biochim. Biophys. Acta* **31**, 573.

Smith, J. D. and R. E. F. Matthews (1957), *Biochem. J.* **66**, 323.

Smith, K. C. (1962), *Biochem. Biophys. Res. Commun.* **8**, 157.

Smith, L. H., Jr., and M. Sullivan (1960), *Biochim. Biophys. Acta* **39**, 554.

Smull, C. E., M. F. Mallette, and E. H. Ludwig (1961), *Biochem. Biophys. Res. Commun.* **5**, 247.

Soldo, A. T., G. A. Godoy, and W. J. van Wagtendonk (1966), *Federation Proc.* **25**, 783.

Somerville, R., K. Ebisuzaki, and G. R. Greenberg (1959), *Proc. Natl. Acad. Sci. U.S.* **45**, 1240.

Sonne, J. C., J. M. Buchanan, and A. M. Delluva (1946), *J. Biol. Chem.* **166**, 395.

Sonne, J. C., J. M. Buchanan, and A. M. Delluva (1948), *J. Biol. Chem.* **173**, 69.

Sonnenberg, B. P. and G. Zubay (1965), *Proc. Natl. Acad. Sci. U.S.* **54**, 415.

Sorensen, L. B. (1959), *Metabolism*, **8**, 687.

Sparrow, A. H. and F. Rosenfeld (1946), *Science* **104**, 245.

Spiegelman, S., I. Haruna, I. B. Holland, G. Beaudreau, and D. Mills (1965), *Proc. Natl. Acad. Sci. U.S.* **54**, 919.

Sporn, B. M., C. W. Dingman, H. L. Phelps, and G. N. Wogan (1966), *Science* **151**, 1539.

Spyrides, G. J. and F. Lipmann (1962), *Proc. Natl. Acad. Sci. U.S.* **48**, 1977.

Srinivasan, P. R. and E. Borek (1963), *Proc. Natl. Acad. Sci. U.S.* **49**, 529.

Srinivasan, P. R., S. Nofal and C. Sussman (1964), *Biochem. Biophys. Res. Commun.* **16**, 82.

Stacey, K. A. (1961), *Biochem. Biophys. Res. Commun.* **5**, 486.

Stacey, K. A., M. Cobbs, S. F. Cousens, and P. Alexander (1958), *Ann. N. Y. Acad. Sci.* **68**, 682.

Stadtman, E. R. (1966), *Advan. Enzymol.* **28**, 41.

Staehelin, T., F. O. Wettstein, and H. Noll (1963), *Science* **139**, 180.

Staehelin, T., F. O. Wettstein, H. Oura, and H. Noll (1964), *Nature* **201**, 264.

Stanley, W. M. (1935), *Science* **81**, 644.

Steele, W. J. (1962), *Proc. Am. Assoc. Cancer Res.* **3**, No. 4, 364.

Steiner, R. F. and R. F. Beers, Jr. (1961), Polynucleotides. Natural and Synthetic Nucleic Acids," Amsterdam-London-New York, Elsevier.

Stent, G. S. (1963), "Molecular Biology of Bacterial Viruses," San Francisco, W. H. Freeman and Co.

Stent, G. S. (1964), *Science* **144**, 816.

Stent, G. S. and S. Brenner (1961), *Proc. Natl. Acad. Sci. U.S.* **47**, 2005.

Stern, H. (1958), *J. Biophys. Biochem. Cytol.* **4**, 157.

Stetten, M. R. and C. L. Fox, Jr. (1945), *J. Biol. Chem.* **161**, 333.

Steudel, H. (1901), *Z. Physiol. Chem.* **32**, 285.

Stevens, L. and L. A. Stocken (1960), *Biochem. Biophys. Res. Commun.* **3**, 155.

Stevens, L. and L. A. Stocken (1962), *Biochem. Biophys. Res. Commun.* **7**, 315.

Stewart, G. A. and E. Farber (1967), *Science* **157**, 67.

Stich, H. F. (1960), *J. Natl. Cancer Inst.* **24**, 1283.

Stich, H. F. (1963), *Exptl. Cell Res., Suppl.* **9**, 277.

Stich, H. F. and M. L. Florian (1958), *Can. J. Biochem. Physiol.* **36**, 855.

Stocken, L. A. (1965), cited by Allfrey *et al.* (1966).

Stocking, C. R. and E. M. Gifford, Jr. (1959), *Biochem. Biophys. Res. Commun.* **1**, 159.

Stokstad, E. L. R., B. L. Hutchings, J. H. Mowat, J. H. Boothe, C. W. Waller, R. R. Angier, J. Semb, and Y. Subba Row (1948), *J. Am. Chem. Soc.* **70,** 5.

Stokstad, E. L. R. and J. Koch (1967), *Physiol. Rev.* **47,** 83.

Stone, J. E. and V. R. Potter (1957), *Cancer Res.* **17,** 800.

Strangeways, T. S. P. and H. E. H. Oakley (1923), *Proc. Roy. Soc. London, Ser. B* **95,** 373.

Stransky, E. (1933), *Biochem. Z.* **266,** 287.

Straus, D. B. and E. Goldwasser (1961), *J. Biol. Chem.* **236,** 849.

Strauss, B., H. Reiter, and T. Searashi (1966), *Radiation Res., Suppl.* **6,** 201.

Strecker, A. (1861), *Ann. Chem.* **118,** 151.

Stretton, A. O. W. and S. Brenner (1965), *J. Mol. Biol.* **12,** 456.

Strominger, J. L. (1955), *Biochim. Biophys. Acta* **17,** 283.

Strominger, J. L. (1960), *Physiol. Rev.* **40,** 55.

Strominger, J. L., H. M. Kalckar, J. Axelrod, and E. S. Maxwell (1954), *J. Am. Chem. Soc.* **76,** 6411.

Stubblefield, E. and G. C. Mueller (1962), *Cancer Res.* **22,** 1091.

Stutz, E. and H. Noll (1967), *Proc. Natl. Acad. Sci. U.S.* **57,** 774.

Stuy, J. H. (1961a), *Radiation Res.* **14,** 56.

Stuy, J. H. (1961b), *Radiation Res.* **15,** 45.

Stuy, J. H. (1962), *Photochem. Photobiol.* **1,** 41.

Su, J. C. and W. Z. Hassid (1960), *J. Biol. Chem.* **235,** PC 36.

Subak-Sharpe, H. and J. Hay (1965), *J. Mol. Biol.* **12,** 924.

Sueoka, N. and T. Kano-Sueoka (1965), in "Developmental and Metabolic Control Mechanisms and Neoplasia," p. 114, Baltimore, Md., The Williams & Wilkins Company.

Sugino, Y. and N. Sugino (1960), *Biochim. Biophys. Acta* **40,** 425.

Sugino, Y., H. Teraoka, and H. Shimono (1966), *J. Biol. Chem.* **241,** 961.

Sugiura, K., G. H. Hitchings, L. F. Cavalieri, and C. C. Stock (1950), *Cancer Res.* **10,** 178.

Sugiyama, T. and H. Fraenkel-Conrat (1961), *Proc. Natl. Acad. Sci. U.S.* **47,** 1393.

Sullivan, M. P., E. C. Beatty, Jr., C. B. Hyman, M. L. Murphy, M. I. Pierce, and N. C. Severo (1962), *Cancer Chemotherapy Rept.* **18,** 83.

Sutherland, T. W. and E. W. Rall (1958), *J. Biol. Chem.* **232,** 1077.

Svensson, I., H. G. Boman, K. G. Eriksson, and K. Kjellin (1963), *J. Mol. Biol.* **7,** 254.

Swartz, M. N., T. A. Trautner, and A. Kornberg (1962), *J. Biol. Chem.* **237,** 1961.

Swez, J. and E. C. Pollard (1966), *Radiation Res.* **29,** 475.

Swift, H. (1955), in "The Nucleic Acids," Vol. II, p. 51, New York, Academic Press.

Swift, H. (1964), in "The Nucleohistones" (J. Bonner and P. Ts'o, editors), p. 169, San Francisco-London-Amsterdam, Holden-Day.

Sztumpf, E. and D. Shugar (1962), *Biochim. Biophys. Acta* **61,** 555.

Szybalski, W. (1962), in "The Molecular Basis of Neoplasia," p. 147, Austin, Texas, University of Texas Press.

Szybalski, W. and V. Bryson (1952), *Microbial Genet. Bull.,* No. 6, 29.

Szybalski, W., N. K. Cohn, and C. Heidelberger (1963), *Federation Proc.* **22,** 532.

Szybalski, W. and V. N. Iyer (1964), *Federation Proc.* **23,** 946.

Szybalski, W. and E. H. Szybalska (1961), *Cancer Chemotherapy Rept.* **11,** 87.

Tabor, H. (1961), *Biochem. Biophys. Res. Commun.* 4, 228.

Tabor, H. and O. Hayaishi (1955), *J. Am. Chem. Soc.* 77, 505.

Tabor, H. and J. C. Rabinowitz (1956), *J. Am. Chem. Soc.* 78, 5705.

Taborsky, G. (1959), *J. Biol. Chem.* 234, 2652.

Takahashi, I. and J. Marmur (1963), *Nature* 197, 794.

Takai, M. (1966), *Biochim. Biophys. Acta* 119, 20.

Talwar, G. P., S. L. Gupta, and F. Gros (1964), *Biochem. J.* 91, 655.

Tamm, C., M. E. Hodes, and E. Chargaff (1952), *J. Biol. Chem.* 195, 49.

Tanaka, N., K. Nagai, H. Yamaguchi, and H. Umezawa (1965), *Biochem. Biophys. Res. Commun.* 21, 328.

Tata, J. R. (1964), *Biochim. Biophys. Acta* 87, 528.

Tata, J. R. (1965), *Nature* 207, 378.

Tata, J. R. (1966a), *Develop. Biol.* 13, 77.

Tata, J. R. (1966b), *Progr. Nucleic Acid. Res. Mol. Biol.* 5, 191.

Tata, J. R. and C. C. Widnell (1966), *Biochem. J.* 98, 604.

Taylor, D., J. P. Greenstein, and A. Hollaender (1947), *Science* 105, 263.

Taylor, J. H. (1963), in "Molecular Genetics," Vol. I (J. H. Taylor, editor), p. 65, New York, Academic Press.

Taylor, J. H., W. F. Haut, and J. Tung (1962), *Proc. Natl. Acad. Sci. U.S.* 48, 190.

Taylor, J. H., P. S. Woods, and W. L. Hughes (1957), *Proc. Natl. Acad. Sci. U.S.* 43, 122.

Taylor, S. H. (1960), *Ann. N. Y. Acad. Sci.* 90, 409.

Tekman, S. and N. Öner (1963), *Nature* 200, 77.

Temin, H. M. (1963), *Virology* 20, 577.

Temin, H. M. (1964a), *Virology* 23, 486.

Temin, H. M. (1964b), *Proc. Natl. Acad. Sci. U.S.* 52, 323.

Temin, H. M. (1966), *Cancer Res.* 26, 212.

Tener, G. M. and H. G. Khorana (1958), *J. Am. Chem. Soc.* 80, 1999.

Terzaghi, B. E., G. Streisinger, and F. W. Stahl (1962), *Proc. Natl. Acad. Sci. U.S.* 48, 1519.

Tessman, I. (1959), *Virology* 7, 263.

Tessman, I. (1962), in "The Molecular Basis of Neoplasia," p. 172, Austin, Texas, University of Texas Press.

Thach, R. E., M. A. Cecere, T. A. Sundararajan, and P. Doty (1965), *Proc. Natl. Acad. Sci. U.S.* 54, 1167.

Theil, E. C. and S. Zamenhof (1963), *J. Biol. Chem.* 238, 3058.

Thomson, J. F., M. S. Carttar, and W. W. Tourtellotte (1954), *Radiation Res.* 1, 165.

Thomson, R. Y., F. C. Heagy, W. C. Hutchison, and J. N. Davidson (1953), *Biochem. J.* 53, 460.

Timonen, S. and E. Therman (1950), *Nature* 166, 995.

Tissières, A. (1959), *J. Mol. Biol.* 1, 365.

Tissières, A., S. Bourgeois, and F. Gros (1963), *J. Mol. Biol.* 7, 100.

Tissières, A. and J. D. Watson (1958), *Nature* 182, 778.

Tissières, A., J. D. Watson, D. Schlessinger, and B. R. Hollingworth (1959), *J. Mol. Biol.* 1, 221.

Toth, B. (1963), *Proc. Soc. Exptl. Biol. Med.* 112, 873.

Touster, O., R. H. Mayberry, and D. B. McCormick (1957), *Biochim. Biophys. Acta* 25, 196.

Touster, O. and D. R. D. Shaw (1962), *Physiol. Rev.* **42**, 181.

Traut, R. R. and R. E. Monro (1964), *J. Mol. Biol.* **10**, 63.

Trautner, T. A., M. N. Swartz, and A. Kornberg (1962), *Proc. Natl. Acad. Sci. U.S.* **48**, 449.

Tremblay, G. C. and S. A. Thayer (1964), *J. Biol. Chem.* **239**, 3321.

Trentin, J. J., Y. Yabe, and G. Taylor (1962), *Science* **137**, 835.

Tsugita, A. and H. Fraenkel-Conrat (1962), *J. Mol. Biol.* **4**, 73.

Tsugita, A., D. T. Gish, J. Young, H. Fraenkel-Conrat, C. A. Knight, and W. M. Stanley (1960), *Proc. Natl. Acad. Sci. U.S.* **46**, 1463.

Tsukada, K. and I. Lieberman (1964a), *J. Biol. Chem.* **239**, 1564.

Tsukada, K. and I. Lieberman (1964b), *J. Biol. Chem.* **239**, 2952.

Tuve, T. W. and C. B. Anfinsen (1960), *J. Biol. Chem.* **235**, 3437.

Tyner, E. P., C. Heidelberger, and G. A. LePage (1953), *Cancer Res.* **13**, 186.

Tzagoloff, H. and W. W. Umbreit (1963), *J. Bact.* **85**, 49.

Udaka, S. and B. Vennesland (1962), *J. Biol. Chem.* **237**, 2018.

Umbarger, H. E. (1961), *Cold Spring Harbor Symp. Quant. Biol.* **26**, 301.

Utter, M. F. (1960), in "The Enzymes" (P. D. Boyer, H. Lardy, and K. Myrbäck, editors), Vol. 2, p. 75, New York, Academic Press.

Uyeda, K. and J. C. Rabinowitz (1964), *Arch. Biochem.* **107**, 419.

Valentin, R. C., R. Bojanowski, E. Gaudy, and R. S. Wolfe (1962), *J. Biol. Chem.* **237**, 2271.

Van Bruggen, E. F. J., P. Borst, G. J. C. M. Ruttenberg, M. Gruber, and A. M. Kroon (1966), *Biochim. Biophys. Acta* **119**, 437.

Van Eckenstein, W. A. and J. J. Blanksma (1913), *Chem. Weeksblad* **10**, 664.

Vanecko, S. and M. Laskowski, Sr. (1961), *J. Biol. Chem.* **236**, 3312.

Van Overbeek, J. (1966), *Science* **152**, 721.

Vendrely, R. (1955), in "The Nucleic Acids," Vol. II, p. 155, New York, Academic Press.

Verachtert, H., S. T. Bass, and R. G. Hansen (1964), *Biochem. Biophys. Res. Commun.* **15**, 158.

Vermund, H., C. P. Barnum, R. A. Huseby, and K. W. Stenstrom (1953), *Cancer Res.* **13**, 633.

Vermund, H., J. Hodgett, and F. J. Ansfield (1961), *Am. J. Roentgenol. Radium Therapy Nucl. Med.* **85**, 559.

Viñuela, E., M. Salas, and S. Ochoa (1967), *Proc. Natl. Acad. Sci. U.S.* **57**, 729.

Visser, D. W., I. Goodmann, and K. Dittmer (1948), *J. Am. Chem. Soc.* **70**, 1926

Vogt, M. and R. Dulbecco (1962), *Virology* **16**, 41.

Vogt, M. and E. Harbers (1967), *Strahlentherapie* **133**, 426.

Volkin, E. (1954), *J. Am. Chem. Soc.* **76**, 5892.

Volkin, E. (1960), *Proc. Natl. Acad. Sci. U.S.* **46**, 1336.

Volkin, E. and A. Ruffilli (1962), *Proc. Natl. Acad. Sci. U.S.* **48**, 2193.

Voll, M. J. and S. H. Goodgal (1961), *Proc. Natl. Acad. Sci. U.S.* **47**, 505.

Von Borstel, R. C. (1955), *Nature* **175**, 342.

Von Euler, H. and G. von Hevesy (1942), *Det. Kgl. Danske Videnskabernes Selskab Biol. Medd.* **17**, No. 8.

Von Euler, H. and G. von Hevesy (1944), *Arkiv. Kemi. Mineral. Geol.* **17A**, No. 30, 1.

Von Hevesy, G. (1948), *Advan. Biol. Med. Phys.* **1**, 409.

Wacker, A. (1959), in "Chemie der Genetik," p. 85, Berlin-Göttingen-Heidelberg, Springer-Verlag.

Wacker, A. (1961), *J. Chim. Phys.* **58**, 1041.

Wacker, A. (1963), *Progr. Nucleic Acid. Res.* **1**, 369.

Wacker, A., H. Dellweg, and D. Jacherts (1962), *J. Mol. Biol.* **4**, 410.

Wacker, A., H. Dellweg, and E. Lodemann (1961a), *Angew. Chem.* **73**, 64.

Wacker, A., H. Dellweg, and D. Weinblum (1961b), *J. Mol. Biol.* **3**, 787.

Wacker, A. and D. Jacherts (1962), *J. Mol. Biol.* **4**, 413.

Wacker, A., E. Lodemann, K. Gauri, and P. Chandra (1966), *J. Mol. Biol.* **18**, 382.

Wacker, A., A. Trebst, D. Jacherts, and F. Weygand (1954), *Z. Naturforsch.* **9b**, 616.

Wacker, A., D. Weinblum, L. Träger, and Z. H. Moustafa (1961), *J. Mol. Biol.* **3**, 790.

Wagner, R. R. (1965), *Am. J. Med.* **38**, 726.

Wagner, R. R. and A. S. Huang (1965), *Proc. Natl. Acad. Sci. U. S.* **54**, 1112.

Wagner, R. R. and A. S. Huang (1966), *Virology* **28**, 1.

Wahba, A. J. and M. Friedkin (1961), *J. Biol. Chem.* **236**, PC 11.

Wahba, A. J., R. S. Miller, C. Basilio, R. S. Gardner, P. Lengyel, and J. F. Speyer (1963), *Proc. Natl. Acad. Sci. U.S.* **49**, 880.

Wahl, R. and L. M. Kozloff (1962), *J. Biol. Chem.* **237**, 1953.

Wakisaka, G., H. Uchino, T. Nakamura, H. Sotobayashi, S. Shirakawa, A. Adachi, and M. Sakurai (1963), *Nature* **198**, 385.

Walerych, W. S., S. Ventakaraman, and B. C. Johnson (1966), *Biochem. Biophys. Res. Commun.* **23**, 368.

Walker, J. B. (1961), *J. Biol. Chem.* **236**, 493.

Walker, J. B. (1962), in "The Molecular Basis of Neoplasia," p. 403, Austin, Texas, University of Texas Press.

Wallach, D. P. and S. Grisolia (1957), *J. Biol. Chem.* **226**, 277.

Waller, J. P. (1964), *J. Mol. Biol.* **10**, 319.

Walwick, E. R. and R. K. Main (1962), *Biochim. Biophys. Acta* **55**, 225.

Wang, T. (1962), *Arch. Biochem.* **97**, 387.

Wang, T. P. and J. O. Lampen (1952a): *J. Biol. Chem.* **194**, 775.

Wang, T. P. and J. O. Lampen (1952b), *J. Biol. Chem.* **194**, 785.

Wanka, F., I. K. Vasil, and H. Stern (1964), *Biochim. Biophys. Acta* **85**, 50.

Wannamaker, L. W. (1958), *J. Exptl. Med.* **107**, 797.

Warburg, O. (1924), *Biochem. Z.* **142**, 317.

Warburg, O. (1930), "The Metabolism of Tumors," London, Constable and Co.

Warburg, O. and W. Christian (1938), *Biochem. Z.* **298**, 150.

Warburg, O., W. Christian, and A. Griese (1935), *Biochem. Z.* **282**, 157.

Ward, D. C., E. Reich, and I. H. Goldberg (1965), *Science* **149**, 1259.

Ward, D. N. and J. D. Spain (1957), *Cancer Res.* **17**, 623.

Waring, M. J. (1965), *Mol. Pharmacol.* **1**, 1.

Warnaar, S. O. and J. A. Cohen (1966), *Biochem. Biophys. Res. Commun.* **24**, 554.

Warner, J. R., A. Rich, and C. E. Hall (1962), *Science* **138**, 1339.

Warner, R. C., H. H. Samuels, M. T. Abbott, and J. S. Krakow (1963), *Proc. Natl. Acad. Sci. U. S.* **49**, 533.

Warren, L. and J. M. Buchanan (1957), *J. Biol. Chem.* **229**, 613.

Warren, L., J. G. Flaks, and J. M. Buchanan (1957), *J. Biol. Chem.* **229**, 627.

Warwick, G. P. and J. J. Roberts (1967), *Nature* **213**, 1206.

Watson, J. D. (1963), *Science* **140**, 17.
Watson, J. D. and F. H. C. Crick (1953a), *Nature* **171**, 737.
Watson, J. D. and F. H. C. Crick (1953b), *Nature* **171**, 964.
Webb, T. E., G. Blöbel, and V. R. Potter (1964), *Cancer Res.* **24**, 122.
Webb, T. E., G. Blöbel, V. R. Potter, and H. P. Morris (1965), *Cancer Res.* **25**, 1219.
Webb, T. E. and H. P. Morris (1966), *Cancer Res.* **26**, 253.
Weber, H. H. (1957), *Ann. Rev. Biochem.* **26**, 667.
Webster, R. E., D. L. Engelhardt, and N. D. Zinder (1966), *Proc. Natl. Acad. Sci. U. S.* **55**, 155.
Wecker, E. (1959), *Z. Naturforsch.* **14b**, 370.
Weckman, B. G. and B. W. Catlin (1957), *J. Bacteriol.* **73**, 747.
Weigert, M. G. and A. Garen (1965), *J. Mol. Biol.* **12**, 448.
Weil, R., A. Bendich, and C. M. Southam (1960), *Proc. Soc. Exptl. Biol. Med.* **104**, 670.
Weil, R., M. R. Michel, and G. K. Ruschmann (1965), *Proc. Natl. Acad. Sci. U.S.* **53**, 1468.
Weiler, E. (1956), *Z. Naturforsch.* **11b**, 31.
Weinbaum, G. and R. J. Suhadolnik (1964), *Biochim. Biophys. Acta* **81**, 236.
Weisberger, A. S. (1963), *Proc. Natl. Acad. Sci. U. S.* **48**, 68.
Weisberger, A. S., S. Armentrout, and S. Wolfe (1963), *Proc. Natl. Acad. Sci. U. S.* **50**, 86.
Weisberger, A. S. and S. Wolfe (1964), *Federation Proc.* **23**, 976.
Weisburger, E. K. and J. H. Weisburger (1958), *Advan. Cancer Res.* **5**, 331.
Weiss, B. and C. C. Richardson (1967), *Proc. Natl. Acad. Sci. U. S.* **57**, 1021.
Weiss, J. J. (1964), *Progr. Nucleic Acid Res. Mol. Biol.* **3**, 103.
Weiss, S. B. and T. Nakamoto (1961), *J. Biol. Chem.* **236**, PC 18.
Weissberger, E. and S. Okada (1961), *Intern. J. Radiation Biol.* **3**, 331.
Weissman, S. M., R. M. S. Smellie, and J. Paul (1960), *Biochim. Biophys. Acta* **45**, 101.
Weissmann, B., P. A. Bromberg, and A. B. Gutman (1957), *J. Biol. Chem.* **224**, 407.
Weissmann, C. and P. Borst (1963), *Science* **142**, 1188.
Weissmann, C., P. Borst, R. H. Burdon, M. A. Billeter, and S. Ochoa (1964), *Proc. Natl. Acad. Sci. U. S.* **51**, 682.
Weissmann, C. and G. Feix (1966), *Proc. Natl. Acad. Sci. U. S.* **55**, 1264.
Weissmann, C. and S. Ochoa (1967), *Progr. Nucleic Acid Res. Mol. Biol.* **6**, 353.
Weissmann, C., L. Simon, and S. Ochoa (1963), *Proc. Natl. Acad. Sci. U. S.* **49**, 407.
Welch, A. D. (1965), *Proc. Natl. Acad. Sci. U. S.* **54**, 1359.
Welch, A. D., R. E. Handschumacher, J. J. Jaffe, S. S. Cardoso, and P. A. Calabresi (1960), *Cancer Chemotherapy Rept.* **9**, 39.
Werkheiser, W. C. (1961), *J. Biol. Chem.* **236**, 888.
Wettstein, F. O., T. Staehelin, and H. Noll (1963), *Nature* **197**, 430.
Wheeler, C. M. and S. Okada (1961), *Intern. J. Radiation Biol.* **3**, 23.
Wheeler, G. P. and J. A. Alexander (1961), *Cancer Res.* **21**, 399.
Wheeler, H. L. and H. F. Merriam (1903), *Am. Chem. J.* **29**, 478.
Wheeler, W. L. and T. B. Johnson (1903), *Am. Chem. J.* **29**, 492.
Whitely, H. R. (1952), *J. Bacteriol.* **63**, 163.
Whitely, H. R. and H. C. Douglas (1951), *J. Bacteriol.* **61**, 605.

Wiest, W. G. and C. Heidelberger (1953), *Cancer Res.* **13**, 246.

Wilczok, T. (1962), *Nature* **196**, 1314.

Wilkins, M. H. F. and G. Zubay (1959), *J. Biophys. Biochem. Cytol.* **5**, 55.

Willett, F. M., L. V. Foye, Jr., M. Roth, and B. E. Hall (1961), *Dis. Chest* **39**, 38.

Williams, E. J., S. C. Sung, and M. Laskowski, Sr. (1961), *J. Biol. Chem.* **236**, 1130.

Williams, R. B., R. P. de Long, and J. J. Jaffe (1952), *Am. J. Pathol.* **28**, 546.

Williams, W. J. and J. M. Buchanan (1953), *J. Biol. Chem.* **202**, 253.

Wilson, C. M. (1967), *J. Biol. Chem.* **242**, 2260.

Wilson, J. D. (1963), *Proc. Natl. Acad. Sci. U. S.* **50**, 93.

Wilson, J. D. and P. M. Loeb (1965), in "Developmental and Metabolic Control Mechanisms and Neoplasia," p. 375, Baltimore, Md., The Williams & Wilkins Company.

Windmueller, H. G. and A. E. Spaeth (1965), *J. Biol. Chem.* **240**, 4398.

Wittmann, H. G. (1960), *Virology* **12**, 609.

Wittmann, H. G. (1961), *Naturwissenschaften* **48**, 729.

Wittmann, H. G. (1962), *Z. Vererbungslehre* **93**, 491.

Wolberg, W. H. (1964), *Cancer Res.* **24**, 1437.

Wong, W. T. and B. S. Schweigert (1957), *Proc. Soc. Exptl. Biol. Med.* **94**, 455.

Wood, H. G. and J. Katz (1958), *J. Biol. Chem.* **232**, 1279.

Woods, D. D. (1940), *Brit. J. Exptl. Pathol.* **21**, 74.

Wool, I. G. and A. J. Munro (1963), *Proc. Natl. Acad. Sci. U. S.* **50**, 918.

Wooley, D. W. (1960), *J. Biol. Chem.* **235**, 3238.

Wright, B. E., M. L. Anderson, and E. C. Herman (1958), *J. Biol. Chem.* **230**, 271.

Wright, L. D., C. S. Miller, H. R. Skeggs, J. W. Huff, L. L. Weed, and D. W. Wilson (1951), *J. Am. Chem. Soc.* **73**, 1898.

Wu, R. and D. W. Wilson (1956), *J. Biol. Chem.* **223**, 195.

Wulff, D. L. and C. S. Rupert (1962), *Biochem. Biophys. Res. Commun.* **7**, 237.

Wulff, V. J., H. Quastler, and F. G. Sherman (1962), *Proc. Natl. Acad. Sci. U. S.* **48**, 1373.

Wyatt, G. R. (1951), *Biochem. J.* **48**, 584.

Wyatt, G. R. and S. S. Cohen (1953), *Biochem. J.* **55**, 774.

Wykes, J. R. and R. M. S. Smellie (1966), *Biochem. J.* **99**, 347.

Wyngaarden, J. B. and J. T. Dunn (1957), *Arch. Biochem.* **70**, 150.

Wyngaarden, J. B. and R. A. Greenland (1963), *J. Biol. Chem.* **238**, 1054.

Wyngaarden, J. B. and D. Stetten (1953), *J. Biol. Chem.* **203**, 9.

Yamasaki, M. and K. Arima (1967), *Biochim. Biophys. Acta* **139**, 202.

Yarmolinsky, M. B. and G. L. de la Haba (1959), *Proc. Natl. Acad. Sci. U. S.* **45**, 1721.

Yates, R. A. and A. B. Pardee (1956a), *J. Biol. Chem.* **221**, 743.

Yates, R. A. and A. B. Pardee (1956b), *J. Biol. Chem.* **221**, 757.

York, J. L. and G. A. LePage (1966), *Can. J. Biochem.* **44**, 19.

Younathan, E. S., J. E. Stone, and T. S. Harris (1964), *J. Biol. Chem.* **239**, 290.

Young, E. G. and W. W. Hawkins (1944), *J. Bacteriol.* **47**, 351.

Yu, T. F. and A. B. Gutman (1964), *Am. J. Med.* **37**, 885.

Zachau, H. G., G. Acs, and F. Lipmann (1958), *Proc. Natl. Acad. Sci. U. S.* **44**, 885.

Zachau, H. G., D. Dütting, and H. Feldmann (1966), *Angew. Chem.* **78**, 392.

Zakrzewski, S. F. (1963), *J. Biol. Chem.* **238**, 1485.

Zalokar, M. (1961), in "Control Mechanisms in Cellular Processes" (D. Bonner, editor), p. 87, New York, The Ronald Press Company.

Zamecnik, P. C. (1966), *Cancer Res.* **26**, 1.

Zamenhof, S. (1956), *Progr. Biophys.* **6**, 85.

Zamenhof, S. and E. Chargaff (1949), *J. Biol. Chem.* **180**, 727.

Zamenhof, S., B. Reiner, R. de Giovanni, and K. Rich (1956), *J. Biol. Chem.* **219**, 165.

Zillig, W., E. Fuchs, and R. Millette (1966), in "Procedures in Nucleic Acid Research" (G. L. Cantoni and D. R. Davies, editors), p. 323, New York and London, Harper & Row, Publishers.

Zillig, W., W. Krone, and M. Albers (1959), *Z. Physiol. Chem.* **317**, 131.

Zillig, W., D. Schachtschabel, and W. Krone (1960), *Z. Physiol. Chem.* **318**, 100.

Zimm, B. H. (1960), *J. Chem. Phys.* **33**, 1349.

Zimm, B. H. (1965), personal communication.

Zimmer, K. G. (1958), *Naturwissenschaften* **45**, 325.

Zimmerman, E. F. and S. A. Greenberg (1965), *Mol. Pharmacol.* **1**, 113.

Zimmerman, M. (1962), *Biochem. Biophys. Res. Commun.* **8**, 169.

Zimmerman, M. (1963), *Federation Proc.* **22**, 291.

Zimmerman, S. B. and A. Kornberg (1961), *J. Biol. Chem.* **236**, 1480.

Zimmermann, F., H. Kröger, U. Hagen, and K. Keck (1964), *Biochim. Biophys. Acta* **87**, 160.

Zinder, N. D. (1958), *Sci. Am.* **199**, No. 5, 38.

Zinder, N. D. and J. Lederberg (1952), *J. Bacteriol.* **64**, 679.

Zittle, C. A. (1946), *J. Biol. Chem.* **166**, 499.

Zubay, G. and M. R. Watson (1959), *J. Biophys. Biochem. Cytol.* **5**, 51.

Index